Grundlehren der mathematischen Wissenschaften 307

A Series of Comprehensive Studies in Mathematics

Albert S. Schwarz

Quantum Field Theory and Topology

With 30 Figures

Springer-Verlag

Berlin Heidelberg New York
London Paris Tokyo
Hong Kong Barcelona
Budapest

Albert S. Schwarz
Department of Mathematics
565 Kerr Hall
University of California
Davis, CA 95616, USA

Translators:

Eugene Yankowsky

Silvio Levy
Geometry Center
1300 South Second St.
Minneapolis, MN 55454, USA

Title of the original Russian edition: Kvantovaya teoriya polya i topologiya. Nauka, Moscow 1989. An expanded version of the last third of the Russian edition is being published separately in English under the title: Topology for Physicists.

Mathematics Subject Classification (1991): 81Txx

ISBN 3-540-54753-3 Springer-Verlag Berlin Heidelberg New York
ISBN 0-387-54753-3 Springer-Verlag New York Berlin Heidelberg

© Springer-Verlag Berlin Heidelberg 1993
Printed in the United States of America

Typesetting: Camera-ready copy produced from the translation editor's output file using a Springer TeX macro package
41/3140-5 4 3 2 1 0   Printed on acid-free paper

Preface

In recent years topology has firmly established itself as an important part of the physicist's mathematical toolkit. It has many applications, first of all in quantum field theory, but increasingly also in other areas of physics—suffice it to say that topological ideas play an important role in one of the suggested explanations of high-temperature superconductivity. Topology is also used in the analysis of another remarkable discovery of recent years, the quantum Hall effect.

The main focus of this book is on the results of quantum field theory that are obtained by topological methods, but some topological aspects of the theory of condensed matter are also discussed. The topological concepts and theorems used in physics are very diverse, and for this reason I have included a substantial amount of purely mathematical information.

This book is aimed at different classes of readers—from students who know only basic calculus, linear algebra and quantum mechanics, to specialists in quantum field theory and mathematicians who want to learn about the physical applications of topology.

Part I can be considered as an introduction to quantum field theory: it discusses the basic Lagrangians used in the theory of elementary particles. Part II is devoted to the applications of topology to quantum field theory. Part III provides the mathematical prerequisites for the understanding of Parts I and II. Readers with a weak background in group theory and topology might profit from skimming Part III before reading the first two parts. For a more extensive introduction to topology, the reader is directed to *Topology for Physicists* [59]. Parts I and II contain some references to that book: the reference T1.3, for example, means Chapter 1, Section 3 of *Topology for Physicists*. In spite of these references, this book tries to be self-contained; the quantum field theorist who has looked through Part III and the Definitions and Notations at the beginning of the book will likely be able to read Part II without trouble. Likewise, mathematicians can use Part I to familiarize themselves with the foundations of quantum field theory, and then proceed with Part II for a study of the physical applications of topology.

Text in smaller print can be skipped without detriment to the main flow of ideas.

Here is a more detailed outline of Part II. Chapter 8, devoted to topologically stable defects in condensed matter, is essentially independent of the rest of the book. Chapters 9–15 and 18 are devoted to topological integrals of motion and topologically nontrivial particles, such as magnetic monopoles. In Chapters 16 and 17 we analyze symmetric gauge fields, and apply the results to the study of magnetic monopoles. Topologically nontrivial strings are the subject of Chapters 19 and 20. Nonlinear fields and multivalued action integrals occupy Chapters 21 and 22, which are closely related. Functional integrals and their applications to quantum theory are discussed in Chapters 23 and 24, the results being used in Chapter 25 to study the quantization of gauge theories. Chapters 26–28 contain mathematical information on elliptic operators and their determinants, to be used in the succeeding chapters. Chapter 29 studies quantum anomalies, and Chapters 30–32 instantons in gauge theories. Finally, Chapter 33 considers functional integrals in theories with fermions, and the results are applied in Chapter 34 to the study of the instanton contribution in quantum chromodynamics.

For a minimum understanding of the applications of topology to quantum field theory, I suggest reading Chapters 9–13, 23–25 and 30. Readers interested in topologically nontrivial particles and strings can then proceed to read Chapters 14, 19 and 21, and as supplementary material Chapters 18 (which depends on 14), 20 (based on 14 and 19) and 22 (dependent on 21). Readers interested in instantons should turn to Chapters 31–34, after having read Chapters 26–29 for mathematical background, if necessary.

The decision to keep this book to a manageable size and to a relatively elementary level meant that many fascinating physical results obtained through topological methods could not be included. In particular, I have not discussed the rapidly developing theory of strings, in spite of the crucial role that topology plays in it.

The questions discussed in this book were discussed with many of my colleagues. I am particularly grateful to L. D. Faddeev, S. P. Novikov and A. M. Polyakov, who made important contributions to the field covered in the book. I take this opportunity to thank my former students, V. A. Fateev, I. V. Frolov, D. B. Fuchs and Yu. S. Tyupkin, and especially M. A. Baranov and A. A. Rosly, who gave me invaluable assistance. I would also like to thank my wife, L. M. Kissina, for her understanding and help.

A. S. Schwarz

Contents

Introduction

Topology is the study of continuous maps. From the point of view of topology, two spaces that can be transformed into each other without tearing or gluing are equivalent. More precisely, a *topological equivalence*, or *homeomorphism*, is a continuous bijection whose inverse, too, is continuous.

For example, every convex, bounded, closed subset of n-dimensional space that is not contained in an $(n-1)$-dimensional subspace is homeomorphic to an n-dimensional ball. The boundary of such a set is homeomorphic to the boundary of an n-ball, that is, an $(n-1)$-dimensional sphere.

For continuity to have a meaning, it is sufficient that there be a concept of distance between any two points in the space. Such a rule for assigning a distance to each pair of points is called a *metric*, and a space equipped with it is a *metric space*. But a space doesn't have to have a metric in order for continuity to make sense; it is enough that there be a well-defined, albeit qualitative, notion of points being close to one another. The existence of this notion, which is generally formalized in terms of *neighborhoods* or *limits*, makes the space into a *topological space*.

Topological spaces are found everywhere in physics. For example, the configuration space and the phase space of a system in classical mechanics are equipped with a natural topology, as is the set of equilibrium states of a system at a given temperature, in statistical physics. In quantum field theory there arise infinite-dimensional topological spaces.

All this opens up possibilities for using topology in physics. Of course, the primary focus of interest to a physicist—quantitative descriptions of physical phenomena—cannot be reduced to topology. But qualitative features can be understood in terms of topology. If a physical system, and consequently the associated topological space, depends on a parameter, it may happen that the space's topology changes abruptly for certain values of the parameter; this change in topology is reflected in qualitative changes in the system's behavior. For instance, critical temperatures (those where a phase transition occurs) are characterized by a change in the topology of the set of equilibrium states.

Physicists are interested not only in topological spaces, but even more so in the topological properties of continuous maps between such spaces. These maps are generally fields of some form—for example, a nonzero vector field defined on a subset of n-dimensional space can be thought of as a map from this set into the set of nonzero vectors.

Most important are the homotopy invariants of continuous maps. A number, or some other datum, associated with a map is called a *homotopy invariant* of the map if it does not change under infinitesimal variations of the map. More precisely, a homotopy invariant is something that does not change under a continuous deformation, or *homotopy*, of the map. A continuous deformation can be thought of as the accumulation of infinitesimal variations.

For example, given a nonzero, continuous vector field on the complement of a disk D in the plane, we can compute an integer called the *rotation number* or *index* of the field, which tells how many times the vector turns as one goes around a simple loop encircling D. More formally, let the vector field $\boldsymbol{\Psi}(x, y)$ have components $(\Psi_1(x, y), \Psi_2(x, y))$, and assume without loss of generality that D contains the origin. We can form the complex-valued function

$$\Psi(r, \varphi) = \Psi_1(r \cos \varphi, r \sin \varphi) + i\Psi_2(r \cos \varphi, r \sin \varphi)$$

and write it in the form $\Psi(r, \varphi) = A(r, \varphi)e^{i\alpha(r,\varphi)}$, where $\alpha(r, \varphi)$ is continuous: this is because the field is nowhere zero (so $e^{i\alpha(r,\varphi)}$ is well-defined and continuous) and the domain of the field avoids the origin (so a branch of the log can be chosen continuously for α). The index n is defined by $2\pi n = \alpha(r, 2\pi) - \alpha(r, 0)$. It is easy to see that the index is a homotopy invariant: by definition, it changes continuously with the field, but being an integer it can only vary discretely. Therefore it cannot change at all under continuous variations of the field.

In particular, the radial field $\boldsymbol{\Psi}(x, y) = (x, y)$ has index $n = 1$, because $\Psi(r, \varphi) = re^{i\varphi}$; this agrees with the intuitive idea that as you go around the origin the field turns around once. The tangential field $\boldsymbol{\Psi}(x, y) = (-y, x)$ also has index $n = 1$, because $\Psi(r, \varphi) = re^{i\varphi+\pi/2}$. The field $\boldsymbol{\Psi}(x, y) = (x^2 - y^2, 2xy)$ has index 2, because $\Psi(r, \varphi) = r^2 e^{2i\varphi}$.

Maps that can be continuously deformed into one another are called *homotopic*, and a continuous family of deformations going from one to the other is a *homotopy* between the two. Homotopic maps are also said to belong to the same *homotopy class*. For fields we often talk about a *topological type* instead of a homotopy class.

By definition, any homotopy invariant has the same value for all maps in a homotopy class. Conversely, it may happen that if a certain homotopy invariant has the same value for two maps, the maps belong to the same homotopy class: we then say that the invariant *characterizes* such maps up to homotopy.

For example, the index is sufficient to characterize the homotopy class of nonzero vector fields defined away from the origin: two vector fields having the same index can be continuously deformed into one another, without ever vanishing. (This is somewhat harder to prove than the fact that the index is a homotopy invariant.) If a nonzero vector field defined outside a disk can be extended continuously to a nonzero field on the whole plane, its index is zero and the field is said to be *topologically trivial*. Generally, n can be interpreted as the algebraic number of singular points that appear when the field is extended to the interior of the disk. (Singular points are those at which the field vanishes as well as points where the field is undefined.)

There are many physical interpretations for the mathematical results just discussed. For instance, the plane with a given vector field may represent the phase space of a system with one degree of freedom, in which case the field determines the dynamics of the system. The topology of the problem then gives information about the equilibrium positions (points where the vector field vanishes). Or the vector field may represent a magnetization field, and the singular points can be interpreted as defects in a ferromagnet. A complex-valued function might represents the wave function of a superconductor (the order parameter), and its singular points vortices in the semiconductor.

In field theory, both classical and quantum, topological invariants can be considered as integrals of motion: if we can assign to each field with a finite energy a number that does not change under continuous variations of the field, this number is an integral of motion, because it does not vary with time as the field changes continuously. In particular, topological integrals of motion can arise in theories that admit a continuum of classical vacuums (a *classical vacuum* is the classical analogue of the ground state). One can capture the asymptotic behavior of the field at infinity by defining a map from a "sphere at infinity" into the space of classical vacuums, and any homotopy invariant of this map is a topological integral of motion for the system. This works whether the classical vacuums form a linear space or a manifold such as a sphere.

An example of a topological integral of motion is the magnetic charge. The magnetic charge of a field in a domain V is defined as $(4\pi)^{-1}$ times the flux of the magnetic field strength $\mathbf{H}$ over the boundary of V. If the field is defined everywhere inside V, the relation div $\mathbf{H} = 0$ implies that the magnetic charge is zero. That is the situation in electromagnetism. But in grand unification theories (theories that account for electromagnetic, weak and strong interactions), the electromagnetic field strength is not defined everywhere, and it is possible to have fields with nonzero magnetic charge. In fact, it turns out that such fields always exist, and one concludes that in grand unification theories there exist particles that carry magnetic charge (*magnetic monopoles*).

The simplest and most important applications of topology to physics have to do with homotopy theory. But another branch of topology, *homology theory*, also plays an important role. Homology theory can be applied either directly (for example, in analyzing multiple integrals arising from Feynman diagrams) or as a technical means for building homotopy invariants. Homology theory is closely linked with the multidimensional generalizations of Green's formula, of Gauss's divergence theorem and of Stokes' theorem. Such generalizations are generally formulated most conveniently in the language of *exterior forms*, that is, sums of antisymmetrized products of differentials.

The basic concepts of homology theory are *cycles*, which can be seen as closed objects (that is, curves, surfaces, etc., having no boundary), and *boundaries*. Boundaries are cycles that are *homologous to zero*, or *homologically trivial*. For example, if Γ is a closed curve in three-space, the complement $\mathbf{R}^3 \setminus \Gamma$ contains one-dimensional cycles that are not homologous to zero—that is, that cannot bound a surface that avoids Γ. This intuitive statement can be proved formally by considering the magnetic field of a current flowing along Γ. The field strength $\mathbf{H}$ satisfies the condition rot $\mathbf{H} = 0$ outside Γ. If a one-dimensional cycle (closed curve) Γ_1 in $\mathbf{R}^3 \setminus \Gamma$ is the boundary of a surface S that lies entirely in $\mathbf{R}^3 \setminus \Gamma$, Stokes' Theorem implies that

$$\oint_{\Gamma_1} \mathbf{H} \cdot d\mathbf{l} = \int_S \text{rot}\, \mathbf{H} \cdot dS.$$

Hence, a cycle Γ_1 for which $\oint_{\Gamma_1} \mathbf{H} \cdot d\mathbf{l}$ does not vanish cannot be homologous to zero in $\mathbf{R}^3 \setminus \Gamma$.

If a cycle that is not homologous to zero undergoes a continuous deformation, the result cannot be homologous to zero. This remark opens the way for the application of homology theory to homotopy theory. For example, if φ is a loop in the domain $\mathbf{R}^3 \setminus \Gamma$ considered above, that is, a map from the circle into the $\mathbf{R}^3 \setminus \Gamma$, Stokes' Theorem implies that the integral $\oint_\varphi \mathbf{H} \cdot d\mathbf{l}$ is a homotopy invariant.

Another important concept from topology that finds an application in physics is that of a *fiber space* or *fibration*. A common situation in both mathematics and physics is to have, for each point b of a space B, some space F_b depending on $b \in B$. If all the F_b are topologically equivalent, we say that the union E of the F_b is a *fiber space* over the *base space* B, and each F_b is called a *fiber*.

Suppose, for example, that B is the configuration space of a mechanical system. Fixing a point in B, we consider all possible sets of values for the generalized velocities; each such set of values is a tangent vector to the configuration space B at the given point. If the system has n degrees of freedom, we get an n-dimensional vector space for each point of B. The union of all such vector spaces is a fiber space, called the *tangent space* of B.

Another example of a fiber space is the space of all gauge fields, each fiber being a class of gauge-equivalent fields, and the base space being set of all classes.

It is often necessary to select in each fiber a single point that depends continuously on the fiber: in other words, to construct a *section* of the fiber space. In both examples above the concept of a section has a physical interpretation: a section of the tangent space is a vector field (or velocity field) on the configuration space; while for the fibering of the space of gauge fields, the construction of a section amounts to choosing a gauge condition. (For nonabelian gauge fields it is not possible to find a gauge condition that singles out exactly one field in each class. This means that the fiber space has no section.)

Besides arising directly from physical applications, fiber spaces play an important technical role in the solution of problems of homotopy theory. The concept of a gauge field, so important in physics, is intimately linked to the concept of a fiber space. Mathematically it is equivalent to the notion of a connection on a principal fiber space (a fiber space whose fiber is a group).

Many other topological concepts, in addition to the ones mentioned above, are used in contemporary physics. Although we cannot cover them all in this brief introduction, we will encounter several of them later on.

It is worth mentioning that the flow of ideas between topology and physics goes both ways. Ideas from quantum field theory have recently been applied to topology and have led, in particular, to new invariants of smooth manifolds and of knots. One such construction is based on the following simple idea.

Consider an action functional on fields defined on a smooth manifold M, and the associated physical quantities (partition function, correlation functions). If the action functional does not depend on the metric on M, these quantities are also independent of the metric. (This statement can be violated in the case of the so-called quantum anomalies, but in many cases one can prove that quantum anomalies do not arise.) For example, one can consider the action functional

(∗) $$S(A) = \frac{1}{2} \int \varepsilon^{\lambda\mu\nu} A_\lambda \partial_\mu A_\nu \, d^3x,$$

where $A_\mu(x)$ is an electromagnetic field on the compact three-dimensional manifold M. This functional does not depend on the metric on M. It is invariant with respect to gauge transformations $A_\mu \to A_\mu + \partial_\mu \lambda$, and therefore to calculate the partition function we have to impose a metric-dependent gauge condition; the answer can be expressed in terms of the determinants of the scalar-field and vector-field Laplacians. Both determinants depend on the choice of the metric, but the partition function is independent of this choice. We thus obtain an invariant of the manifold M, which turns out to be the well-known Ray–Singer torsion (a smooth version of the Reidemeister torsion).

One can generalize the action (∗) in many ways, obtaining new invariants. In particular, for gauge fields A_μ taking values in the Lie algebra of a compact Lie group, one can construct an analog of (∗) leading to invariants of three-dimensional manifolds closely related to the Jones polynomial of a knot.

Definitions and Notations

Set Theory

The set of points α satisfying the condition (or conditions) Y is denoted by $\{\alpha \mid Y\}$.

If f is a map from A into B, the *image* of A is the set $f(A) = \{f(a) \mid a \in A\}$, and the *inverse image* of a point $x \in B$ is $f^{-1}(x) = \{a \mid a \in A, f(a) = x\}$.

A *transformation* is a bijective map between two sets. (A *map* between topological spaces is always assumed to be continuous.)

The *Cartesian product* of two sets A and B is

$$A \times B = \{(x, y) \mid x \in A, y \in B\}.$$

Linear Algebra

$\mathbf{R}^n$ denotes the n-dimensional real vector space of n-tuples of real numbers, and $\mathbf{C}^n$ the n-dimensional complex vector space of n-tuples of complex numbers.

E^* denotes the dual space to a vector space E, that is, the space of linear functionals on E. If $f \in E^*$ is a linear functional on E, the value of f on x is denoted by $f(x) = \langle f, x \rangle$, and is called the *scalar product* of f and x.

If A is a linear operator from E_1 into E_2, the adjoint $A^\dagger$ of A is a linear operator from E_2^* into E_1^*, defined as follows: $\langle A^\dagger f, x \rangle = \langle f, Ax \rangle$ for $f \in E_2^*$ and $x \in E_1$. If E is a Hilbert space, there is a canonical identification of E with E^*.

The *image* of an operator $A : E \to F$ is $\operatorname{Im} A = \{Ax \mid x \in E\}$.

The *kernel* of an operator $A : E \to F$ is $\operatorname{Ker} A = \{x \mid x \in F, Ax = 0\}$.

The dimension of a vector space E is denoted by $\dim E$.

The *nullity* of an operator A is $l(A) = \dim \operatorname{Ker} A$.

$A_{ik} = \langle i|A|k \rangle$ stands for a matrix entry of the operator A acting on E.

The *trace* of an operator A is $\operatorname{tr} A = \sum \langle i|A|i \rangle = \sum \lambda_i$, where the λ_i are the eigenvalues of A.

Group Theory

$G_1 = G_2$ indicates that the groups G_1 and G_2 are isomorphic. The same notation is used for the (topological) isomorphism of topological groups.

$G_1 \approx G_2$ indicates that the topological groups G_1 and G_2 are locally isomorphic.

A *(left) action* of a group G on a space X is a family of transformations φ_g, for $g \in G$, such that $\varphi_{g_1 g_2} = \varphi_{g_1} \varphi_{g_2}$. A *right action* is similar, but it satisfies $\varphi_{g_2 g_1} = \varphi_{g_1} \varphi_{g_2}$.

Given an action of G on X, the *orbit* of a point $x \in X$ is the set $N_x = \{\varphi_g(x) \mid g \in G\}$, and the *stabilizer* of a point $x \in X$ is the subgroup $H_x = \{g \mid \varphi_g(x) = x, g \in G\}$. An action is *free* is the stabilizer of every point is trivial, that is, $\varphi_g(x) = x$ only if $g = 1$. The set of orbits is denoted by X/G and is called the *quotient* space of X by G.

If H is a subgroup of G, the *right action* of H on G is given by the formula $\varphi_h(g) = gh$, where $g \in G$ and $h \in H$. The quotient space G/H is the set of all orbits of this action (right cosets). If H is a normal subgroup (that is, a subgroup invariant under inner automorphisms $\alpha_g h = ghg^{-1}$), the quotient G/H inherits a group structure, and is called the *quotient group* of G by H.

GL$(n, \mathbf{R})$ and GL$(n, \mathbf{C})$ denote the groups of invertible real and complex n-by-n matrices.

$U(n) = \{a \mid a^\dagger a = a a^\dagger = 1\}$ denotes the group of unitary matrices of order n.

SU$(n) = \{a \mid a^\dagger a = a a^\dagger = 1, \det a = 1\}$ denotes the group of unimodular unitary matrices of order n.

$O(n) = \{a \mid a^T a = a a^T = 1\}$ denotes the group of orthogonal matrices of order n.

SO$(n) = \{a \mid a^T a = a a^T = 1, \det a = 1\}$ denotes the rotation group of order n.

The Lie algebras corresponding to these groups are denoted by $\mathfrak{gl}(n, \mathbf{R})$, $\mathfrak{gl}(n, \mathbf{C})$, $\mathfrak{u}(n)$, $\mathfrak{su}(n)$, $\mathfrak{o}(n)$ and $\mathfrak{so}(n)$. In general, $\mathcal{G}$ will denote the Lie algebra of a group G (Chapter 36).

The adjoint representation of a group G (Chapter 41) is denoted by $\tau_g(x) = gxg^{-1}$, for $g \in G$ and $x \in \mathcal{G}$. The adjoint representation of a Lie algebra $\mathcal{G}$ is denoted by $\sigma_a(x) = [a, x]$, for $a, x \in \mathcal{G}$.

Homotopy Theory

If $A \subset X$ and $B \subset Y$, a map $f : X \to Y$ is said to be a map from the pair (X, A) into the pair (Y, B) if $f(A) \subset B$. If A and B consist of a single point, we talk of maps of pointed spaces of basepoint-preserving maps.

Two maps $f_0, f_1 : X \to Y$ are *homotopic* (as maps from X to Y) if there is a continuous family of maps $f_t : X \to Y$ connecting f_0 and f_1. (In other words, there is a map $F : [0, 1] \times X \to Y$ such that f_0 and f_1 equal the restriction of

F to $\{0\} \times X$ and $\{1\} \times X$, respectively.) Such a family is a *homotopy* between f_0 and f_1. A similar definition applies for maps between pairs of spaces.

The set of homotopy classes of maps from X into Y is denoted by $\{X, Y\}$, and the set of homotopy classes of maps from (X, A) into (Y, B) is denoted by $\{(X, A); (Y, B)\}$.

$S^n = \{x \mid x \in \mathbf{R}^{n+1}, \|x\| = 1\}$ denotes the unit sphere of dimension n, and $s = (-1, 0, \ldots, 0) \in S^n$ its south pole.

We denote by $\pi_n(X, x) = \{(S^n, s); (X, x_0)\}$ the set of homotopy classes of basepoint-preserving maps from the sphere S^n with basepoint s into a space X with basepoint x_0. For $n \geq 1$ the set $\pi_n(X, x_0)$ has a group structure, and is called the *n-th homotopy group* of X (T8.1). The relative homotopy group of the pair (X, A) is denoted by $\pi_n(X, A)$.

Manifolds

A *smooth* map is one whose coordinate functions are differentiable infinitely many times. A *smooth manifold* M is a space that can be covered with coordinate patches (local coordinate systems) such that the change-of-coordinate map between any two overlapping patches is smooth.

An exterior form of degree k, or *k-form*, on M is given in local coordinates by the formula $\omega = \omega_{i_1 \ldots i_k}(x) \, dx^{i_1} \wedge \cdots \wedge dx^{i_k}$, where $\wedge$ is the exterior product of differentials $(dx^i \wedge dx^j = -dx^j \wedge dx^i)$ (T5.1 and T6.3). The *exterior derivative* of ω is $d\omega = d\omega_{i_1 \ldots i_k} \wedge dx^{i_1} \wedge \cdots \wedge dx^{i_k}$. A k-form ω is *closed* if $d\omega = 0$; it is *exact* if there exists a $(k-1)$-form σ such that $\omega = d\sigma$.

The *k-th cohomology group* of M, denoted by $H^k(M) = Z^k(M)/B^k(M)$, is the quotient of the space $Z^k(M)$ of closed k-forms by the space $B^k(M)$ of exact k-forms (T5.2 and T6.3).

The *k-th homology group* of M, denoted by $H_k(M) = Z_k(M)/B_k(M)$, is the quotient of the group of *cycles*, or k-dimensional closed surfaces, by the group of *boundaries*, or closed surfaces that bound a $(k+1)$-dimensional surface (T5.2 and T6.1).

The homology and cohomology groups with coefficients in the abelian group A (T6.1 and T6.2) are denoted by $H_k(M, A)$ and $H^k(M, A)$, respectively.

Fibrations

A *fiber space* or *fibration* (E, B, F, p) is a map $p : E \to B$ such that the inverse image of every point in B is homeomorphic to F. We call B the *base space*, F the *fiber*, p the *projection* and E the *total space* (T9.1). Sometimes E itself is called a fibration. A Cartesian product $E = B \times F$ is a *trivial fibration* with projection $p(b, f) = b \in B$.

A *section* of a fibration (E, B, F, p) is a map $q : B \to E$ such that $q(b) \in p^{-1}(b) = F_b$ for all $b \in B$ (T9.2).

If a group G acts freely on E, the fibration $(E, E/G, G, p)$ is called a *principal fibration* (T9.3).

The *exact homotopy sequence* of a fibration is

$$\cdots \to \pi_n(F, e_0) \to \pi_n(E, e_0) \to \pi_n(B, b_0) \to \pi_{n-1}(F, e_0) \to \cdots ,$$

where e_0 and b_0 are base points in E and B (T11.2). (An *exact sequence* of homeomorphisms is one in which the image of each homeomorphisms coincides with the kernel of the next.)

If G acts on a space F and (E, B, G, p) is a principal fibration, the *associated F-fibration* is the fibration obtained by patching together the products $U_i \times F$ in the same way that E is patched together from the products $U_i \times G$ (T9.4 and T15.1).

Miscellaneous

All functions, maps, manifolds and sections of fibrations are assumed smooth unless we state otherwise. (However, all the assertions remain valid if instead of infinite differentiability we assume differentiability up to a certain order, which depends on the problem. Usually continuous differentiability or just continuity is enough.)

$A_\mu(x)$ denotes a *gauge field* (a covector field with values in the Lie algebra of a gauge group G). The *strength* of the gauge field A_μ is denoted by $\mathcal{F}_{\mu\nu} = \partial_\mu A_\nu - \partial_\nu A_\mu + [A_\mu, A_\nu]$.

The *covariant derivative* of a field φ that transforms according to some representation T of the gauge group G is $\nabla_\mu \varphi = (\partial_\mu + t(A_\mu))\varphi$, where t is the representation of the Lie algebra $\mathcal{G}$ corresponding to T (see Chapter 4 and T15.1).

We adopt throughout the convention that repeated indices are to be summed over. We denote by boldface letters the spatial components of a space-time vector.

The line element of Minkowski space-time is written

$$dx_\mu dx^\mu = (dx^0)^2 - (d\mathbf{x})^2,$$

where $x^0 = t$. The speed of light c is taken to be 1.

Part I

The Basic Lagrangians of
Quantum Field Theory

1. The Simplest Lagrangians

Classical field theory is founded upon the action functional

(1.1)
$$S = \int \mathcal{L}\, dx,$$

where $\mathcal{L}$, the *action density* or *Lagrangian*, is a function of the field variables and of their derivatives, and the integral is over all spatial variables and time. The equations of motion are obtained from the principle of least action: the variation of S must vanish under fixed boundary conditions.

We will consider the simplest Lagrangians that are invariant under the Lorentz group. We start with the Lagrangian of a free scalar field $\varphi(x)$:

(1.2)
$$\mathcal{L} = \tfrac{1}{2}\partial_\mu\varphi\,\partial^\mu\varphi - \tfrac{1}{2}a^2\varphi^2.$$

The corresponding equation of motion, called the *Klein–Gordon equation*, has the form

$$\Box\varphi + a^2\varphi = 0,$$

where $\Box = \partial_\mu\partial_\mu$. The solutions to this equation are linear combinations of plane waves

(1.3)
$$\exp(-ikx) = \exp(-ik^0 x^0 + i\mathbf{k}\cdot\mathbf{x}),$$

with $k^2 = (k^0)^2 - \mathbf{k}^2 = a^2$.

Upon quantization, the Lagrangian (1.2) yields a theory describing particles of mass $\hbar a$; the plane wave (1.3) corresponds to a particle with energy $E = \hbar k^0$ and momentum $\mathbf{p} = \hbar\mathbf{k}$, so that $E^2 - \mathbf{p}^2 = m^2 = \hbar^2 a^2$. Notice that (1.2) does not contain Planck's constant $\hbar$; this constant appears in the mass of the particle only because it occurs in the canonical commutation relations

$$[\hat{\pi}(\mathbf{x}), \hat{\varphi}(\mathbf{x}')] = -i\hbar\delta(\mathbf{x} - \mathbf{x}'),$$

which are postulated in process of quantization. The same remark is true of all the other Lagrangians about to be discussed.

The Klein–Gordon equation is often considered the result of quantization of the Lagrangian of a free relativistic particle. (For this reason the process of quantization of the Klein–Gordon equation is sometimes called second quantization.) In this approach, the Planck constant occurs in the Klein–Gordon equation proper. We do not adhere to this point of view, however. From now on we set $\hbar = 1$.

An electromagnetic field can be described by a potential $A_\mu(x)$, which transforms as a vector under Lorenz transformations. If two fields $A'_\mu(x)$ and $A_\mu(x)$ differ by the gradient of a scalar function λ, that is, if $A'_\mu = A_\mu + \partial_\mu \lambda$, the fields are physically equivalent. The transformation $A_\mu \to A_\mu + \partial_\mu \lambda$ is known as a *gauge transformation*. The tensor $F_{\mu\nu} = \partial_\mu A_\nu - \partial_\nu A_\mu$ is called the *field-strength tensor*; it does not change under gauge transformations. The components F_{0i} of this tensor are identified with the components E_i of the electric field **E**, and the components F_{ij}, for $i, j = 1, 2, 3$, are related to the components H_k of the magnetic field **H** by

$$F_{ij} = \varepsilon_{ijk} H_k.$$

The Lagrangian of an electromagnetic field can then be written in the form

(1.4) $$\mathcal{L} = -\tfrac{1}{4} F_{\mu\nu} F^{\mu\nu}.$$

Taking advantage of gauge invariance, we can impose on the vector potential A_μ additional restrictions such that in each class of physically equivalent fields there exists at least one field satisfying these restrictions. These are known as *gauge conditions*. For instance, we can impose the *Lorentz gauge* condition $\partial_\mu A^\mu = 0$ or the *Coulomb gauge* conditions $A^0 = 0$ and div $\mathbf{A} = 0$. The Coulomb gauge singles out exactly one field in each class of physically equivalent fields (if one requires that fields fall off at infinity); the Lorentz gauge does not possess this property.

The equations of motion corresponding to the Lagrangian (1.4) are given by $\partial_\mu F^{\mu\nu} = 0$, and coincide with the Maxwell equations in a vacuum. In the Coulomb gauge, to each wave vector **k** there correspond two linearly independent plane waves with frequency $k^0 = |\mathbf{k}|$, both satisfying the equations of motion. (A plane wave is characterized by a wave vector **k** and a polarization vector orthogonal to the wave vector.) This implies that, upon quantization, the Lagrangian (1.4) gives a theory describing massless particles (photons). For each value of the momentum the photon can be in two independent states.

Adding to (1.4) the "mass term" $\tfrac{1}{2} a^2 A_\mu A^\mu$, we obtain the Lagrangian

(1.5) $$\mathcal{L} = \tfrac{1}{2}(\partial_\mu A_\nu \, \partial^\mu A^\nu - \partial_\mu A_\nu \, \partial^\nu A^\mu) + \tfrac{1}{2} a^2 A_\mu A^\mu,$$

which describes a massive vector field. This Lagrangian is no longer gauge invariant. Upon quantization, it gives a theory that describes vector particles of mass a; for a given value of the momentum **p** a particle can have three independent states, because for a fixed value of the wave vector there are three independent plane waves satisfying the equations of motion.

We now turn to fields that transform according to two-valued representations of the Lorentz group.

When quantizing such fields one must use canonical anticommutation relations. In fact, strictly speaking, even before quantization these field values should be considered anticommuting. For our purposes, however, this subtlety is almost everywhere inessential.

We start with a two-component spinor field φ^α, that is, a field that transforms according to a two-dimensional complex representation of the Lorentz group. (We do not include reflections in the Lorentz group.) From two spinors, say φ^α and χ^α, we can set up a scalar $\varepsilon_{\alpha\beta}\varphi^\alpha\chi^\beta = \varphi\chi$ and a vector $\varphi^\alpha\sigma^\mu_{\alpha\dot\alpha}\bar\chi^{\dot\alpha} = \varphi\sigma^\mu\bar\chi$, where σ^0 stands for the identity matrix, and σ^1, σ^2 and σ^3 for the Pauli matrices. (The spinor $\overline{\chi^\alpha} = \bar\chi^{\dot\alpha}$, the complex conjugate of χ^α, transforms as a dotted spinor.) Thus, we can write the Lagrangian of φ^α as

(1.6) $\quad \mathcal{L} = \mathrm{Re}(i\varphi\sigma^\mu\,\partial_\mu\bar\varphi + 2a\varphi\varphi) = \dfrac{i}{2}(\varphi\sigma^\mu\,\partial_\mu\bar\varphi - \partial_\mu\varphi\sigma^\mu\bar\varphi) + a(\varphi\varphi + \bar\varphi\bar\varphi).$

The corresponding equations of motion,

$$i\partial_\mu\varphi\sigma^\mu - a\bar\varphi = 0,$$

have solutions in the form of plane waves:

$$\varphi = u\exp(-ikx),$$

with $k^2 = a^2$. Hence, if we quantize Lagrangian (1.6), we get a theory that describes particles of mass $m = a$. Since in the quantization process we employ the canonical anticommutation relations

$$[\varphi^\alpha(\mathbf{x}), \varphi^\beta(\mathbf{x}')]_+ = 0,$$
$$[\varphi^\alpha(\mathbf{x}), \bar\varphi^\beta(\mathbf{x}')]_+ = \delta^{\alpha\dot\beta}\delta(\mathbf{x} - \mathbf{x}'),$$

these particles are fermions.

In quantum field theory, charged particles are described by complex-valued fields: the corresponding Lagrangian must remain unchanged when the field is multiplied by a complex number of absolute value 1. If $a \neq 0$, the Lagrangian (1.6) is not invariant under the substitution $\varphi^\alpha \to e^{i\lambda}\varphi^\alpha$; hence, it describes neutral massive fermions, or *Majorana neutrinos*. For $a = 0$ it describes *Weyl neutrinos*, which are massless.

To construct a Lagrangian that will describe charged fermions we must use two spinor fields, φ^α and χ^α. The resulting *Dirac Lagrangian* has the form

(1.7) $\quad\quad\quad \mathcal{L} = i(\partial_\mu\varphi\,\sigma^\mu\bar\varphi + \partial_\mu\chi\,\sigma^\mu\bar\chi) - m(\varphi\chi + \bar\chi\bar\varphi).$

By introducing the bispinor

(1.8) $\quad\quad\quad\quad\quad\quad \psi = \begin{pmatrix} \varphi^\alpha \\ \bar\chi_{\dot\alpha} \end{pmatrix},$

where $\bar\chi_{\dot\alpha} = \varepsilon_{\dot\alpha\dot\beta}\bar\chi^{\dot\beta}$, we can rewrite the Dirac Lagrangian as

(1.9) $\quad\quad\quad\quad\quad\quad \mathcal{L} = i\bar\psi\gamma^\mu\partial_\mu\psi - m\bar\psi\psi,$

where

(1.10) $\bar{\psi} = \psi^+\gamma^0, \qquad \gamma^0 = \begin{pmatrix} 0 & 1 \\ 1 & 0 \end{pmatrix}, \qquad \gamma = \begin{pmatrix} 0 & -\sigma \\ \sigma & 0 \end{pmatrix}.$

The matrices in (1.10) satisfy

(1.11) $\gamma^\mu\gamma^\nu + \gamma^\nu\gamma^\mu = 2g^{\mu\nu}.$

Four-dimensional matrices γ^μ that satisfy (1.11) are known as *Dirac matrices*. If in (1.9) we replace matrices (1.10) by arbitrary Dirac matrices, we obtain an equivalent Lagrangian.

A two-dimensional complex-valued representation of the Lorentz group can also be seen as a four-dimensional real-valued representation of the same group. In other words, a two-component complex-valued spinor φ^α can be considered as a four-component real-valued spinor, or *Majorana spinor*:

$$\psi = \begin{pmatrix} \mathrm{Re}\,\varphi^1 \\ \mathrm{Re}\,\varphi^2 \\ \mathrm{Im}\,\varphi^1 \\ \mathrm{Im}\,\varphi^2 \end{pmatrix}.$$

In terms of Majorana spinors, we can rewrite the Lagrangian (1.6) in the form (1.9), where the γ^μ are real-valued Dirac matrices.

2. Quadratic Lagrangians

We consider the simplest Lagrangians describing multicomponent fields. We start with the Lagrangian

$$(2.1) \qquad \mathcal{L} = \frac{1}{2}\sum_i \partial_\mu \varphi^i \, \partial^\mu \varphi^i - \frac{1}{2}\sum_{i,j} k_{ij}\varphi^i \varphi^j,$$

which describes an n-component scalar field $\varphi = (\varphi^1, \ldots, \varphi^n)$. One can always diagonalize the quadratic form $\sum_{i,j} k_{ij}\varphi^i \varphi^j$ by means of an orthogonal transformation $\varphi^i \to \sum_j a^i_j \varphi^j$, where (a^i_j) is an orthogonal matrix. This substitution does not affect the kinetic part $\frac{1}{2}\sum_i \partial_\mu \varphi^i \, \partial^\mu \varphi^i$ of $\mathcal{L}$, so (2.1) can be reduced to the simpler form

$$(2.2) \qquad \mathcal{L} = \frac{1}{2}\sum_i (\partial_\mu \varphi^i \, \partial^\mu \varphi^i - \nu_i \varphi^i \varphi^i),$$

where the ν_i are the eigenvalues of the symmetric matrix (k_{ij}). Thus, when all the eigenvalues are nonnegative, the Lagrangian (2.1) describes scalar particles of mass $m_i = \sqrt{\nu_i}$. If there is at least one negative eigenvalue, the Hamiltonian is not bounded below, so the theory cannot be interpreted in terms of particles. (One sometimes says that the theory describes particles of imaginary mass, or *tachyons*, but it must be added that such particles do not exist in nature.)

The quadratic Lagrangian describing a multicomponent vector field,

$$(2.3) \qquad \mathcal{L} = -\frac{1}{4}\sum_i (\partial_\mu A^i_\nu - \partial_\nu A^i_\mu)(\partial^\mu A^{\nu i} - \partial^\nu A^{\mu i}) + \frac{1}{2}\sum_{i,j} k_{ij} A^i_\mu A^{\mu j},$$

can be studied in a similar manner. If the symmetric matrix (k_{ij}) is nonnegative definite and has eigenvalues ν_i, we conclude by diagonalizing that the Lagrangian describes particles of mass $\sqrt{\nu_i}$.

Next we consider the quadratic Lagrangian describing n spinor fields,

$$(2.4) \qquad \mathcal{L} = \frac{i}{2}\sum_j (\chi_j \sigma^\mu \partial_\mu \bar{\chi}_j - \partial_\mu \chi_j \sigma^\mu \bar{\chi}_j) + \sum_{i,j} (M_{ij}\chi_i \chi_j + \bar{M}_{ij}\bar{\chi}_i \bar{\chi}_j),$$

where the M_{ij} form a symmetric complex matrix (the mass matrix). By applying a transformation $\chi_i \to U_{ij}\chi^j$, where $U = (U_{ij})$ is a unitary matrix, we can reduce (2.4) to the form

$$(2.5) \qquad \mathcal{L} = \frac{i}{2} \sum_j (\chi_j \sigma^\mu \partial_\mu \bar{\chi}_j - \partial_\mu \chi_j \sigma^\mu \bar{\chi}_j) + \sum_j m_j (\chi_j \chi_j + \bar{\chi}_j \bar{\chi}_j),$$

where the m_j are the square roots of the eigenvalues of $M^\dagger M$. We show the existence of a diagonalizing matrix in the next paragraph; here we just observe that the transformation $\chi_i \to U_{ij} \chi_j$ does not change the first sum in (2.4), and that it has the effect of replacing the mass matrix $M = (M_{ij})$ by $U^T M U$. Thus $M^\dagger M$ becomes $U^\dagger M^\dagger U^{T\dagger} U^T M U = U^{-1} M^\dagger M U$, so its eigenvalues do not change, and we conclude that the m_j can be interpreted as masses.

We must still show that the quadratic form $\mathcal{M}(\chi) = \sum_{i,j} M_{ij} \chi_i \chi_j$ on $\mathbf{C}^n$ can be reduced to the form $\sum_i m_i \chi_i \chi_i$ by a unitary transformation. We identify $\mathbf{C}^n$ with $\mathbf{R}^{2n}$, via the correspondence $(u_1 + iv_1, \ldots, u_n + iv_n) \mapsto (u_1, v_1, \ldots, u_n, v_n)$, and we consider the quadratic form

$$\begin{aligned} \mathcal{N}(x) &= \mathcal{M}(x) + \overline{\mathcal{M}(x)} \\ &= M_{kj}(u_k + iv_k)(u_j + iv_j) + M_{kj}(u_k - iv_k)(u_j - iv_j) \end{aligned}$$

on $\mathbf{R}^{2n}$. Multiplication by i in $\mathbf{C}^n$ gives in $\mathbf{R}^{2n}$ a linear operator J satisfying $J^2 = -1$, and taking $(u_1, v_1, \ldots, u_n, v_n) \in \mathbf{R}^{2n}$ to $(-v_1, u_1, \ldots, -v_n, u_n)$. The standard basis vectors in $\mathbf{R}^{2n}$ are $e_1, Je_1, \ldots, e_n, Je_n$, where $\{e_1, \ldots, e_n\}$ is the standard basis of $\mathbf{C}^n$. Since $\mathcal{M}(ix) = -\mathcal{M}(x)$, we have

$$(2.6) \qquad\qquad\qquad \mathcal{N}(Jx) = -\mathcal{N}(x).$$

The standard Hermitian scalar product in $\mathbf{C}^n$ corresponds to the standard scalar product in $\mathbf{R}^{2n}$, and unitary operators in $\mathbf{C}^n$ are orthogonal operators in $\mathbf{R}^{2n}$ that commute with J. Using the scalar product in $\mathbf{R}^{2n}$, we can write the quadratic form $\mathcal{N}(x)$ as $\mathcal{N}(x) = \langle Nx, x \rangle$, where N is a self-adjoint operator in $\mathbf{R}^{2n}$. From (2.6) it follows that $NJ = -JN$, which means that for each eigenvector x of N with an eigenvalue λ there is an eigenvector Jx with eigenvalue $-\lambda$. A standard reasoning then implies that there exists a complete orthonormal set of eigenvectors of N, consisting of the vectors $x_1, Jx_1, \ldots, x_n, Jx_n$, corresponding to eigenvalues $\lambda_1, -\lambda_1, \ldots, \lambda_n, -\lambda_n$, with each λ_i nonnegative. The orthogonal operator that maps the standard basis vectors $e_1, Je_1, \ldots, e_n, Je_n$ to the eigenvectors $x_1, Jx_1, \ldots, x_n, Jx_n$ of N commutes with J, and therefore represents the desired unitary transformation in $\mathbf{C}^n$.

3. Internal Symmetries

By a *symmetry* we mean a transformation of fields that leaves the action functional unchanged. A symmetry is *internal* if it affects only the values of the fields and not the coordinate variables.

Take, for instance, the Lagrangian of a system of n *massless scalar fields* $\varphi^1(x), \ldots, \varphi^n(x)$, or, equivalently, of an *n-component scalar field* $\varphi(x) = (\varphi^1(x), \ldots, \varphi^n(x))$:

$$(3.1) \qquad \mathcal{L}_b = \frac{1}{2} \sum_{i=1}^{n} \partial_\mu \varphi^i \, \partial^\mu \varphi^i = \tfrac{1}{2} \langle \partial_\mu \varphi, \partial^\mu \varphi \rangle.$$

(The angle brackets stand for the standard scalar product in $\mathbf{R}^n$.) This Lagrangian, and hence the corresponding action functional, do not change under a transformation of the type

$$(3.2) \qquad \varphi(x) \mapsto \varphi'(x) = A\varphi(x),$$

where $A = (a_j^i)$ is an orthogonal matrix. Thus (3.1) possesses the group of internal symmetries $O(n)$. In Chapter 2 we used this fact to reduce (2.1) to (2.2).

The Lagrangian

$$(3.3) \qquad \mathcal{L} = \mathcal{L}_b - \tfrac{1}{2} m^2 \sum_{i=1}^{n} (\varphi^i)^2,$$

which describes n noninteracting scalar fields of equal mass, and the Lagrangian

$$(3.4) \qquad \mathcal{L} = \mathcal{L}_b - \lambda \left(\sum_{i=1}^{n} (\varphi^i)^2 - a^2 \right)^2,$$

also possess the group of internal symmetries $O(n)$.

More generally, the Lagrangian of an n-component scalar field invariant under Lorentz transformations can be written as

$$(3.5) \qquad \mathcal{L} = \tfrac{1}{2} \langle \partial_\mu \varphi, \partial^\mu \varphi \rangle - U(\varphi),$$

where $U(\varphi) = U(\varphi^1, \ldots, \varphi^n)$ is a function on $\mathbf{R}^n$. (This encompasses (3.3) and (3.4) as particular cases, with $U(\varphi) = \tfrac{1}{2} m^2 \langle \varphi, \varphi \rangle$ and $U(\varphi) = \lambda (\langle \varphi, \varphi \rangle - a^2)^2$, respectively.) This Lagrangian has internal symmetry group $G \subset O(n)$ if $U(\varphi)$

is a G-invariant function on $\mathbf{R}^n$, that is, if $U(g\varphi) = U(\varphi)$ for all $g \in G$ and $\varphi \in \mathbf{R}^n$.

In the examples above, the action functional was defined on fields taking values in $\mathbf{R}^n$. We can also speak of an internal symmetry group when the action functional is defined on M-valued fields, where M is a manifold. Any homeomorphism of M gives rise to a transformation of fields; a group G acting on M is an *internal symmetry group* if the action functional remains unchanged under the field transformations corresponding to the elements of G. In particular, if an n-component scalar field $\varphi(x)$ can take on values only on the unit sphere $\langle \varphi, \varphi \rangle = 1$, the Lagrangian

$$(3.6) \qquad \mathcal{L} = \tfrac{1}{2} \langle \partial_\mu \varphi, \partial^\mu \varphi \rangle$$

has internal symmetry group $O(n)$.

Next, we can speak of internal symmetries not only of with scalar fields, but also of fields that transform by other representations of the Lorentz group. For instance, we have seen that the Dirac Lagrangian (1.9) is invariant under the transformation

$$(3.7) \qquad \psi(x) \mapsto \psi'(x) = e^{i\alpha}\psi(x),$$

and so has internal symmetry group $U(1)$. To the symmetry (3.7) corresponds an integral of motion $Q = \int d^3 x \, \bar{\psi}\gamma^0\psi$, which has the physical meaning of charge.

The Lagrangian

$$\mathcal{L}_f = \frac{i}{2} \sum_{j=1}^{m} (\xi_j \sigma^m \partial_\mu \bar{\xi}_j - \partial_\mu \xi_j \sigma^\mu \bar{\xi}_j),$$

where the ξ_i are (two-component) spinor fields, is obviously invariant under the transformation $\xi_i \to u_{ij}\xi_j$, where (u_{ij}) is a unitary matrix. Thus, the internal symmetry group for Lagrangian (3.8) can be taken as $U(m)$. This fact was used when we analyzed Lagrangian (2.4).

The Lorentz-invariant Lagrangian describing n scalar fields $\varphi^1, \ldots, \varphi^n$ and m spinor fields $\xi_1, \ldots, \xi_m$ can be written in the form

$$(3.9) \qquad \mathcal{L} = \mathcal{L}_b + \mathcal{L}_f + \mathcal{L}_{\text{int}},$$

where $\mathcal{L}_b$ and $\mathcal{L}_f$ are defined by (3.1) and (3.8) and

$$(3.10) \qquad \mathcal{L}_{\text{int}} = \sum_{i,j,k} \Gamma_{ijk}\xi_i\xi_j\varphi^k - U(\varphi).$$

If we impose the additional condition that the quantum field corresponding to (3.10) be renormalizable, we must assume that $U(\varphi)$ is a polynomial of degree no higher than four:

$$(3.11) \qquad U(\varphi) = a_i\varphi^i + b_{ij}\varphi^i\varphi^j + c_{ijk}\varphi^i\varphi^j\varphi^k + d_{ijkl}\varphi^i\varphi^j\varphi^k\varphi^l.$$

Note that the coefficients Γ_{ijk}, b_{ij}, ... , in (3.10) and (3.11) can be assumed to satisfy the symmetry conditions

$$\Gamma_{ijk} = -\Gamma_{jik}, \quad b_{ij} = b_{ji}, \quad \ldots$$

If $\mathcal{L}_{\text{int}} = 0$, the Lagrangian (3.9) has internal symmetry group $O(n) \times U(m)$, consisting of orthogonal transformations of the scalar fields and unitary transformations of the spinor fields. If $\mathcal{L}_{\text{int}} \neq 0$, the symmetry group is the subgroup of $O(n) \times U(m)$ that leaves $\mathcal{L}_{\text{int}}$ invariant.

The question often arises of how to describe Lagrangians for which the internal symmetry group is isomorphic to a given group G, such as $\mathrm{SU}(2)$, for instance.

First, G must be realized as a subgroup of $O(n) \times U(m)$, that is, we must construct an isomorphism from G into $O(n) \times U(m)$. Suppose we have an n-dimensional real and an m-dimensional complex representation of G, that is, homomorphisms $T_1 : G \to O(n)$ and $T_2 : G \to U(m)$. Then we can construct a homomorphism $G \to O(n) \times U(m)$, by taking $g \in G$ to $(T_1(g), T_2(g))$. Thus we associate with g an internal symmetry of the Lagrangian $\mathcal{L}_b + \mathcal{L}_f$, in which scalar fields transform according to T_1 and spinor fields according to T_2.

It may occur that for some $g \neq 1$ both images $T_1(g)$ and $T_2(g)$ are the identity, that is, the associated internal symmetry is trivial. In this case the homomorphism from G into $O(n) \times U(m)$ is not an isomorphism.

If the representations T_1 and T_2 are fixed, one can establish, using group-theory considerations, what form the Lagrangian $\mathcal{L}_{\text{int}}$ must have in order to be invariant by all the transformations $T_1(g)$ and $T_2(g)$. In other words, if we know how the scalar and spinor fields φ and ξ transform, we can find out all G-invariant expressions in these variables.

For example, take $G = \mathrm{SO}(3)$, and let the scalar field transform by a three-dimensional irreducible representation of $\mathrm{SO}(3)$, and the spinor field by the direct sum of two three-dimensional irreducible representations. In other words, the scalar fields $\varphi = (\varphi^1, \varphi^2, \varphi^3)$ form a three-dimensional vector, and the spinor fields form two three-dimensional vectors, $\mathbf{L}_1 = (L_1^1, L_1^2, L_1^3)$ and $\mathbf{L}_2 = (L_2^1, L_2^2, L_2^3)$. From two vectors, say $\mathbf{A}_1$ and $\mathbf{A}_2$, one can construct a unique (to within a factor) scalar $(\mathbf{A}_1, \mathbf{A}_2) = \sum_a A_1^a A_2^a$ that depends linearly on $\mathbf{A}_1$ and $\mathbf{A}_2$; from three vectors $\mathbf{A}_1, \mathbf{A}_2, \mathbf{A}_3$, one can construct a unique scalar $\varepsilon_{abc} A_1^a A_2^b A_3^c$ that depends linearly on $\mathbf{A}_1, \mathbf{A}_2, \mathbf{A}_3$ (this scalar will be antisymmetric in the three vectors); and finally, there is a unique scalar $(\mathbf{A}, \mathbf{A})^2$ that is a fourth-degree polynomial in the components of a vector $\mathbf{A}$. Using these facts, one can easily write the most general form of $\mathcal{L}_{\text{int}}$ for this case:

$$\mathcal{L}_{\text{int}} = a\varphi^2 + \sum_i b_i(\varphi, \mathbf{L}_i) + \sum_{i,j} c_{ij}(\mathbf{L}_i, \mathbf{L}_j) + d\varepsilon_{abc}\varphi^a L_1^b L_2^c + f\varphi^4 .$$

In general, to establish the most general form that a G-invariant expression of type (3.11) can take, we must find out how many times the trivial (scalar) representation of G appears in the decomposition of $T_1^{(2)}$, $T_1^{(3)}$ and $T_1^{(4)}$ into irreducible representations, where $T_1^{(k)}$ denotes the k-th symmetric tensor power of T_1. To specify all the ways to construct the remaining terms in $\mathcal{L}_{\text{int}}$, one must solve a similar problem for the representations $T_2^{(2)}$ and $T_1 \otimes T_2^{(2)}$.

Lagrangians having symmetry group $U(1)$ are easily characterized. Indeed, every complex representation of $U(1)$ can be decomposed into one-dimensional irreducible representations. This means that, after applying a unitary transformation, we can ensure that the spinor fields $\xi_1, \ldots, \xi_m$ transform as

$$(3.12) \qquad \xi_j \mapsto e^{in_j\alpha}\xi_j$$

under the action of $e^{i\alpha} \in U(1)$, where $\alpha \in \mathbf{R}$ and each n_j is an integer, called the $U(1)$-*charge* of ξ_j.

A quadratic Lagrangian that contains spinor fields and is invariant with respect to $U(1)$ can be written as

$$(3.13) \qquad \mathcal{L} = \mathcal{L}_f + \left(\sum_{i,j} M_{ij}\xi_i\xi_j + \text{c.c.} \right),$$

where $M_{ij} \neq 0$ only if $n_i + n_j = 0$. This can be decomposed as a sum of Lagrangians $\mathcal{L}_n$, where each $\mathcal{L}_n$ involves only fields whose $U(1)$-charge has absolute value n.

To reduce $\mathcal{L}_n$ to its simplest form, we use complex conjugate fields (that is, *dotted* spinors) with positive $U(1)$-charges, instead of spinor fields with negative $U(1)$-charges. Thus we have, for $n > 0$:

$$(3.14) \quad \mathcal{L}_n = \frac{i}{2}\sum(\xi_j\sigma^\mu\,\partial_\mu\bar{\xi}_j - \partial_\mu\xi_j\,\sigma^\mu\bar{\xi}_j)$$
$$+ \frac{i}{2}\sum(\bar{\chi}_k\sigma^\mu\,\partial_\mu\chi_k - \partial_\mu\bar{\chi}_k\,\sigma^\mu\chi_k) + \sum(a_{jk}\xi_j\bar{\chi}_k + \text{c.c.}),$$

where the ξ_j are spinors and the χ_j are dotted spinors.

Using well-known algebraic facts, one can find a unitary transformation that takes the matrix (a_{jk}) into a diagonal matrix with non-zero elements (the unitary transformation being performed on ξ_j and χ_k separately).

To prove this, look at (a_{jk}) as representing a linear operator A from a complex vector space E into a complex vector space E'. We are looking for orthonormal bases $\{e_j\}$ for E and $\{e'_k\}$ for E' such that A, expressed with respect to these bases, is diagonal. For E we select an orthonormal basis consisting of eigenvectors of the nonnegative self-adjoint operator $A^\dagger A$. If $A^\dagger Ae_j = \lambda_j e_j$, with λ_j positive, we set $e'_j = \lambda_j^{-1/2}Ae_j$. These new vectors are orthogonal to each other and normalized, because

$$\langle e'_i, e'_j \rangle = \frac{1}{\sqrt{\lambda_i\lambda_j}}\langle Ae_i, Ae_j \rangle = \frac{1}{\sqrt{\lambda_i\lambda_j}}\langle A^\dagger Ae_i, e_j \rangle = \frac{1}{\sqrt{\lambda_i\lambda_j}}\lambda_i\delta_{ij} = \delta_{ij}.$$

We obtain the necessary orthonormal basis for E' by adjoining to the e'_j a sufficient number of normalized vectors orthogonal to the e'_j and among themselves.

If (a_{ij}) is a square matrix, that is, if the original Lagrangian contained an equal number of fields with $U(1)$-charges equal to n and $-n$, we conclude that

$\mathcal{L}_n$ can be represented as a sum of several Dirac Lagrangians. But if (a_{ij}) is not square, the reduced form of Lagrangian (3.14) will comprise, in addition to Dirac Lagrangians, other Lagrangians describing massless (Weyl) neutrinos.

Thus, if a fermion has some type of $U(1)$-charge (for example, an electric, baryonic or leptonic charge), it is described either by the Dirac equation (if massive) or by the Weyl equation (if massless). A fermion may be thought of as being a massive Majorana neutrino only if it carries no $U(1)$-charge. This explains the special role played by the Dirac and Weyl equations in elementary particle physics.

Every real-valued representation of $U(1)$ breaks down into a direct sum of one-dimensional trivial and two-dimensional irreducible representations. Moreover, two-dimensional real representations of $U(1)$ are in obvious correspondence with one-dimensional complex representations. Thus, given a real-valued representation of $U(1)$ describing the transformation of scalar fields, we can choose new fields φ^j that transform by the rule

$$\varphi^j \mapsto \begin{cases} e^{in_j\alpha}\varphi^j & \text{for } j = 1, \ldots, r, \\ \varphi^j & \text{for } j > r \end{cases}$$

under the action of $e^{i\alpha} \in U(1)$. Each new field φ^j, for $1 \leq j \leq r$, should be thought of as a complex scalar field, and n_j is called its $U(1)$-charge, just as in the case of a fermion field. The remaining scalar fields are real-valued; they don't change under $U(1)$, and their $U(1)$-charge is zero.

A Lorentz-invariant Lagrangian that contains spinor and scalar fields and is invariant under $U(1)$ can be expressed by a formula similar to (3.13); one must only make sure that each term in the Lagrangian contains a product of fields whose $U(1)$-charges add up to zero.

4. Gauge Fields

As noted before, the Dirac Lagrangian (1.9) is invariant under the transformation $\psi'(x) = U\psi(x)$, where $U = e^{i\alpha}$ (in other words, a transformation in $U(1)$, the internal symmetry group of this Lagrangian). It is not invariant under transformations of the form $\psi'(x) = U(x)\psi(x)$, where $U = e^{i\alpha(x)}$ is a function taking values on the unit circle. However, one can consider the Lagrangian of a bispinor field interacting with an electromagnetic field $A_\mu(x)$; this Lagrangian is already invariant under the transformation $\psi'(x) = U(x)\psi(x)$ if the transformation law for the electromagnetic field is chosen as

$$A'_\mu(x) = A_\mu(x) + \frac{1}{e}\partial_\mu\alpha(x),$$

where e is the electric charge.

There exists a simple general rule that makes it possible to switch on the interaction with the electromagnetic field: namely, we must replace ∂_μ with $\nabla_\mu = \partial_\mu - ieA_\mu$ and add $-\frac{1}{4}F_{\mu\nu}F^{\mu\nu}$ to the Lagrangian, where $F_{\mu\nu} = \partial_\mu A_\nu - \partial_\nu A_\mu$ is the electromagnetic field strength tensor. This rule can always be applied if we are dealing with a complex-valued field and the Lagrangian does not change as a result of multiplying the field by $e^{i\alpha}$. For example, the Lagrangian of a complex-valued scalar field $\varphi(x)$ interacting with an electromagnetic field $A_\mu(x)$ has the form

$$\mathcal{L} = \tfrac{1}{2}(\partial_\mu\varphi - ieA_\mu\varphi)(\partial^\mu\varphi^* + ieA^\mu\varphi^*) - \tfrac{1}{2}m^2\varphi\varphi^* - \tfrac{1}{4}F_{\mu\nu}F^{\mu\nu}.$$

Now let G be a group of n-dimensional matrices, and consider a Lagrangian having internal symmetry group G, that is, invariant under symmetries of type $\varphi'(x) = g\varphi(x)$, where $\varphi(x)$ is an n-component field and $g \in G$. Starting from $\mathcal{L}$, we construct a new Lagrangian $\tilde{\mathcal{L}}$ invariant under transformations of a broader class,

(4.1) $$\varphi'(x) = g(x)\varphi(x),$$

where $g(x)$ is a G-valued function. More precisely, in full analogy with the standard procedure for introducing an electromagnetic field, we must replace ∂_μ with $\nabla_\mu = \partial_\mu + eA_\mu(x)$, where $A_\mu(x)$ is a matrix-valued vector field and ∇_μ is the covariant derivative. The transformation law for $A_\mu(x)$ must be selected in such a way that

(4.2) $$\nabla'_\mu\varphi'(x) = g(x)\nabla_\mu\varphi(x),$$

with $\nabla'_\mu = \partial_\mu + eA'_\mu(x)$. One easily checks that the transformation law must be

$$(4.3) \qquad A'_\mu = gA_\mu g^{-1} - \frac{1}{e}\partial_\mu g\, g^{-1}.$$

Condition (4.2) implies that the new Lagrangian $\tilde{\mathcal{L}}$ is invariant under the transformations (4.1) and (4.3).

Note that we need not take arbitrary matrix-valued functions as candidates for $A_\mu(x)$, but only functions taking values in the Lie algebra $\mathcal{G}$ of G. For example, if $G = U(n)$ is the group of all unitary matrices, $A_\mu(x)$ takes on values in the set of anti-Hermitian matrices. Equation (4.3) should also be interpreted as taking place in the Lie algebra $\mathcal{G}$; indeed, if $g(x)$ is a function with values in G, the matrix $\partial_\mu g(x)\, g^{-1}(x)$ belongs to $\mathcal{G}$, and if $a \in \mathcal{G}$, we have $gag^{-1} \in \mathcal{G}$. We can easily verify directly that (4.3) transforms an anti-Hermitian field $A_\mu(x)$ into another such field.

Thus, starting from a Lagrangian $\mathcal{L}$ having internal symmetry group G, and replacing ∂_μ with $\nabla_\mu = \partial_\mu + eA_\mu$, where $A_\mu(x)$ is a vector field taking values in the Lie algebra $\mathcal{G}$, we have constructed a Lagrangian $\tilde{\mathcal{L}}$ invariant under the *local gauge transformations* (4.1) and (4.3). $A_\mu(x)$ is known as a *gauge field*, or *Yang–Mills field*, and represents a generalization of the electromagnetic field. The electromagnetic field can be considered, to within a factor of $-i$, as a gauge field for the group $U(1)$.

However, if the gauge field $A_\mu(x)$ is to be dynamic, like the electromagnetic field, we must add to $\tilde{\mathcal{L}}$ an expression that contains the derivatives of $A_\mu(x)$ and is invariant under gauge transformations. To construct such an expression we note that the commutator of two covariant derivatives can be written as

$$(4.4) \qquad [\nabla_\mu, \nabla_\nu]\varphi = e\mathcal{F}_{\mu\nu}\varphi,$$

where $\mathcal{F}_{\mu\nu} = \partial_\mu A_\nu - \partial_\nu A_\mu + e[A_\mu, A_\nu]$ is known as the strength of the gauge field $A_\mu(x)$. Equations (4.3) and (4.4) imply that $\mathcal{F}'_{\mu\nu} = g\mathcal{F}_{\mu\nu}g^{-1}$, so that

$$\mathcal{L}_{\mathrm{YM}} = \tfrac{1}{4}\operatorname{tr}\mathcal{F}_{\mu\nu}\mathcal{F}^{\mu\nu}$$

is gauge-invariant. Physically, $\mathcal{L}_{\mathrm{YM}}$ stands for the Lagrangian of the gauge field $A_\mu(x)$. It must be added to $\tilde{\mathcal{L}}$ so that we obtain a Lagrangian $\hat{\mathcal{L}}$ describing the field $\varphi(x)$ and its interaction with the field $A_\mu(x)$. One says that the Lagrangian $\hat{\mathcal{L}}$ is obtained from $\mathcal{L}$ by *localization* of the internal symmetry group G, which is called the *gauge group*.

It is convenient to include the factor e, the "charge", in the gauge field. Then the covariant derivative becomes $\nabla_\mu = \partial_\mu + A_\mu$, the gauge field strength becomes

$$F_{\mu\nu} = \partial_\mu A_\nu - \partial_\nu A_\mu + [A_\mu, A_\nu],$$

and an additional factor e^{-2} appears in the expression for $\mathcal{L}_{\mathrm{YM}}$.

We now look at a simple but important example that illustrates the construction above. Let $\mathcal{L}$ be the Lagrangian (1.9), describing an n-component bispinor field. As we know, $\mathcal{L}$ is invariant under the transformation $\psi'(x) = g\psi(x)$, for $g \in U(n)$. By localizing the $U(n)$-symmetry, we get

$$(4.5) \qquad \hat{\mathcal{L}} = i\bar{\psi}\gamma^\mu(\partial_\mu + A_\mu)\psi - m\bar{\psi}\psi + \frac{1}{4e^2}\operatorname{tr}\mathcal{F}_{\mu\nu}\mathcal{F}^{\mu\nu},$$

where A_μ belongs to the Lie algebra of $U(n)$, that is, it is anti-Hermitian.

It is not necessary to start with the full internal symmetry group of the Lagrangian when localizing; we can use any subgroup. For instance, for Lagrangian (1.9) we could limit ourselves to localizing the $SU(n)$ symmetry. Then in (4.5) we must assume that the matrices A_μ belong to the Lie algebra of $SU(n)$, that is, that they satisfy $\operatorname{tr} A_\mu = 0$ in addition to being anti-Hermitian.

So far we have assumed that the elements of the internal symmetry group G are matrices acting on the space V where the multicomponent field φ takes its values. As already noted, it is often convenient to assume instead that G is an abstract group, and that we have a representation T of G in V, with the Lagrangian $\mathcal{L}$ being invariant under transformations $\varphi(x) \mapsto \varphi'(x) = T(g)\varphi(x)$. Then the procedure for constructing $\hat{\mathcal{L}}$ changes in the following manner:

By a covariant derivative we now understand the expression $\nabla_\mu = \partial_\mu + t(A_\mu)$, where t is the representation of the Lie algebra $\mathcal{G}$ of G corresponding to T. The gauge-field Lagrangian must be written as

$$\mathcal{L}_{\text{YM}} = -\frac{1}{4e^2}\langle\mathcal{F}_{\mu\nu}, \mathcal{F}^{\mu\nu}\rangle,$$

where the angle brackets stand for the invariant scalar product in $\mathcal{G}$, which means that $\langle\tau_g a, \tau_g b\rangle = \langle a, b\rangle$, for $a, b \in \mathcal{G}$, $g \in G$, and τ_g the adjoint representation of G. (For a matrix group we have $\tau_g a = gag^{-1}$, and the invariant scalar product can be taken as $\langle a, b\rangle = -\operatorname{tr} ab$.) It follows from the transformation law $\mathcal{F}'_{\mu\nu}(x) = \tau_{g(x)}\mathcal{F}_{\mu\nu}(x)$ that $\hat{\mathcal{L}}$ is invariant under the local gauge transformations

$$\varphi'(x) = T(g(x))\varphi(x),$$
$$A'_\mu(x) = \tau_{g(x)}A_\mu(x) - \partial_\mu g(x)g^{-1}(x).$$

When G is a direct product of two groups G_1 and G_2, its Lie algebra is the direct sum of the two Lie algebras $\mathcal{G}_1$ and $\mathcal{G}_2$. This makes it possible, knowing the invariant scalar products $\langle\,,\,\rangle_1$ and $\langle\,,\,\rangle_2$ in $\mathcal{G}_1$ and $\mathcal{G}_2$, to set up a two-parameter family of scalar products in $\mathcal{G}$, namely,

$$\langle x, y\rangle = \lambda_1\langle x_1, y_1\rangle_1 + \lambda_2\langle x_2, y_2\rangle_2,$$

where x_1 and x_2 are the projections of $x \in \mathcal{G}$ onto $\mathcal{G}_1$ and $\mathcal{G}_2$, and similarly for y_1 and y_2. Accordingly, the Lagrangian of the gauge field with values in $\mathcal{G} = \mathcal{G}_1\dot{+}\mathcal{G}_2$ contains two coupling constants:

$$\mathcal{L}_{\text{YM}} = -\frac{1}{4e_1^2}\langle(\mathcal{F}_1)_{\mu\nu}, (\mathcal{F}_1)^{\mu\nu}\rangle_1 - \frac{1}{4e_2^2}\langle(\mathcal{F}_2)_{\mu\nu}, (\mathcal{F}_2)^{\mu\nu}\rangle_2.$$

In general, the Lie algebra of a compact Lie group G breaks up into a direct sum of simple Lie algebras; the number of parameters on which the scalar product in $\mathcal{G}$ depends, and consequently the number of coupling constants in the gauge-field Lagrangian, is equal to the number of elements in this decomposition.

5. Particles Corresponding to Nonquadratic Lagrangians

We consider the Lagrangian

(5.1)
$$\mathcal{L} = \frac{1}{2}\sum_i \partial_\mu\varphi^i\,\partial^\mu\varphi^i - W(\varphi),$$

describing a multicomponent scalar field, and put it in the form

(5.2)
$$\mathcal{L} = \mathcal{L}_0 - V(\varphi) = \frac{1}{2}\sum_i \partial_\mu\varphi^i\,\partial^\mu\varphi^i - \frac{1}{2}\sum_{i,j} k_{ij}\varphi^i\varphi^j - V(\varphi),$$

where V is a polynomial in φ containing terms of order three and higher. In Chapter 2 we showed that if the matrix (k_{ij}) is nonnegative definite, $\mathcal{L}_0$ describes particles whose squared masses are the eigenvalues of this matrix. The term $V(\varphi)$ can be taken into account by perturbation techniques; although it does not change the spectrum qualitatively, it nevertheless introduces corrections to the masses of the particles.

If (k_{ij}) is not nonnegative definite, $\mathcal{L}_0$ cannot serve as a meaningful initial approximation for the perturbation approach. In this case it is reasonable to replace the fields $\varphi^i(x)$ by new fields $\chi^i(x) = \varphi^i(x) - c^i$, where the constants c^i are chosen so that the quadratic part of $\mathcal{L}$ in the new fields does become nonnegative definite, and so allows an interpretation in terms of particles. It is enough to choose for $c = (c^1, \ldots, c^n)$ a point where $W(\varphi)$ is minimal; such a point is known as a *classical vacuum*. For example, if $W(\varphi) = a\varphi^2 + b\varphi^4$, with $a < 0$, we take $c = \pm\sqrt{-a/2b}$. Setting $\chi(x) = \varphi(x) - c$, we get

(5.3)
$$\mathcal{L} = \tfrac{1}{2}\partial_\mu\chi\,\partial^\mu\chi + 2a\chi^2 \mp \sqrt{-8ab}\,\chi^3 - b\chi^4 + \frac{a^2}{4b},$$

which, to order zero, describes particles of mass $m^2 = -4a$.

There may be many classical vacuums: this phenomenon is usually related to symmetry breaking. Symmetry breaking arises when not all transformations that leave the Lagrangian invariant map a classical vacuum into itself. For example, the polynomial $W(\varphi) = a\varphi^2 + b\varphi^4$ is invariant under the transformation $\varphi \mapsto -\varphi$, but for $a < 0$ this is a broken symmetry, since it permutes the classical vacuums $\pm\sqrt{-a/2b}$.

As in the example of (5.3), so in the general case we can find the particle spectrum by looking at the quadratic part of the expansion of the Lagrangian in powers of the deviation of the field from a classical vacuum. But if the theory contains gauge fields, we must impose gauge conditions in advance, which lift the gauge freedom in the definition of a classical vacuum. Consider, for example, the Lagrangian

$$(5.4) \qquad \mathcal{L} = |\partial_\mu \varphi - ieA_\mu \varphi|^2 - \lambda(|\varphi|^2 - a^2)^2 - \tfrac{1}{4}F_{\mu\nu}^2,$$

which describes a complex scalar field φ interacting with the electromagnetic field A_μ. Assuming $\lambda > 0$, the condition for a classical vacuum is $|\varphi| = a$, that is, $\varphi = ae^{i\alpha}$. All these vacuums are gauge-equivalent, so we can impose a gauge condition that lifts the degeneracy. One straightforward condition is

$$(5.5) \qquad \operatorname{Im}\varphi = 0, \qquad \operatorname{Re}\varphi > 0.$$

Then (5.4) becomes

$$(5.6) \qquad \mathcal{L} = (\partial_\mu \chi)^2 - \lambda\chi^2(\chi + 2a)^2 - \tfrac{1}{4}F_{\mu\nu}^2 + e^2 A_\mu^2(\chi + a)^2,$$

where $\chi(x) = \varphi(x) - a$. We see that, to order zero, $\mathcal{L}$ describes a scalar particle of mass $m = 2a\sqrt{\lambda}$ and a vector particle of mass $\sqrt{2}ea$. By contrast, for $\lambda = 0$ the Lagrangian (5.4) describes two scalar particles, with charges $\pm e$, and one massless vector particle, the photon. For $\lambda > 0$ the $U(1)$-symmetry breaks down, the vector particle becomes massive, and there remains only one scalar particle; this is a manifestation of the Higgs effect. This effect also occurs in non-abelian gauge theories; in Chapter 6 we discuss this in greater detail, using the Weinberg–Salam model.

The procedure described above for finding the particle spectrum by choosing an appropriate initial approximation and employing perturbation techniques is based on the assumption that the corrections do not radically alter the spectrum of the system. This, however, is not always the case. In quantum chromodynamics (QCD) (Section 1.6), the procedure leads to the prediction that fermion and gauge fields correspond to particles—quarks and gluons. These particles, however, are not observed in experiments as free particles. Nevertheless, the study of quarks and gluons has proved to be extremely fruitful at high momenta. The explanation is that, in QCD, the higher the momentum, the better the perturbation-theory approximation (asymptotic freedom).

6. Lagrangians of Strong, Weak and Electromagnetic Interactions

Hadrons can be considered as consisting of quarks. Each quark is characterized by color—yellow, blue or red—and flavor. At present six flavors are known, and are commonly called up (u), down (d), strange (s), charmed (c), bottom (b) and top (t). Quarks are described by a multicomponent bispinor field ψ_a^i, where the superscript denotes the color (and therefore takes three values) and the subscript the flavor (six values).

The Lagrangian describing free quarks has the form

$$(6.1) \qquad \mathcal{L} = \sum_{k,a}(i\bar{\psi}_a^k \gamma^\mu \partial_\mu \psi_a^k - m_a \bar{\psi}_a^k \psi_a^k).$$

Since the mass of a quark depends only on its flavor, not on its color, this Lagrangian is invariant under rotations in "color space," that is, transformations $\psi_a^i \mapsto \psi_a^{i\prime} = g_j^i \psi_a^j$, where (g_j^i) is a unitary 3×3 matrix. Thus, (6.1) is invariant under $U(3)$, and *a fortiori* under SU(3). It is postulated that strong interactions are described by the Lagrangian $\mathcal{L}_{QCD}$ obtained from (6.1) by localizing the SU(3) symmetry (color symmetry), namely

$$(6.2)$$
$$\mathcal{L}_{QCD} = \sum_{n,k,a}(i\bar{\psi}_a^k \gamma^\mu(\partial_\mu \delta_n^k + (A_\mu)_n^k)\psi_a^n - m_a \bar{\psi}_a^k \psi_a^k) + \frac{1}{4g^2}\sum_{k,n}(\mathcal{F}_{\mu\nu})_n^k(\mathcal{F}^{\mu\nu})_k^n$$
$$= \sum_a i\bar{\psi}_a \gamma^\mu \nabla_\mu \psi_a - \sum_a m_a \bar{\psi}_a \psi_a + \frac{1}{4g^2}\operatorname{tr}\mathcal{F}_{\mu\nu}\mathcal{F}^{\mu\nu}.$$

Here A_μ takes values in the Lie algebra of SU(3), that is, $(A_\mu)_n^m = -\overline{(A_\mu)_m^n}$ and $\sum_n(A_\mu)_n^n = 0$, and

$$(6.3) \qquad (F_{\mu\nu})_n^m = \partial_\mu(A_\nu)_n^m - \partial_\nu(A_\mu)_n^m + \sum_k((A_\mu)_k^m(A_\nu)_n^k - (A_\nu)_k^m(A_\mu)_n^k).$$

The particles corresponding to the field A_μ are known as *gluons*. Thus, the central hypothesis of QCD is that the interaction between quarks is carried by eight vector particles, the gluons.

Often, instead of working with bispinor fields ψ_a^i, it is convenient to use spinor fields L_a^i and R_i^a such that ψ_a^i decomposes into L_a^i and the dotted spinor complex conjugate to R_i^a. Under the action of the color group SU(3), the spinor L_a^i transforms like ψ_a^i (that is, like a vector), but R_i^a transforms like a complex conjugate quantity. (For the group $SU(n)$, the complex conjugate of a vector can also be seen as a covector, that is, a vector with subscripts instead of superscripts.)

We now turn to the *Weinberg–Salam theory*, which unifies weak and electromagnetic interactions. We start its construction by describing bosons. We take two complex scalar fields φ^1 and φ^2, with the Lagrangian

$$(6.4) \qquad \begin{aligned} \mathcal{L}_S &= \partial_\mu \varphi^1 \, \overline{\partial^\mu \varphi^1} + \partial_\mu \varphi^2 \, \overline{\partial^\mu \varphi^2} - \lambda(|\varphi^1|^2 + |\varphi^2|^2 - \eta^2)^2 \\ &= \langle \partial_\mu \varphi, \partial^\mu \varphi \rangle - \lambda(\langle \varphi, \varphi \rangle - \eta^2)^2. \end{aligned}$$

We consider the following action of $U(2)$: a unitary matrix $u \in U(2)$ transforms a vector $\varphi = (\varphi^1, \varphi^2)$ into ψ', with $\psi'^i = u^i_j \varphi^j$. (In other words, φ is an $SU(2)$-doublet and carries a $U(1)$-charge of $+1$.) Clearly, $\mathcal{L}_S$ is invariant under this action.

Now, using the localization procedure, we switch on the interaction of φ with a gauge field taking values in the Lie algebra of $U(2)$ (the algebra $u(2)$ of all anti-Hermitian matrices). The resulting Lagrangian is

$$(6.5) \qquad \tilde{\mathcal{L}} = \langle \nabla_\mu \varphi, \nabla^\mu \varphi \rangle - \lambda(\langle \varphi, \varphi \rangle - \eta^2)^2 + \mathcal{L}_{\text{YM}},$$

where $\nabla_\mu \varphi = \partial_\mu \varphi + A_\mu \varphi$, with A_μ an anti-Hermitian matrix, and the field $\varphi = (\psi^1, \psi^2)$ is written as a column. Now $U(2)$ is not simple, since $U(2) \approx U(1) \times SU(2)$; in terms of Lie algebras, this says that every element of $u(2)$ (an anti-Hermitian matrix) is the sum of a scalar matrix with a traceless matrix. Therefore, the Lagrangian of the gauge field with values in $u(2)$ contains two coupling constants, g_1 and g_2:

$$(6.6) \qquad \mathcal{L}_{\text{YM}} = \frac{1}{4g_1^2} f_{\mu\nu} f^{\mu\nu} + \frac{1}{2g_2^2} \operatorname{tr} G_{\mu\nu} G^{\mu\nu},$$

where $f_{\mu\nu}$ is the scalar part and $G_{\mu\nu}$ the traceless part of $\mathcal{F}_{\mu\nu}$:

$$(\mathcal{F}_{\mu\nu})^m_n = \tfrac{1}{2} f_{\mu\nu} \delta^m_n + (G_{\mu\nu})^m_n.$$

It is convenient to take the matrices $\tfrac{1}{2}i$, $\tfrac{1}{2}i\tau^1$, $\tfrac{1}{2}i\tau^2$ and $\tfrac{1}{2}i\tau^3$ as generators of $U(2)$, where the τ^i are the Pauli matrices. The first matrix generates the subgroup of scalar matrices, and the other three generate $SU(2)$. If we write A_μ in terms of these generators,

$$A_\mu = \tfrac{1}{2}ib_\mu + \tfrac{1}{2}i\mathbf{c}_\mu \boldsymbol{\tau},$$

the Lagrangian (6.5) becomes

$$\begin{aligned} \tilde{\mathcal{L}} = {} & |(\partial_\mu + \tfrac{1}{2}ib_\mu + \tfrac{1}{2}i\mathbf{c}_\mu \boldsymbol{\tau})\varphi|^2 - \lambda(|\varphi^1|^2 + |\varphi^2|^2 - \eta^2)^2 \\ & - \frac{1}{4g_1^2}(\partial_\mu b_\nu - \partial_\nu b_\mu)^2 - \frac{1}{4g_2^2}(\partial_\mu \mathbf{c}_\nu - \partial_\nu \mathbf{c}_\mu + [\mathbf{c}_\mu, \mathbf{c}_\nu])^2. \end{aligned}$$

In the theory described by Lagrangian (6.5) there is spontaneous symmetry breaking. Indeed, the minimum of the function $U(\varphi) = \lambda(\langle \varphi, \varphi \rangle - \eta^2)^2$ is attained at $\langle \varphi, \varphi \rangle = \eta^2$. Choosing $\varphi_0 = \binom{0}{\eta}$ as a classical vacuum, we see that the group of symmetries that do not alter the classical vacuum (the unbroken symmetry group) consists of matrices of the form $\left(\begin{smallmatrix} a & 0 \\ 0 & 1 \end{smallmatrix}\right)$, for $|a| = 1$, and therefore is isomorphic to $U(1)$. As a generator of the unbroken symmetry group we can take $\frac{1}{2}i(1 + \tau^3)$. Bearing in mind that fields related by a gauge transformation are physically equivalent, we conclude that we can impose the conditions $\varphi^1 = 0$ and $\operatorname{Im} \varphi^2 = 0$ to single out one field in each gauge equivalence class.

Expanding Lagrangian (6.5) in powers of the deviation from the classical vacuum φ_0, and retaining only the quadratic terms, we obtain

$$(6.7) \quad \mathcal{L}_{\mathrm{qu}} = \left(\frac{\partial \chi}{\partial x_\mu}\right)^2 - 4\lambda\chi^2\eta^2 + \tfrac{1}{4}(c_\mu^3 - b_\mu)^2\eta^2 + \tfrac{1}{2}\eta^2 c_\mu^- c_\mu^+ + \mathcal{L}_{\mathrm{YM}}$$

$$= (\partial_\mu\chi)^2 - 4\lambda\eta^2\chi^2 + \tfrac{1}{4}(c_\mu^3 - b_\mu)^2\eta^2 + \tfrac{1}{2}\eta^2 c_\mu^- c_\mu^+$$

$$- \frac{1}{4g_1^2}(\partial_\mu b_\nu - \partial_\nu b_\mu)^2 - \frac{1}{4g_2^2}(\partial_\mu \mathbf{c}_\nu - \partial_\nu \mathbf{c}_\mu)^2,$$

where $c_\mu^\pm = \frac{1}{\sqrt{2}}(c_\mu^1 \pm ic_\mu^2)$ and $\chi = \varphi^2 - \eta$.

To find the masses of the particles described by the quadratic Lagrangian (6.7), we must diagonalize it. This can be done by introducing the following fields:

$$W_\mu^\pm = \frac{1}{\sqrt{2}g_2}(c_\mu^1 \pm ic_\mu^2) = \frac{1}{g_2}c_\mu^\pm,$$

$$Z_\mu = \frac{c_\mu^3 - b_\mu}{\sqrt{g_1^2 + g_2^2}},$$

$$A_\mu = \frac{g_1 g_2^{-1}c_\mu^3 + g_2 g_1^{-1}b_\mu}{\sqrt{g_1^2 + g_2^2}}.$$

Now (6.7) takes the form

$$(6.8) \quad \mathcal{L}_{\mathrm{qu}} = (\partial_\mu\chi)^2 - 4\lambda^2\eta^2\chi^2 - \tfrac{1}{2}(\partial_\mu W_\nu^- - \partial_\nu W_\mu^-)(\partial_\mu W_\nu^+ - \partial_\nu W_\mu^+)$$

$$+ \tfrac{1}{2}g_2^2\eta^2 W_\mu^- W_\mu^+ - \tfrac{1}{4}(\partial_\mu A_\nu - \partial_\nu A_\mu)^2$$

$$- \tfrac{1}{4}(\partial_\mu Z_\nu - \partial_\nu Z_\mu)^2 + \tfrac{1}{4}\eta^2(g_1^2 + g_2^2)Z_\mu^2.$$

Thus, the Lagrangian (6.7) describes a scalar particle of mass $2\lambda\eta$, a massless vector particle (photon), and three massive vector particles, one with mass $\frac{1}{\sqrt{2}}\eta\sqrt{g_1^2 + g_2^2}$ and two with mass $\frac{1}{\sqrt{2}}\eta g_2$.

Notice that there are two $U(1)$-charges involved here: one, associated with the generator $\frac{1}{2}i(1 + \tau^3)$ of the unbroken symmetry group, is the usual electric charge; the other is associated with the $U(1)$ from the direct-factor decomposition of $U(2) = U(1) \times \mathrm{SU}(2)$, and is called the *hypercharge*. The photon and the particle corresponding to the field Z_μ, known as the Z-boson, are electrically neutral. The vector particles corresponding to the fields $W_\mu^\pm$, known as W-bosons, have electric charge ± 1.

From (6.8) and (6.5) it follows that the electromagnetic coupling constant is $e = g_1 g_2 / \sqrt{g_1^2 + g_2^2}$. Often one considers, instead of the two coupling constants g_1

and g_2, the electromagnetic coupling constant e and the so-called *Weinberg angle* θ_W, defined by $e = g_1 \cos\theta_W = g_2 \sin\theta_W$; it has been determined experimentally that $\theta_W \approx 20°$.

We now examine how to incorporate fermions in the Weinberg–Salam model. To describe the representation of $U(2) \approx \mathrm{SU}(2) \times U(1)$, according to which fermion fields transform, it is convenient to study the behavior of these fields under the action of $\mathrm{SU}(2)$ and $U(1)$ separately. We recall that irreducible representations of $\mathrm{SU}(2)$ are characterized by the dimension, and those of $U(1)$ by the $U(1)$-charge, an integer.

It turns out that leptons are described in the Weinberg–Salam model by spinor fields that are $\mathrm{SU}(2)$-doublets or singlets. The $U(1)$-charges of these fields (hypercharges) are $Y_{\mathrm{doubl}} = -1$ and $Y_{\mathrm{sing}} = 2$. Considering that there is one doublet l^a and one singlet r in the model, the most general Lorentz-invariant Lagrangian having internal symmetry group $U(2)$ and describing the interaction of the spinor fields l^a and r with the scalar field φ^a is

$$(6.9) \qquad \mathcal{L}_{\mathrm{int}} = f(l^a r \bar\varphi_a + \bar l_a \bar r \varphi^a),$$

where $\bar\varphi_a$, $\bar l_a$ and $\bar r$ are the fields complex conjugate to φ^a, l^a and r. Here we have assumed the interaction Lagrangian to be quadratic in the spinor fields and linear in the scalar fields: see (3.10). The total Lagrangian of the spinor and scalar fields is

$$(6.10) \qquad \mathcal{L} = \mathcal{L}_f + \mathcal{L}_b + \mathcal{L}_{\mathrm{int}} + \mathcal{L}_{\mathrm{self\text{-}int}},$$

where $\mathcal{L}_f$ and $\mathcal{L}_b$ are the massless spinor and scalar field Lagrangians, and $\mathcal{L}_{\mathrm{self\text{-}int}} = -\lambda(\varphi^2 - \eta^2)^2$ is the self-interaction Lagrangian of the scalar field. The Lagrangian in the one-generation Weinberg–Salam model can be obtained from (6.10) by localizing the $U(2)$ internal symmetry group, that is, by replacing the derivatives with covariant derivatives and adding the gauge-field Lagrangian (6.6). Note that the covariant derivative of l^a can be written as

$$(\nabla_\mu l)^a = ((\partial_\mu + \tfrac{1}{2} i\boldsymbol\tau \mathbf{c}_\mu - \tfrac{1}{2} i b_\mu) l)^a,$$

and the covariant derivative of r as

$$\nabla_\mu r = (\partial_\mu + i b_\mu) r.$$

As before, we impose the gauge conditions $\varphi^1 = 0$ and $\mathrm{Im}\,\varphi^2 = 0$. Expanding the Weinberg–Salam Lagrangian in powers of the deviation $\chi = \varphi^2 - \eta$ of the field from the classical vacuum, and retaining only the quadratic terms, we obtain the particle spectrum. The boson part of the spectrum was described earlier. To establish the fermion spectrum, we notice that the mass term is

$$(6.11) \qquad f\eta l^2 r + \text{c.c.}$$

This implies that the field l^1 describes a massless particle (identified with the electron neutrino), while l^2 and r together define a bispinor field corresponding to a Dirac particle of mass $f\eta$ (identified with the electron).

We have described the Weinberg–Salam model for a single generation of leptons: the electron and electron neutrino. Other leptons—the muon and muon neutrino, and the antiparticles of all of these—can be incorporated in a straightforward way. For this we must assume that the model contains spinor fields l_1^a, l_2^a and l_3^a, for $a = 1, 2$, which are doublets in SU(2), and spinor fields r_1, r_2 and r_3, which are singlets in SU(2). The $U(1)$-charges of these fields must be assumed equal to -1 and $+2$, respectively. The Lagrangian giving the interaction of the spinor fields l_k^a and r_a with the scalar field φ^a must be chosen as

$$(6.12) \qquad \mathcal{L}_I = \sum_{i=1}^{3} f_i (l_i^a r_i \bar{\varphi}_a + \text{c.c.}).$$

By repeating the reasoning and the procedure above, we obtain a Lagrangian that describes three types of massless particles (neutrinos) and three types of charged particles with masses f_1, f_2 and f_3.

Formula (6.12) does not give the most general Lagrangian for the interaction of the fields l_k^a, r_k and φ^a. We can also consider an interaction of the type

$$(6.13) \qquad \mathcal{L}_I = \sum_{i,k=1}^{3} f^{ik} l_i^a r_k \bar{\varphi}_a + \text{c.c.},$$

where f^{ik} is an arbitrary complex matrix. However, a unitary transformation of variables reduces (6.13) to (6.12) because, as shown in Chapter 3, the matrix f^{ik} can be diagonalized by multiplication on the right and on the left by unitary matrices. This means that the Weinberg–Salam model does not mix different generations of leptons.

This restriction is lifted if we make the neutrino massive. We can do this by introducing spinor fields s_1, s_2 and s_3 that are singlets in SU(2) and have $U(1)$-charge equal to zero.

We now turn to theories that account simultaneously for strong, weak and electromagnetic interactions. In such a theory the gauge group must contain the group SU(3) of strong interactions and the Weinberg–Salam group SU(2)×$U(1)$. Thus SU(3) × SU(2) × $U(1)$ is the smallest possible gauge group. It breaks down to SU(3) × $U(1)$ via a doublet of complex-valued scalar fields φ^1 and φ^2, which transforms like a two-dimensional vector under transformations belonging to $U(2) \approx$ SU(2) × $U(1)$, and does not change under transformations belonging to SU(3). Hence, the boson part of the SU(3) × SU(2) × $U(1)$ theory differs from the corresponding part in the model of electroweak interaction only by the presence of eight gauge fields corresponding to SU(3)—the gluon fields.

The fermion part of the the SU(3) × SU(2) × $U(1)$ model contains first of all leptons, fields that are present already in the Weinberg–Salam model. They do not change under transformations in SU(3), that is, they are SU(3) singlets. There are also quark fields, which transform by the vector or covector representation of SU(3) (representations **3** and **3̄**.) If quarks are represented by bispinor fields ψ_a^i, all of them transform according to the vector representation of SU(3); but for us it is more convenient to work with the spinor fields L_a^i and R_i^a.

Quark fields may be SU(2)-doublets or singlets: those that transform by the vector representation of SU(3) are SU(2)-doublets and have a $U(1)$-charge of $\frac{1}{3}$, while those that transform by the covector representation are SU(2)-singlets (that is, SU(2)-invariant) and their $U(1)$-charge may be $-\frac{4}{3}$ or $\frac{2}{3}$. To summarize, quark fields transform according to one of the following representations: $(\mathbf{3}, \mathbf{2}, \frac{1}{3})$, $(\bar{\mathbf{3}}, \mathbf{1}, \frac{2}{3})$, or $(\bar{\mathbf{3}}, \mathbf{1}, -\frac{4}{3})$, where the first number gives the transformation law with respect to SU(3), the second the transformation law with respect to SU(2), and the third the $U(1)$-charge.

The fact that the $U(1)$-charges of quarks are fractional and multiples of $\frac{1}{3}$ is due to convention; one could always make all charges integral by renormalizing the generator of $U(1)$.

In the minimal set of quark fields necessary to explain the existing experimental data, each of the multiplets $(\mathbf{3}, \mathbf{2}, \frac{1}{3})$, $(\bar{\mathbf{3}}, \mathbf{1}, \frac{2}{3})$ and $(\bar{\mathbf{3}}, \mathbf{1}, -\frac{4}{3})$ appears three times. In other words, there are three generations of quarks. Thus, the set of quark fields consists of the spinors $L_1^{a\alpha}$, $L_2^{a\alpha}$, $L_3^{a\alpha}$, R_{a1}, R_{a2}, R_{a3}, $\tilde{R}_{a1}$, $\tilde{R}_{a2}$ and $\tilde{R}_{a3}$, where $a = 1, 2, 3$ is the SU(3)-index, or color, $\alpha = 1, 2$ is the SU(2)-index, and the $U(1)$-charges of the fields L, R and $\tilde{R}$ are $\frac{1}{3}$, $-\frac{4}{3}$ and $\frac{2}{3}$.

To construct the Lagrangian of the SU(3) $\times$ SU(2) $\times$ $U(1)$ model we must write the most general Lagrangian $\mathcal{L}$ that is invariant under SU(3)$\times$SU(2)$\times U(1)$ and describes the interactions of all the spinor fields in the model and the fields φ^α, and then switch on the interaction with the gauge fields, that is, localize the SU(3) $\times$ SU(2) $\times$ $U(1)$ symmetry. The part of $\mathcal{L}$ responsible for the lepton fields and for φ was constructed above, so we need only describe the part responsible for the interaction of the quark fields with φ.

Invariance under SU(3) $\times$ SU(2) $\times$ $U(1)$ reduces the possibilities to $L^{a\alpha} R_a \varphi_\alpha$ and $L^{a\alpha} \tilde{R}_a \bar{\varphi}_\alpha$ (and their complex conjugates), where $\varphi_\alpha = \varepsilon_{\alpha\beta}\varphi^\beta$ and $\bar{\varphi}_\alpha = \overline{\varphi^\alpha}$. (Notice that $\bar{\varphi}$ transforms by the same representation of SU(2) as φ, but has opposite $U(1)$-charge.) Therefore the desired interaction Lagrangian is

$$(6.14) \qquad \mathcal{L}_{\text{int}} = \sum_{i,k}(A_{ik} L_i^{a\alpha} R_{ak}\varphi_\alpha + B_{ik} L_i^{a\alpha} \tilde{R}_{ak}\bar{\varphi}_\alpha) + \text{c.c.}$$

Using unitary transformations

$$L_i^{a\alpha} \to U_i^k L_k^{a\alpha}, \qquad R_{ai} \to V_i^k R_{ak}, \qquad \tilde{R}_{ai} \to W_i^k \tilde{R}_{ak},$$

which do not alter the free part of the Lagrangian, we can simplify the matrices A_{ik} and B_{ik}. As discussed in Chapter 2, A_{ik} can be diagonalized, so it takes the form $A_{ij} = a_i \delta_{ij}$. Next, by transforming the fields $\tilde{R}$ appropriately, we can make the matrix B_{ij} Hermitian; this is because any matrix can be written as the product of a Hermitian matrix and a unitary one. But even after doing this, we can still simplify B_{ij} by performing transformations of the type $L_k^{a\alpha} \to u_k L_k^{a\alpha}$, $R_{ak} \to u_k^{-1} R_{ak}$, and $\tilde{R}_{ak} \to u_k^{-1} \tilde{R}_{ak}$, where the u_k are complex numbers of absolute value 1.

For the case of two generations, we can then make the matrix (B_{ij}) real—the diagonal entries are already real because the matrix is Hermitian, so there is only one other complex entry $B_{12} = \bar{B}_{21}$ that needs to be made real, and this is possible because we have two independent parameters at our disposal. In the case of three generations B cannot generally be made real, because it includes three complex parameters, $B_{12} = \bar{B}_{21}$, $B_{13} = \bar{B}_{31}$ and $B_{23} = \bar{B}_{32}$.

To analyze what particles are described by this model, we must, as usual, expand the Lagrangian in a power series in the deviation from the classical vacuum, retain only the quadratic part, and diagonalize it. For the boson part and for lepton fields this was done when we examined the Weinberg–Salam model. To see what particles correspond to the quark fields $L_k^{a\alpha}$, R_{ak} and $\tilde{R}_{ak}$, we must diagonalize the relevant mass term, which is obtained from (6.14) by replacing the field φ^α with its vacuum value $(0, \eta)$, and which has the form

$$(6.15) \qquad \mathcal{M} = \sum_{i,k=1}^{3} \left(A_{ik} L_i^{a1} R_{ak} \eta + B_{ik} L_i^{a2} \tilde{R}_{ak} \eta \right) + \text{c.c.}$$

For the case of two generations, this equation becomes (replacing the summation limit by 2):

$$(6.16) \quad \mathcal{M} = a_1 L_1^{a1} R_{a1} \eta + a_2 L_2^{a1} R_{a2} \eta + b_{11} L_1^{a2} \tilde{R}_{a1} \eta$$
$$+ b_{12} (L_1^{a2} \tilde{R}_{a2} + L_2^{a2} \tilde{R}_{a1}) \eta + b_{22} L_2^{a2} \tilde{R}_{a2} \eta + \text{c.c.}$$

(We assume that the matrix $A_{ik} = a_i \delta_{ik}$ is diagonal and that B_{ik} is Hermitian and real.) We combine the spinors L_1^{a1}, L_1^{a2}, L_2^{a1} and L_2^{a2} with the spinors complex conjugate to R_{a1}, $\tilde{R}_{a1}$, R_{a2} and $\tilde{R}_{a2}$ to form the bispinors u^a, d'^a, c^a and s'^a. Then (6.16) becomes

$$\mathcal{M} = a_1 \eta \bar{u} u + a_2 \eta \bar{c} c + b_{11} \eta \bar{d}' d' + b_{12} \eta (\bar{d}' s' + \bar{s}' d') + b_{22} \eta \bar{s}' s'.$$

Next we change from d' and s' to the bispinors d and s given by

$$d = d' \cos \theta_C - s' \sin \theta_C, \qquad s = d' \sin \theta_C - s' \cos \theta_C,$$

where θ_C is selected in such a way that the mass term, expressed in terms of u, c, d and s, is diagonal. Now the fields u, c, d and s correspond to the physical quarks. The angle θ_C, which characterizes generation mixing, is known as the Cabibbo angle, and experiments place it at $\theta_C \approx 13°$.

7. Grand Unifications

The $SU(3) \times SU(2) \times U(1)$ model describes strong, weak and electromagnetic interactions, but it cannot be considered a unified theory for all these interactions, because each of the groups $SU(3)$, $SU(2)$ and $U(1)$ has its own coupling constant. A true unification of all three interactions is achieved in the so-called *grand unification* theories, which, although duplicating much of the Weinberg–Salam and $SU(3) \times SU(2) \times U(1)$ models, involve a simple gauge group and therefore a single coupling constant.

The possibility of having only one coupling constant arises because in quantum field theory the effective coupling constant depends on the momentum. More precisely, in studying the scattering matrix, even to the lowest order of approximation one cannot use the "bare" coupling constant that occurs in the Lagrangian; one must replace it by a corrected number that depends on the characteristic momentum of the particle participating in the scattering process.

Within the perturbation-theory framework, the semiclassical approximation corresponds to using tree diagrams. (An expansion in powers of $\hbar$ corresponds to grouping perturbation diagrams by the number of closed loops.) However, the use of only tree diagrams with bare vertices proves to be insufficient, although one often can make do with tree diagrams containing "heavy" vertices, that is, vertices in which the bare coupling constant is replaced with a vertex function depending on the momenta of the incoming lines. If all the momenta in the diagram have the same order of magnitude, say p, we can assume that each vertex has the same number (depending on p, of course). This means we can employ the semiclassical approximation—that is, tree diagrams, or, for the next order, diagrams with one loop—assuming that the coupling constant depends on p. In Chapter 25 we will study this dependence in greater detail.

In the $SU(3) \times SU(2) \times U(1)$ model, the dependence of the effective coupling constants on the momentum is such that, for a certain value of the momentum (roughly 10^{15} GeV), these constants approach each other. It is therefore assumed that the true gauge group is larger than $SU(3) \times SU(2) \times U(1)$, but that below a certain energy the symmetry breaks down to $SU(3) \times SU(2) \times U(1)$ (and below energies of about 10^2 GeV it breaks down to $SU(3) \times U(1)$).

There are many choices for the gauge group G underlying a grand unification theory. The smallest admissible group is $SU(5)$, which gives rise to the model we discuss now.

The boson part of this model consists of two multicomponent scalar fields, φ and χ, which transform according to the adjoint (24-dimensional) and vector (five-dimensional) representations of SU(5). Thus, φ can be considered as a traceless Hermitian matrix, and χ as a column vector. Of course, the boson part of the model also incorporates vector gauge fields, which take on values in the Lie algebra of SU(5).

The interaction potential of scalar fields is selected in such a way that the classical vacuum can be chosen in the form $\varphi = a\varphi_0$ and $\chi = b\chi_0$, where

$$
(7.1) \qquad \varphi_0 = \begin{pmatrix} -\frac{2}{3} & & & & \\ & -\frac{2}{3} & & \text{\Large 0} & \\ & & -\frac{2}{3} & & \\ & \text{\Large 0} & & 1 & \\ & & & & 1 \end{pmatrix}, \qquad \chi_0 = \begin{pmatrix} 0 \\ 0 \\ 0 \\ 0 \\ 1 \end{pmatrix},
$$

and a and b are real. The unbroken symmetry group corresponding to this choice of the classical vacuum is the subgroup of SU(5) consisting of block-diagonal matrices of the form

$$
(7.2) \qquad \begin{pmatrix} K & & \\ & l & \\ & & 1 \end{pmatrix},
$$

where K is a 3×3 matrix. Obviously, K is a unitary matrix, $|l| = 1$ and $l \det K = 1$; in other words, $K = e^{-i\alpha}A$ and $l = e^{3i\alpha}$, for $A \in$ SU(3). This implies that the unbroken symmetry group H is locally isomorphic to SU(3) $\times$ U(1). (If $\alpha = 2\pi n/3$ and $A = e^{2\pi n i/3}$, we have $K = 1$ and $l = 1$; hence H is the quotient of SU(3) $\times$ U(1) by a subgroup of order three.)

However, the constants $a \sim 10^{15}$ GeV and $b \sim 10^2$ GeV are selected in such a way that at high energies the symmetry breaking associated with the field χ becomes inessential. At these energies the unbroken symmetry group becomes bigger: it equals H_0 of all matrices $h \in$ SU(5) that commute with φ_0, that is, all matrices of the from

$$
(7.3) \qquad h = \begin{pmatrix} M & 0 \\ 0 & N \end{pmatrix},
$$

where M is a 3×3 matrix and N is a 2×2 matrix. The group H_0 is locally isomorphic to SU(3) $\times$ SU(2) $\times$ U(1), since $h = h_0 e^{i\alpha\varphi_0}$, where $h_0 = \begin{pmatrix} M_0 & \\ & N_0 \end{pmatrix}$ for unimodular 3×3 and 2×2 matrices M_0 and N_0, and φ_0 is defined by (7.1), that is, is a generator of U(1).

The fermion part of the SU(5) model consists of the spinor fields ψ_a and κ^{ab}, for $a, b = 1, 2, \ldots, 5$. Under the action of SU(5) the field φ_a transforms as a covector (a vector with subscripts instead of superscripts), and κ^{ab} as an anti-symmetric tensor. In other words, ψ_a transforms by the representation $\bar{\mathbf{5}}$ and κ^{ab} by $\mathbf{10}$. We can easily decompose these representations in SU(3) $\times$ SU(2) $\times$ U(1), as follows:

The field ψ_a breaks up into the irreducible components ψ_α and ψ_τ. (We reserve the indexes α, β, γ for the first three coordinates and σ, τ for the last two.) Clearly, ψ_α is an SU(3)-triplet (more precisely, it transforms according to the covector representation $\bar{\mathbf{3}}$), an SU(2)-singlet, and has $U(1)$-charge (hypercharge) $\frac{2}{3}$. Meanwhile, ψ_τ is an SU(3)-singlet, an SU(2) doublet, and has $U(1)$-charge -1.

Next, the field κ^{ab} has irreducible components $\kappa^{\alpha\beta} = \varepsilon^{\alpha\beta\gamma}\rho_\gamma$, $\kappa^{\alpha\tau}$ and $\kappa^{\tau\sigma} = \varepsilon^{\tau\sigma}\nu$. It is clear that ρ_γ is an SU(3) covector, an SU(2) singlet and has $U(1)$-charge $-\frac{4}{3}$; while $\kappa^{\alpha\tau}$ is an SU(3) vector, an SU(2) doublet and has $U(1)$-charge $\frac{1}{3}$. Finally, ν is a singleton with respect to both SU(3) and SU(2), and has $U(1)$-charge 2.

We see that the representations of $\mathrm{SU}(3) \times \mathrm{SU}(2) \times U(1)$ obtained as a result of decomposing the representations of SU(5) that appear in the SU(5) model coincide with the representations by which quarks and leptons transform in the $\mathrm{SU}(3) \times \mathrm{SU}(2) \times U(1)$ model. From this we can conclude that, at energies E in the range from 100 to 10^{15} GeV, the SU(5) model reduces to the $\mathrm{SU}(3) \times \mathrm{SU}(2) \times U(1)$ model. Further symmetry breaking via the field χ, as we saw above, decreases the symmetry group to $\mathrm{SU}(3) \times U(1)$.

Part II

Topological Methods in
Quantum Field Theory

8. Topologically Stable Defects

We now discuss applications of topology to the classification of defects, or violations of local equilibrium, in condensed media. These applications derive from the fact that, in many important cases, the state of thermodynamic equilibrium is degenerate at temperatures below a certain critical temperature.

For example, if the temperature of a ferromagnetic material is less than the so-called *Curie point* T_c, spontaneous magnetization sets in and the equilibrium magnetization vector $\mathbf{M}$ can take different directions—in fact, any direction, if the material is isotropic. Only the length of $\mathbf{M}$ is determined by the temperature, $|\mathbf{M}| = M(T)$. Thus, for $T < T_c$, the space of possible equilibrium states of an isotropic ferromagnetic material, or *degeneracy space*, is the two-dimensional sphere S^2.

In an anisotropic ferromagnetic material of the easy-plane type, $\mathbf{M}$ must lie on a certain plane, and its length is, as before, given by the temperature. In this case the degeneracy space is a circle S^1. Finally, for a ferromagnetic material of the easy-axis type, $\mathbf{M}$ can point in two opposite directions; the degeneracy space has two points, and can be thought of as a zero-dimensional sphere S^0.

In general, the *degeneracy space* R of a system is the set of equilibrium states at a fixed temperature T. This is a topological space, because there is a notion of equilibrium states being close to each other. One can formalize the notion of closeness, but we will not do it here because in each concrete situation it will be clear what it means.

Usually the structure of the degeneracy space can be studied by applying the Landau theory of second-order phase transitions. Recall that an equilibrium state is a state with minimum free energy. The Landau theory assumes that an equilibrium state can be found by minimizing the free energy not over the set of all states, but over a set of states defined by a finite number of *order parameters*. For example, the magnetization vector $\mathbf{M}$ is an order parameter for a ferromagnetic material. In the isotropic case, the symmetry makes it possible to conclude that the free energy is a function of $|\mathbf{M}|^2$ and T. For $T \geq T_c$ this minimum is attained for $\mathbf{M} = 0$, while for $T < T_c$ it is attained for values of $\mathbf{M}$ such that $|\mathbf{M}|^2 = M^2(T) > 0$. For a uniaxial crystal the free energy's anisotropic term has the form KM_z^2, in appropriate coordinates. If K is positive, the minimum in the potential energy occurs for $M_z = 0$; since the free energy is invariant under rotations about the z-axis, for $T < T_c$ the set of minima is a circle in the xy-plane, and the material is of the easy-plane type. If K is negative, the free energy attains its minimum for $\mathbf{M}$ along the z-axis itself, and the material is of the easy-axis type.

Usually the degeneracy of an equilibrium state is connected with symmetry breaking. For instance, the energy functional for an isotropic ferromagnetic material is invariant under spatial rotations, that is, elements of SO(3). But equilibrium states are characterized by the vector $\mathbf{M}$, and therefore are not invariant under SO(3), but only under a proper subgroup, isomorphic to SO(2).

In general, we denote by G the group of transformations that leave the energy functional invariant, and by H the subgroup of transformations that leave invariant a particular equilibrium state $e \in R$. Thus H is the *stabilizer* of e (in mathematical terminology), or the *unbroken symmetry group* of e (in physical terminology). Any transformation in G takes any equilibrium state into another, because equilibrium states are defined as those that minimize the free energy, which is invariant under G.

A common situation occurs when any two equilibrium states can be obtained from one another by the action of a transformation in G: we say then that G acts *transitively*. Physically, this means that the degeneracy of an equilibrium state is determined solely by symmetry properties. If this is the case, R can be identified with the quotient space G/H (see Chapter 40).

We now consider a body, or region thereof, that is in *local thermodynamic equilibrium*. In this case we can still talk about the body's temperature, but the temperature changes from point to point. If the equilibrium state is degenerate, it is not only the temperature that depends on the point, but also the other parameters characterizing the equilibrium state—for example, the magnetization vector of a ferromagnetic material. It is natural to assume that this dependence is continuous.

We first show that, possessing information on the local equilibrium state in a certain set, we can sometimes conclude that outside this set local equilibrium must be violated. Take the simplest case, that of an easy-axis ferromagnetic material. Then the spontaneous magnetization vector is always parallel to the axis, so it can be described by a scalar times a fixed unit vector along the axis. If there are two points at which the magnetization has opposite directions, and we can connect them by a curve lying inside the region of interest, we deduce that somewhere along this curve the magnetization vector vanishes (since a continuous scalar function that changes sign inside an interval must have a zero in that interval).

We conclude that either there is a point on the curve with temperature $T \geq T_c$, or there is a point where local equilibrium is violated. If $T < T_c$ everywhere, there must be in fact a whole surface where local equilibrium is violated: this is because the two original points can be connected by a two-parameter family of disjoint curves, and along each curve there is at least one point where local equilibrium is violated, as shown in Figure 1. Defects of this kind are known as *domain walls*. Of course, equilibrium may be violated on a set of higher dimension—in a domain wall of non-zero thickness, for example.

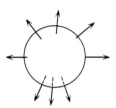

Figure 1   **Figure 2**

We now turn to the case of an isotropic ferromagnetic material. We assume that local equilibrium with $T < T_c$ occurs on a two-dimensional sphere. The magnetization vector field on this sphere is non-zero everywhere, so we get a map $S^2 \to \mathbf{R}^3 \setminus 0$ from the sphere to $\mathbf{R}^3$ minus the origin.

If local equilibrium with $T < T_c$ prevails inside the sphere as well, we have a non-zero vector field in the whole ball D^3 bounded by S^2. Thus we get a map $D^3 \to \mathbf{R}^3 \setminus 0$, which is continuous and extends the map on the sphere, discussed in the previous paragraph. This means the map on the sphere is not arbitrary; for example, the "porcupine" map of Figure 2, where the magnetization points outward everywhere, cannot be extended continuously to the ball.

More precisely, a map $S^2 \to \mathbf{R}^3 \setminus 0$ can be extended to a map $D^3 \to \mathbf{R}^3 \setminus 0$ if and only if it is null-homotopic (see T1.3), that is, if and only if it can be continuously deformed into a constant map (a constant vector field), always avoiding the zero vector. A field that is not null-homotopic, like the one in Figure 2, cannot be extended to an everywhere non-zero field on D^3 (see T1.3), and therefore local equilibrium with $T < T_c$ must fail somewhere inside the ball. The violation can occur at a single point—this is what is called a *point defect*— but it can also occur in a larger set, possibly an entire domain, connected or not.

To every map $f : S^k \to \mathbf{R}^{k+1} \setminus 0$ we can assign an integer $n(f)$, called its *degree*, which remains fixed under continuous changes in f. Two maps f_0 and f_1 are homotopic, that is, can be deformed continuously into one another, if and only if $n(f_0) = n(f_1)$. (This is proved in T2.3 for maps $S^k \to S^k$; the result for maps $S^k \to \mathbf{R}^{k+1} \setminus 0$ follows because S^k and $\mathbf{R}^{k+1} \setminus 0$ are homotopically equivalent: see T1.3.) Thus, the set $\{S^k, \mathbf{R}^{k+1} \setminus 0\}$ of homotopy classes of maps $S^k \to \mathbf{R}^{k+1} \setminus 0$ is in one-to-one correspondence with the set of integers. In particular, a map $f : S^k \to \mathbf{R}^{k+1} \setminus 0$ is null-homotopic if and only if it has index zero.

Using these results, we can easily classify defects in isotropic ferromagnetic materials. Recall that a defect is a point, or larger set, where local equilibrium is violated. If we have a defect that is entirely surrounded by a sphere S^2, so that on the sphere local equilibrium does hold, we can look at the magnetization vector field $\mathbf{M}(x)$ on the sphere as a map from S^2 into $\mathbf{R}^3 \setminus 0$. We then define the *index* of the defect as the degree of this map. A defect whose index is non-zero is said to be *topologically nontrivial*.

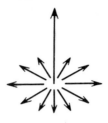

Figure 3

Finally, we discuss the case of an easy-plane ferromagnetic material. Consider a circle where local equilibrium with $T < T_c$ has established itself, and look at the magnetization vector at each point of this circle. Since the magnetization vector must lie on the easy plane, this gives a map from the circle S^1 into $\mathbf{R}^2 \backslash 0$. If the map is not null-homotopic, it cannot be extended to a map from the disk D^2 bounded by S^1 into $\mathbf{R}^2 \backslash 0$, and there must be a defect somewhere in the disk. Furthermore, the disk can be deformed into any other surface that spans the circle and is homeomorphic to a disk; on each such surface there must be a point where local equilibrium is violated or $T < T_c$. There is a one-parameter family of disjoint topological disks of this type, so local equilibrium is violated at least along a one-dimensional family of points, that is, a curve. We thus get a stringlike topologically nontrivial defect.

Again, equilibrium may be violated not only along a curve, but also on a set of higher dimension. As a rule, equilibrium is violated in a small neighborhood of a curve. For example, the magnetization vectors can behave as in Figure 3: outside a certain neighborhood of the line there is local equilibrium (the magnetization vector lies in a horizontal plane, the easy plane); along the line, the vector is vertical, and in a neighborhood of the line it gradually passes from begin vertical to being horizontal.

The defects whose existence is guaranteed by the above reasoning are stable, that is, they cannot vanish with time. Consider, for definiteness, a topologically nontrivial defect in an isotropic ferromagnetic material. On a sphere S^2 surrounding the defect, the magnetization vector $\mathbf{M}(x)$ determines a map $S^2 : \mathbf{R}^3 \backslash 0$ that is not null-homotopic. If this map evolves continuously, and local equilibrium is not violated on the sphere as time goes by, the map must remain homotopic to what it was at the beginning, and therefore can never become null-homotopic. Thus there continues to be a defect inside the sphere (whose position, of course, may change). Not only is the defect preserved, but also the number characterizing its topological type, since this number, being an integer, cannot change under continuous transformations. A similar reasoning holds true for the other cases considered here.

More generally, if some region U is in local equilibrium, we have a map from U to the degeneracy space R (assuming for simplicity that the temperature is uniform). We consider the restriction $f : S^2 \to R$ of this map to a sphere lying entirely within U. If f is not null-homotopic, it cannot be continued to the ball D^3 bounded by S^2, so somewhere inside the ball there are points where local equilibrium does not hold (a defect). In particular, the ball is not entirely contained in U. As time goes by, so long as local equilibrium always holds on S^2, the homotopy class of $f : S^2 \to R$ remains the same, because f is changing continuously. The homotopy class of f characterizes the topological type of the defect, so the topological type of the defect also remains the same.

As we have mentioned, the topological type of a defect in an isotropic ferromagnetic material is an integer, since in this case $R = S^2$ and the homotopy class of a map $f : S^2 \to R$ is characterized by an integer, the degree (T2.3).

The topological type of a defect is also independent of the choice of a surrounding sphere. More precisely, let S_1^2 and S_2^2 be two spheres that can be continuously deformed into one another within the region U where local equilibrium prevails. In order to say that the maps $f_1 : S_1 \to R$ and $f_2 : S_2 \to R$ defined on the two spheres have the same homotopy class, we must somehow identify the set $\{S_2^2, R\}$ of homotopy classes of maps $S_2^2 \to R$ with the set $\{S_1^2, R\}$. To do this we take a model sphere S^2 and homeomorphisms $\varphi_1 : S^2 \to S_1^2$ and $\varphi_2 : S^2 \to S_1^2$, so that φ_1 and φ_2 are homotopic as maps $S^2 \to U$ (this is possible because S_1^2 and S_2^2 can be deformed into one another within U). Then we identify $\{S_1^2, R\}$ with $\{S^2, R\}$, by the one-to-one correspondence that associates with the homotopy class of a map $f : S_1^2 \to R$ the class of the map $f \circ \varphi : S^2 \to R$. We identify $\{S_2^2, R\}$ with $\{S^2, R\}$ in the same way. With this identification, it is clear that the homotopy classes of the maps $S_1^2 \to R$ and $S_2^2 \to R$ coincide, since the composite maps $S^2 \to R$ are homotopic.

If R is *simply connected*, that is, if every map $S^1 \to R$ is null-homotopic, one can define an operation of addition on the set $\{S^k, R\}$ of homotopy classes of maps from S^k into R (T7.1). This makes $\{S^k, R\}$ into a commutative group $\pi_k(R)$, called the *k-th homotopy group* of R. If R is connected but not simply connected, addition in $\{S^k, R\}$ is not well-defined; we can think of it as being multivalued (T7.2). But the homotopy group $\pi_k(R)$ is still well-defined for any $k \geq 1$, and is commutative for $k \geq 2$ (T8.1). In this case $\{S^k, R\}$ can be identified with the quotient $\pi_k(R)/\pi_1(R)$, that is, the set of orbits under the natural action of $\pi_1(R)$ on $\pi_k(R)$ (T8.2). The first homotopy group $\pi_1(R)$, called the *fundamental group* of R, is usually not commutative, so we write the corresponding operation multiplicatively rather than additively; this operation can be thought of as concatenation of loops (maps $S^1 \to R$).

Let U be a domain in $\mathbf{R}^{k+1}$, bounded outside by a k-sphere S^k and inside by two k-spheres S_1^k and S_2^k, and let f, f_1 and f_2 be maps from S^k, S_1^k and S_2^k into R. If there is a map $F : U \to R$ that coincides with f on S^k, with f_1 on S_1^k, and with f_2 on S_2^k, the homotopy class $[f]$ of f is equal to the sum (or one of the possible values of the sum) of $[f_1]$ and $[f_2]$ (T7.3). If you wish, you can consider this as a definition for addition in $\{S^k, R\}$.

What is the physical meaning of this discussion? Suppose there is local equilibrium in the domain $U \subset \mathbf{R}^3$, so that we have a map F from U to the degeneracy space R. Denote the defect inside S_1^2 by K_1, the defect inside S_2^2 by K_2, and the union of the two by $K = K_1 \cup K_2$. The homotopy class $[F|_{S^2}]$ of the restriction of F to S^2 gives the topological type of K, and the homotopy

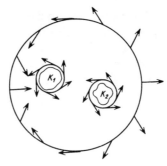

Figure 4

classes $[F|_{S_1^2}]$ and $[F|_{S_2^2}]$ give the topological types of K_1 and K_2. We see that the topological type of the composite defect K is equal to the sum of the topological types of K_1 and K_2 (Figure 4), or, more precisely, one of the possible values of this sum, since the addition operation is multivalued in general.

This fact can yield information on the dynamics of point defects. For one thing, it implies that when a point defect splits, the sum of the topological types of the new defects is equal to the topological type of the disintegrated one.

If R is simply connected, addition in $\{S^2, R\}$ is well-defined, so the topological type of a composite defect is uniquely determined by the types of the components. This is also true when R is not simply connected, but U is. (Roughly speaking, the simple connectivity of U means that there are no stringlike defects.) Indeed, if U is simply connected, $\{S^2, U\}$ and $\pi_2(U)$ coincide, so a sphere S^2 surrounding the defect determines an element $\sigma \in \pi_2(U)$. We also have a map $U \to R$, which gives rise to a homomorphism $\pi_2(U) \to \pi_2(R)$. The image of $\sigma \in \pi_2(U)$ under this homomorphism gives the topological type of the defect. The element of $\pi_2(R)$ corresponding to the composite defect equals the sum of the elements corresponding to the component defects.

We see that topologically nontrivial point defects can appear whenever R is not *aspherical* in dimension two, that is, whenever there exist maps $S^2 \to R$ that are not null-homotopic. Topologically nontrivial stringlike defects can appear if R is not simply connected, that is, if there are maps of $S^1 \to R$ that are not null-homotopic. Finally, wall-like topologically nontrivial defects can appear if R is not connected.

Suppose we have a stringlike defect, so that local equilibrium is violated along a smooth curve or in a small neighborhood of the curve. Draw a small circle S^1 around the curve, contained entirely in the local equilibrium region U (Figure 5). As usual, we have a map $f : U \to R$, and the topological type of the defect is given by the homotopy class of the restriction $f|_{S^1} : S^1 \to R$. As discussed above, the set $\{S^1, R\}$ of homotopy classes of such maps can be identified with the quotient of the action of $\pi_1(R)$ on itself by *inner automorphisms* $x \mapsto gxg^{-1}$—in other words, with the set of conjugacy classes of $\pi_1(R)$. Using the same reasoning as for point defects, we see that if the stringlike defect splits into two, the topological type of the old defect equals the product of the

Figure 5

topological types of the new ones (or, more precisely, one of possible values for the product, when the product is multivalued).

In a nematic liquid crystal, the order parameter is the traceless symmetric tensor $\varepsilon_{\alpha\beta}$, and the symmetry group G is the group SO(3) of space rotations. In all examples observed in nature, the free energy attains its minimum when two eigenvalues of $\varepsilon_{\alpha\beta}$ coincide; let λ be their value. The eigenspace of the remaining eigenvalue -2λ is called the crystal's *axis*. The unbroken symmetry group H is isomorphic to the orthogonal group $O(2)$ preserving the axis, which is also the group preserving the eigenspace with eigenvalue λ (a plane orthogonal to the axis).

Thus, in equilibrium, the system is completely characterized by the direction of the axis; opposite directions are equivalent. The degeneracy space H is the space of directions in $\mathbf{R}^3$, which is, by definition, the projective plane RP^2. This space can also be thought of as the sphere S^2 with antipodal points identified: S^2 is the space of unit vectors in $\mathbf{R}^3$, but unit vectors that differ only by a sign define the same direction. This gives another proof that $R = RP^2$: we have $R = G/H = SO(3)/O(2)$, and $SO(3)/SO(2)$ is homeomorphic to S^2 (Chapter 40), so $R = SO(3)/O(2) = S^2/\mathbf{Z}_2 = RP^2$.

Stringlike defects in a nematic crystal are known as *disclinations*. By the preceding discussion, the topological type of a disclination is an element of $\{S^1, RP^2\}$. It turns out that RP^2 is not simply connected; its fundamental group is $\pi_1(RP^2) = \mathbf{Z}_2$ (this follows from the representation of RP^2 in the form $S^2/\mathbf{Z}_2$ and from the results of T3.2), and since $\mathbf{Z}_2$ is commutative all conjugacy classes have a single element, so we can identify $\pi_1(RP^2)$ with $\{S^1, RP^2\}$. Thus, all maps $S^1 \to RP^2$ that are not null-homotopic are homotopic to one another, and there is a single type of topologically nontrivial disclination. Figures 6 and 7 show a cross-section of the direction field for a nontrivial and a trivial disclination, respectively. The disclination is perpendicular to the plane of the page.

When studying the type of a disclination, it is useful to know that a loop in RP^2 is null-homotopic if and only if it is covered by a loop in S^2 (T3.2). This is equivalent to saying that a direction field on S^1 is topologically trivial if and only if we can make it into a unit vector field by assigning arrows to the directions in a consistent way.

The homotopy group $\pi_2(RP^2)$ is isomorphic to $\pi_2(S^2) = \mathbf{Z}$ (T8.2). If the region U of local equilibrium is simply connected, the topological type of a point

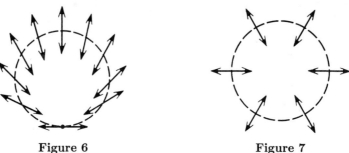

Figure 6   **Figure 7**

defect is an element of $\pi_2(RP^2)$, and hence an integer. When defects merge, these integers are added. But if U is not simply connected—for example, if stringlike defects are present—point defects are classified by $\{S^2, RP^2\}$, rather than by $\pi_2(RP^2)$. Recalling that $\{S^2, RP^2\}$ is the quotient of $\pi_2(RP^2)$ by the action of $\pi_1(RP^2)$, and that the nontrivial element of $\pi_1(RP^2)$ takes a class in $\pi_2(RP^2)$ to its negative (T7.2), we see that the elements of $\{S^2, RP^2\}$ are integers up to sign. The result of merging two point defects with topological numbers m and n is not unique; the resulting point defect can have topological number $m+n$ or $|m-n|$, because all these numbers are only defined up to sign. Thus two defects with the same topological type can annihilate or reinforce each other, depending on the path they follow in approaching each other.

Theoretically, in addition to the *uniaxial* nematic crystals considered above, there can exist *biaxial* nematic crystals, although they have never been observed in nature. In a biaxial crystal, the free energy attains its minimum for a tensor $\varepsilon_{\alpha\beta}$ all of whose eigenvalues are distinct; by a change of coordinates, we can assume that the eigenspaces are the coordinate axes. Then the stabilizer H consists of those rotations of $\mathbf{R}^3$ that map each eigenspace into itself. This group, denoted by K_4 and sometimes called the *Klein group*, has four elements; each nontrivial transformation in K_4 leaves one coordinate unchanged and reverses the signs of the other two. The degeneracy space is homeomorphic to $R = SO(3)/K_4$, and the quotient map $SO(3) \to R$ is a fourfold covering map. This implies that $\pi_2(R) = \pi_2(SO(3)) = 0$ (T8.2), and therefore biaxial nematic crystals cannot have topologically stable point defects.

To compute $\pi_1(R)$, we use the fact that $SU(2)$ is a twofold cover of $SO(3)$. The covering map p is actually a homomorphism, and can be defined as follows: an element of $SU(2)$ is a matrix $\begin{pmatrix} z & -\bar{w} \\ w & \bar{z} \end{pmatrix}$, where $|z|^2 + |w|^2 = 1$, so $SU(2)$ can be identified with the three-sphere S^3. Writing $z = x + iy$ and $w = t + iu$, we map $\begin{pmatrix} z & -\bar{w} \\ w & \bar{z} \end{pmatrix}$ to the element of $SO(3)$ that fixes the line spanned by the vector $(y, t, u) \in \mathbf{R}^3$, and rotates space around this line through an angle $2\arctan(\sqrt{y^2 + t^2 + u^2}/x)$. It is easy to see that two elements of $SU(2)$ map to the same rotation if and only if they differ by a factor of 1 or -1, so that p is a twofold cover.

Since $R = SO(3)/H$, we also have $R = SU(2)/\tilde{H}$, where $\tilde{H} = p^{-1}(H)$ has eight elements. The nontrivial elements of H are rotations by 180° around the coordinate axes, so $\tilde{H}$ consists of the matrices ± 1, $\pm \tilde{h}_1$, $\pm \tilde{h}_2$ and $\pm \tilde{h}_3$, with $\pm \tilde{h}_j = \exp(pi\sigma_j/2) = i\sigma_j$, where σ_1, σ_2 and σ_3 are the Pauli matrices. Since $\mathrm{SU}(2) = S^3$ is simply connected, we conclude that $\pi_1(R) = \tilde{H}$ (T3.2). The anticommutation relations $\tilde{h}_i \tilde{h}_j \tilde{h}_i^{-1} = -\tilde{h}_j$, derived from the same relations for the Pauli matrices, imply that $\tilde{h}_i$ and $-\tilde{h}_i$ are conjugate in $\tilde{H}$. Thus $\tilde{H}$ has five conjugacy classes: two one-element classes, $\{1\}$ and $\{-1\}$, and three two-element classes $\{\tilde{h}_j, -\tilde{h}_j\}$, for $j = 1, 2, 3$.

Thus, in a biaxial nematic crystal, there exist four types of topologically nontrivial stringlike defects. To see how defects combine, we look at the group law in $\pi_1(R)$: given the topological types of two defects, that is, two elements of $\{S^1, R\}$, we take all possible representatives for them in $\pi_1(R)$, and list the possible conjugacy classes of the product of these elements. If we multiply a class $\{\tilde{h}_j, -\tilde{h}_j\}$ by itself, the result is either $\{1\}$ or $\{-1\}$, since $1 = \tilde{h}_j \tilde{h}_j$ and $-1 = -\tilde{h}_j \tilde{h}_j$. In all other cases, multiplication is single-valued: for example, multiplying $\{\tilde{h}_i, -\tilde{h}_i\}$ by $\{\tilde{h}_j, -\tilde{h}_j\}$ gives $\{\tilde{h}_k, -\tilde{h}_k\}$, for $k \neq i, j$, since $\tilde{h}_i \tilde{h}_j = \varepsilon_{ijk}\tilde{h}_k$ and the choice of sign does not affect the class. Thus, if two merging strings have different topological types, the topological type of the resulting string can be predicted unambiguously; but if they have the same type, and the type is not 1 or -1, the result may have type 1 (a topologically trivial string) or type -1.

We now explain how to compute $\pi_1(R)$ and $\pi_2(R)$ in the general case when $R = G/H$, that is, when the degeneracy of the equilibrium state is caused solely by symmetry breaking. We assume, for simplicity, that G is connected and simply connected; if G is not simply connected, we can replace it by its universal cover, and H by its inverse image under the covering map. This is what we did in the preceding discussion, replacing $G = \mathrm{SO}(3)$ by $\mathrm{SU}(2)$ and $H = K_4 \subset \mathrm{SO}(3)$ by the order-eight group $\tilde{H} \subset \mathrm{SU}(2)$.

Under these conditions, $\pi_2(G/H)$ is isomorphic to $\pi_1(H)$ and $\pi_1(G/H)$ is isomorphic to $\pi_0(H) = H/H_{\mathrm{con}}$, where H_{con} is the maximal connected subgroup of H, that is, the set of elements of H that can be connected to the identity by a continuous path, also called *continuous unbroken symmetries*; see T14.1. (A symmetry outside H_{con} is called *discrete*.) Thus, topologically nontrivial point defects exist if and only if the group H of unbroken symmetries is not simply connected, and topologically nontrivial stringlike defects exist if and only if H is disconnected.

Consider the example of superfluid quantum liquids. For superfluid ^{4}He, the only broken symmetry is the gauge invariance. The group of global gauge transformations is isomorphic to $U(1)$: under a gauge transformation the wave function is multiplied by $e^{i\alpha}$. Thus the degeneracy space R is homeomorphic to S^1. This also follows directly from Landau theory: the order parameter is a complex number Ψ, and, below the critical temperature, equilibrium is achieved on the circle $|\Psi| = c$, where c is a positive constant. Since $\pi_2(S^1) = 0$ and $\pi_1(S^1) = \mathbf{Z}$, there are no topologically nontrivial point defects in ^{4}He, while stringlike defects can be characterized by a single integer. Physically, the stringlike defects take the form of vortices in the superfluid. Indeed, the order parameter in a state of

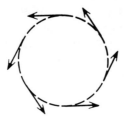

Figure 8

local equilibrium is a complex-valued function $\Psi(\mathbf{x})$ of the coordinate $\mathbf{x}$. The topological number n of a stringlike defect is $(2\pi)^{-1}$ times the phase variation of $\Psi(\mathbf{x})$ as one goes once around the defect. If we denote the phase by $\Phi(\mathbf{x})$, the superfluid velocity $\mathbf{v}_s$ is given by $\mathbf{v}_s = (\hbar/m)\nabla\Phi$, so that

$$n = \frac{1}{2\pi} \oint_\Gamma \nabla\Phi(\mathbf{x})\, d\mathbf{x} = \frac{m}{2\pi\hbar} \oint_\Gamma \mathbf{v}_s\, d\mathbf{x},$$

where Γ is a path going around the defect once. An example of an vortex with topological number 1 can easily be given in polar coordinates: $\Psi(r, \varphi, z) = \alpha(r)e^{i\varphi}$, where $\alpha(r)$ is a real-valued function. In this case the superfluid velocity is tangent to horizontal circles centered along the z-axis; Figure 8 shows a horizontal section.

The topological analysis of defects in a superconductor is entirely analogous to that of superfluid ^{4}He.

We discuss briefly topologically stable defects in ^{3}He. The order parameter in this case is the rank-2 complex-valued tensor A_{ik}, where i can be thought of as a spin index and k as a coordinate index. If we ignore spin-orbit coupling, the free energy is invariant under the group $G = \mathrm{SO}(3) \times \mathrm{SO}(3) \times U(1)$. An element $(V, W, e^{i\alpha})$ of this group, where $V = (V_{il})$ and $W = (W_{kr})$ are rotations in spin space and coordinate space, respectively, transforms A_{ik} into the tensor $A'_{ik} = e^{i\alpha}V_{il}W_{kr}A_{lr}$.

In the B-phase space of ^{3}He, one equilibrium state is given by $A_{ik} = \lambda\delta_{ik}$, where λ is a scalar related to the energy gap Δ by the formula $\lambda = \Delta/\sqrt{3}$. Every other equilibrium state can be obtained from this one by the action of an element of G, and therefore is given by

(8.1)
$$A_{ik} = \lambda e^{i\alpha} V_{ik},$$

where (V_{ik}) is a rotation matrix. The unbroken symmetry group H consists of triples $(V, W, e^{i\alpha})$ with $\alpha = 0$ and $V = W$, and is isomorphic to $\mathrm{SO}(3)$. Thus the degeneracy space R is homeomorphic to $\mathrm{SO}(3) \times \mathrm{SO}(3) \times U(1)/\mathrm{SO}(3) \times U(1)$. Using the fact that the k-th homotopy group of a product of two spaces is the direct sum of the direct sum of the k-th homotopy groups of the factors, and the equations

$$\pi_2(\mathrm{SO}(3)) = 0, \quad \pi_2(U(1)) = 0, \quad \pi_1(\mathrm{SO}(3)) = \mathbf{Z}_2, \quad \pi_1(U(1)) = \mathbf{Z},$$

we see that $\pi_2(R) = 0$ and $\pi_1(R) = \mathbf{Z}_2 \dotplus \mathbf{Z}$. Since $\pi_1(R)$ is commutative, it can be identified with $\{S^1, R\}$. Thus, there are no topologically stable point defects, and stringlike defects are classified by an integer and a residue modulo 2.

In the A-phase the equilibrium state can be selected in such a way that

$$(8.2) \qquad A_{ik} = \frac{\Delta}{\sqrt{2}} \begin{pmatrix} 0 & 0 & 0 \\ 0 & 0 & 0 \\ 1 & i & 0 \end{pmatrix}.$$

This tensor is invariant under rotations about the z-axis in spin space, and it gets multiplied by $e^{-i\alpha}$ under a rotation through an angle α about the z-axis in coordinate space. Therefore the unbroken symmetry group H contains all transformations of the form $(V, W, e^{i\alpha})$, with V any rotation about the z-axis and W a rotation through an angle α about the z-axis; the group of such transformations is isomorphic to $U(1) \times U(1)$. In fact, it turns out that there are no other continuous unbroken symmetries, so $U(1) \times U(1) = H_{\mathrm{con}}$, the maximal connected subgroup of H.

In addition, H contains discrete symmetries, namely, transformations of the form $(V, W, e^{i\beta})$, where V and W are rotations through π about horizontal axes that meet at an angle β (V and W map the z-axis to itself, reversing its direction). All discrete symmetries can be obtained from one such symmetry by multiplication by a transformation in $H_{\mathrm{con}} = U(1) \times U(1)$, that is, H/H_{con} has two elements, and therefore equals $\mathbf{Z}_2$. On the other hand,

$$G/H_{\mathrm{con}} = \mathrm{SO}(3) \times \mathrm{SO}(3) \times U(1)/U(1) \times U(1) = S^2 \times \mathrm{SO}(3),$$

so that $R = (G/H_{\mathrm{con}})/(H/H_{\mathrm{con}}) = S^2 \times \mathrm{SO}(3)/\mathbf{Z}_2$. Thus R has the twofold cover $S^2 \times \mathrm{SO}(3)$; this space, in turn, has the twofold cover $S^2 \times \mathrm{SU}(2)$, which is simply connected. Therefore R has $S^2 \times \mathrm{SU}(2)$ as a fourfold cover, and $\pi_1(R)$ has four elements. In fact it is easy to see that $\pi_1(R) = \mathbf{Z}_4$, so that $\{S^1, R\}$ can be identified with $\pi_1(R)$, we conclude and stringlike defects are classified by a residue modulo 4.

As to point defects, we have

$$\pi_2(R) = \pi_2(S^2 \times \mathrm{SU}(2)) = \pi_2(S^2) \dotplus \pi_2(\mathrm{SU}(2)) = \mathbf{Z},$$

so the topological type of a point defect is characterized by an integer.

Interesting effects occur when a weak field is applied to a system with defects. Consider, for example, a uniaxial nematic crystal placed in a magnetic field. The degeneracy of equilibrium states is lifted, because the crystal axis tends to align itself with the field. If the crystal contains a point defect of the type depicted in Figure 6, the result is that the axis direction field aligns itself with the field throughout the entire crystal, with the exception of a small tube originating at the defect: see Figure 9, which shows lines tangent to the direction field.

Generally, if we apply a weak field to the system, or allow for a weak interaction, such as spin-orbit coupling in ^{3}He, we single out in the degeneracy space R a subspace $\tilde{R}$ consisting of those states that remain in equilibrium in the new situation. Now to

Figure 9

each point in local equilibrium we must associate a point in $\tilde{R}$, rather than in R. However, the weakness of the field implies that the free energies of the new and old equilibrium states differ but little, so it is reasonable to consider also states that are in local equilibrium in the old sense, or *partial local equilibrium*. If a region is in partial local equilibrium, we have a map from this region to R as before.

Just as topological considerations can imply that certain violations of local equilibrium cannot disappear, they show that, in some cases, one cannot get rid of partial local equilibrium. Consider a sphere S^2 on which total local equilibrium holds everywhere, and such that partial local equilibrium holds inside the ball D^3 bounded by S^2. This gives rise to a map of pairs $(D^3, S^2) \to (R, \tilde{R})$, that is, a map $D^3 \to R$ such that S^2 is taken entirely inside $\tilde{R}$ (T11.3). Of course, there may be points in the interior of the ball that are mapped into $\tilde{R} \subset R$; this means there are points in the ball where total local equilibrium holds, which, of course, does not contradict our assumption that partial local equilibrium holds.

We assume that the system evolves in such a way that total local equilibrium always holds on S^2, and partial local equilibrium always holds in D^3. Then the map of pairs $(D^3, S^2) \to (R, \tilde{R})$ changes continuously with time, and therefore its *relative homotopy class*, that is, its class as a map of pairs, cannot change. In particular, if the map $(D^3, S^2) \to (R, \tilde{R})$ is not null-homotopic, total local equilibrium cannot be established everywhere inside D^3; if it did, the map would have been homotopic to a map that takes the entire ball D^3 into $\tilde{R}$, and would, by definition, be homotopically trivial.

Thus, a point or region where total local equilibrium is violated can be assigned an element in the set $\{(D^3, S^2); (R, \tilde{R})\}$ of relative homotopy classes. If $\tilde{R}$ is connected and simply connected, $\{(D^3, S^2); (R, \tilde{R})\}$ can be identified with the relative homotopy group $\pi_3(R, \tilde{R})$. It $\tilde{R}$ is connected but not simply connected, $\{(D^3, S^2); (R, \tilde{R})\}$ is the quotient of $\pi_3(R, \tilde{R})$ by the action of $\pi_1(\tilde{R})$ (T11.3). If $\tilde{R}$ has a single point, that is, if the perturbation lifts the degeneracy completely, the relative homotopy group $\pi_3(R, \tilde{R})$ coincides with the absolute homotopy group $\pi_3(R)$. In general, $\pi_3(R, \tilde{R})$ can be computed from the homotopy exact sequence

$$\cdots \to \pi_3(\tilde{R}) \xrightarrow{i_*} \pi_3(R) \to \pi_3(R, \tilde{R}) \xrightarrow{\partial} \pi_2(\tilde{R})$$
$$\xrightarrow{i_*} \pi_2(R) \to \pi_2(R, \tilde{R}) \xrightarrow{\partial} \pi_1(\tilde{R}) \xrightarrow{i_*} \pi_1(R).$$

(see T11.3).

If total local equilibrium is violated along a string, or a tube surrounding the string, we can take a circle S^1 going around the string and a disk D^2 bounded by

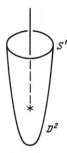

Figure 10

the circle. Assuming that total local equilibrium holds on the circle, and that partial local equilibrium holds on the disk, we obtain a map $(D^2, S^1) \to (R, \tilde{R})$. The relative homotopy class of this map does not change with time, and it determines the topological type of the string. Thus, in this case the topological type is an element of the set $\{(D^2, S^1); (R, \tilde{R})\}$. If $\tilde{R}$ is connected and simply connected, this set coincides with the relative homotopy group $\pi_2(R, \tilde{R})$.

It may happen that the string along which total equilibrium is violated ends at a point defect (Figure 9). This imposes conditions on the topological type of the defect: the corresponding map from $(D^2, S^1) \to (R, \tilde{R})$ cannot be arbitrary, but must give rise to a null-homotopic map $S^1 \to \tilde{R}$, because S^1 can be spanned by a surface homeomorphic to a disk and lying entirely within the region of total equilibrium (Figure 10).

9. Topological Integrals of Motion

Consider a one-dimensional particle moving in a field with potential $V(x)$, and suppose that the potential has an infinite spike at a point $x = a$, as in Figure 11. The particle cannot penetrate an infinitely high potential barrier: if it starts to the left of point a, for example, it remains to the left of that point for all time. This is true for both classical and quantum particles.

To state this in topological language, let U be the set of positions that the particle can occupy. If U is disconnected, the particle must remain in the same connected component where it starts. (Two points belong to the same connected component if they can be connected by a path.) In Figure 11, U is disconnected: it consists of the two intervals $(-\infty, a)$ and (a, ∞).

Now consider a classical system with n degrees of freedom, described by the Hamiltonian $H = T + V$, where the kinetic energy T is a quadratic function of the momenta and the potential energy $V(x^1, \ldots, x^n)$ may become infinite for certain configurations of the system. Assume that the domain U where V is finite is disconnected, and has connected components $U_1, \ldots, U_k$. Here again, it the particle starts in component U_i, it will remain in the same component for all time; in other words, the component where the particle lies is an integral of motion.

A similar statement is true if we quantize the system. If a wave function $\psi(x)$ is zero outside of U_i, the same will be true for the wave function obtained by solving the appropriate time-dependent Schrödinger equation with initial condition $\psi(x)$.

The domain U where the potential energy is finite is the system's configuration space. Instead of the configuration space, we can also consider the phase space, that is, the set of points $(p_1, \ldots, p_n, x^1, \ldots, x^n)$ where the Hamiltonian is finite. Again, a particle cannot leave the connected component of the phase

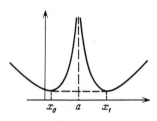

Figure 11

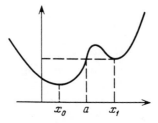

Figure 12

space where it starts. For the system just discussed, the phase space is simply $\mathbf{R}^n \times U$, because the kinetic component of the Hamiltonian is always finite. Therefore the phase space has the same number of components as the configuration space.

We now discuss the ground state and the weakly excited states of the system just discussed. Suppose the minimum of $V(x)$ is attained at a point x_0. The Taylor series at x_0 is

$$(9.1) \qquad V(x) = V(x_0) + \tfrac{1}{2}k_{ij}\xi^i\xi^j + W(\xi),$$

where $\xi = x - x_0$ and $W(\xi) = o(|\xi|^2)$. Assume the quadratic form $k_{ij}\xi^i\xi^j$ is non-degenerate. Neglecting the term $W(\xi)$, we get the Hamiltonian of a multi-dimensional harmonic oscillator, whose energy levels are given by

$$(9.2) \qquad V(x_0) + \hbar \sum (n_i + \tfrac{1}{2})\omega_i,$$

where the ω_i are the normal modes (proper frequencies) and the n_i are non-negative integers. Perturbation theory tells us that the correction to the energy levels once we take into account the influence of $W(\xi)$ is of order $(N\hbar)^{3/2}$, where $N = \max n_i$. This implies that the energy of weakly excited states is determined in the semiclassical approximation by (9.2). In particular, the energy E_0 of the ground state in the semiclassical approximation is $V(x_0)$, or, more precisely, $E_0 = V_0 + \tfrac{1}{2}\hbar \sum \omega_i + O(\hbar^{3/2})$.

There is an obvious lower bound for the ground-state energy: $E_0 \geq V(x_0)$. To obtain an upper bound, one can use the variational principle, taking as a trial function the wave function of the ground state for the harmonic oscillator. A similar estimate can be carried out even when the quadratic form $k_{ij}\xi^i\xi^j$ is degenerate, and in this case, too, $E_0 \approx V(x_0)$ in the semiclassical approximation, with an error of the order of $\hbar$.

In finding the ground state of the system we expanded the potential $V(x)$ in powers of $\xi = x - x_0$, where x_0 is the point where V has its absolute minimum. We could instead have used a point x_1 where V has a *local* minimum. The states obtained in this manner are, in general, quasi-stationary. For example, for the potential depicted in Figure 12, the ground-state wave function of an oscillator with potential $\tfrac{1}{2}V''(x_1)(x - x_1)^2$ represents a quasi-stationary state, whose lifetime is determined by the semiclassical transmission coefficient

$$D = \exp\left(-\frac{2}{\hbar}\int_a^{x_1}\sqrt{2m(V(x)-E)}\,dx\right),$$

which accounts for the transmission through the barrier between the two minima at x_0 and x_1. Thus the lifetime increases exponentially with the width of the barrier and the square root of its height.

The situation changes if the potential barrier is infinitely high and x_1 is the absolute minimum of one of the other components, as in Figure 11. Then the quantum states concentrated around x_1 are stationary, rather than quasi-stationary. Repeating the above reasoning, we conclude that in the semiclassical approximation the lowest possible energy of a quantum state whose wave function is zero outside the connected component of x_1 is given by $V(x_1)$.

To summarize, we have a topological integral of motion whenever the phase space of a system—the set of (p, x) for which the Hamiltonian $H(p, x)$ is finite—is disconnected. The integral of motion is the connected component of the phase state occupied by the system; it remains the same for all time. Topological integrals of motion arise for classical and quantum systems alike; in the quantum case they are often called *topological quantum numbers*.

In the semiclassical approximation, if the phase space has connected components $U_1, \ldots, U_k$ and the minimum of the Hamiltonian on U_i is E_i, the minimal energy of a state with topological quantum number i is E_i. In particular, the ground-state energy is the absolute minimum of H, that is, $\min\{E_1, \ldots, E_k\}$. To obtain the energy values of weakly excited states, we need only expand the Hamiltonian in a power series in the neighborhood of that point, keeping only the quadratic terms.

Topological integrals of motion arise not only in systems with a finite number of degrees of freedom, but in field theory as well. Consider a theory that describes a scalar field $\varphi(x) = \varphi(t, \mathbf{x})$ in one dimension, and assume that the action integral is

$$(9.3) \qquad S = \int \mathcal{L}\,d\mathbf{x}\,dt$$

$$= \frac{1}{2}\int\left(\left(\frac{\partial\varphi}{\partial t}\right)^2 - \left(\frac{\partial\varphi}{\partial \mathbf{x}}\right)^2\right)d\mathbf{x}\,dt - \lambda\int(\varphi^2 - a^2)^2\,d\mathbf{x}\,dt,$$

where $\mathcal{L}$ is the Lagrangian. Then the energy functional has the form

$$(9.4) \qquad E = \frac{1}{2}\int \pi^2(\mathbf{x})\,d\mathbf{x} + \frac{1}{2}\int\left(\frac{\partial\varphi}{\partial\mathbf{x}}\right)^2 d\mathbf{x} + \lambda\int(\varphi^2 - a^2)^2\,d\mathbf{x},$$

where $\pi(\mathbf{x}) = \partial\mathcal{L}/\partial\dot{\varphi}(\mathbf{x}) = \dot{\varphi}(\mathbf{x})$ is the generalized momentum corresponding to the field $\varphi(\mathbf{x})$. The phase space consists of all pairs $(\pi(\mathbf{x}), \varphi(\mathbf{x}))$ for which the energy (9.4) is finite.

Notice that in this case the energy functional can be written in the form $E = T + V$, with $T = \frac{1}{2}\int \pi^2 \partial\mathbf{x}$ and

$$(9.5) \qquad V = V[\varphi] = \frac{1}{2}\int\left(\frac{\partial\varphi}{\partial\mathbf{x}}\right)^2 d\mathbf{x} + \lambda\int(\varphi^2 - a^2)^2\,d\mathbf{x}.$$

The term T, which is quadratic in the generalized momenta $\pi(\mathbf{x})$, represents the kinetic energy, and V, which depends only on the generalized coordinates $\varphi(\mathbf{x})$, represents the potential energy. The configuration space can be interpreted as the set of fields $\varphi(\mathbf{x})$ with $V[\varphi] < \infty$.

Both phase space and configuration space are disconnected. Indeed, in order for (9.5) to be finite, we must have $\lim_{|x| \to \pm\infty} |\varphi(x)| = a$. (By itself, the finiteness of the second term in (9.5) does not imply this, because the integral of a function may be finite even if the function does not tend to 0 as $x \to \pm\infty$. But if the limits exist, they are necessarily equal to a. One can prove the existence of the limits by using the finiteness of the first term in (9.5). We omit the proof because, in physical problems, we can clearly assume that the functions behave nicely at infinity. From now on, we will always assume that a function of n variables tends to zero at infinity if its integral over $\mathbf{R}^n$ is finite.)

Thus, both phase space and configuration space consist of four components, characterized by the field's behavior at infinity. The possibilities are:

(1) $\varphi(+\infty) = \varphi(-\infty) = a$;   (2) $\varphi(+\infty) = -\varphi(-\infty) = a$;

(3) $\varphi(+\infty) = -\varphi(-\infty) = -a$;   (4) $\varphi(+\infty) = \varphi(-\infty) = -a$.

A field satisfying one of these conditions cannot be continuously transformed into a field satisfying another in such a way that all intermediate fields have a finite energy. Thus, fields of different types are effectively separated by an infinitely high potential barrier. On the other hand, fields having the same behavior at infinity can be connected by a continuous family of fields with finite energy. Thus the sets in configuration space determined by these four conditions are indeed the four connected components of this space. The corresponding sets in phase space are likewise connected and form different components; this can be seen, for example, from the fact that any point $(\pi(\mathbf{x}), \varphi(\mathbf{x}))$ in phase space can be joined to the point $(0, \varphi(\mathbf{x}))$ by a continuous curve, such as $(\tau\pi(\mathbf{x}), \varphi(\mathbf{x}))$ for $0 \leq \tau \leq 1$.

The minimum of the energy functional (9.4) is zero, and is reached when $\pi(x) \equiv 0$ and $\varphi(x) \equiv a$ or $\varphi(x) \equiv -a$. Thus, the energy functional achieves its minimum at two points in the phase space separated by an infinitely high energy barrier. These points are known as *classical vacuums*. By convention, we will normalize the energy of the classical vacuums to zero. In all the examples that will concern us, the classical vacuums are translation-invariant fields.

Classical vacuums are linked with the ground states of quantum field theory, in the semiclassical approximation. For example, let's apply the semiclassical approximation to the theory obtained by quantization of the action integral (9.3). Since the classical vacuums are separated by an infinitely high energy barrier, each corresponds to a ground state of the system. We thus have spontaneous breaking of the symmetry $\varphi(\mathbf{x}) \to -\varphi(\mathbf{x})$. As in the case of finitely many degrees of freedom, in order to find the energies of weakly excited states we must expand the action integral (or the energy functional) in a power series in the deviation from a classical vacuum, or energy-minimizing field. Thus, instead of the field $\varphi(\mathbf{x})$ we must consider either $\xi(\mathbf{x}) = \varphi(\mathbf{x}) - a$ or $\xi'(\mathbf{x}) = \varphi(\mathbf{x}) + a$.

In terms of $\xi(\mathbf{x})$, the action becomes $S = S_{\text{qu}} + S_1$, where

$$S_{\text{qu}} = \frac{1}{2} \int \left(\left(\frac{\partial \xi}{\partial t} \right)^2 - \left(\frac{\partial \xi}{\partial \mathbf{x}} \right)^2 - 8\lambda a^2 \xi^2 \right) d\mathbf{x}\, dt,$$

$$S_1 = -4\lambda a \int \xi^3\, d\mathbf{x}\, dt - \lambda \int \xi^4 d\mathbf{x}\, dt.$$

Quantization of the action integral S_{qu} results in scalar particles of mass $m = 2a\hbar\sqrt{2\lambda}$. The terms of order greater than two, subsumed under S_1, yield small corrections to the energy of the particles. A similar reasoning can be applied if we expand in powers of the deviation $\xi'(\mathbf{x})$ of the field from the other classical vacuum.

We now consider the minimum of the energy functional on the components of the phase space where $\varphi(-\infty) \neq \varphi(+\infty)$. Assume, for definiteness, that $\varphi(-\infty) = -a$ and $\varphi(+\infty) = a$. It can easily be checked that the minimum is

(9.6) $$M = \frac{4\sqrt{2}}{3} \sqrt{\lambda}\, a^3,$$

which is achieved for the field

(9.7) $$\varphi(\mathbf{x}) = a \tanh(a\sqrt{2\lambda}\,(\mathbf{x} - \mathbf{c})),$$

where $\mathbf{c} \in \mathbf{R}$ is arbitrary. Thus, in the semiclassical approximation, the lowest energy of quantum states in these two components of the phase space is M, and this is also the mass of the quantum particles corresponding to the fields (9.7). We will not proof this last statement; we merely note that it is true in general.

A field like (9.7), which represents a local minimum of the energy functional, is called a *soliton*. Solitons cannot be translation-invariant, so the minima are degenerate: if a minimum is attained for $\varphi(\mathbf{x})$, it is also attained for $\varphi(\mathbf{x} - \mathbf{c})$, because the energy functional is translation-invariant.

A soliton is a time-independent solution to the classical equations of motion. In the relativistically invariant case that interests us here, Lorentz transformations can be used to transform a time-independent solution into one that depends on time according to the law $s(\mathbf{x} - \mathbf{v}t)$, where $\mathbf{v}$ is a constant. We talk then of a soliton with with velocity $\mathbf{v}$.

This terminology differs from the one used in mathematics, where a solution to the equations of motion that has the form $s(\mathbf{x} - \mathbf{v}t)$ is called a solitary wave with velocity $\mathbf{v}$, and the term "soliton" is used for solitary waves in completely integrable systems.

A soliton in a classical problem resembles a particle, in that it is a *localized* time-invariant solution of the equations of motion. The reason it is localized is that a soliton, like all fields considered here, has a finite energy, and therefore it differs considerably from a classical vacuum only in a finite region of space.

It turns out that, in the semiclassical approximation, a soliton corresponds to a quantum particle whose mass equals the soliton's energy. In general, this particle is not stable, but in many interesting cases its stability is guaranteed by topological conservation laws. For example, if the soliton yields an absolute minimum for the energy functional among all fields of a given topological type (that is, among all fields in a given component of the phase space), it is stable, because the topological type cannot change with time, as discussed at the beginning of this chapter. This is the case for the field (9.7), and is true in multidimensional, as well as one-dimensional, field theories.

Throughout this book we use a system of units in which $\hbar = 1$. This is caused by the desire not to deviate from the standard notation, in which the commutation relation for the generalized momenta $\pi(\mathbf{x})$ and the generalized coordinates $\varphi(\mathbf{x})$ is given by $[\pi(\mathbf{x}), \varphi(\mathbf{x}')] = i^{-1}\delta(\mathbf{x} - \mathbf{x}')$. If we lift the condition $\hbar = 1$, we must carry out quantization by using the commutation relations $[\pi(\mathbf{x}), \varphi(\mathbf{x}')] = \hbar i^{-1}\delta(\mathbf{x} - \mathbf{x}')$. Then the mass of particles corresponding the quadratic part of the action turns out to be of order $\hbar$ (in the model above, for example, it equals $2a\hbar\sqrt{2\lambda}$). On the other hand, the mass of the particle corresponding to a soliton is determined by the classical energy of the soliton, and therefore does not contain $\hbar$. Thus, as $\hbar \to 0$, the ratio between the masses of the soliton and of the standard particle tends to infinity. It can be demonstrated that, when it makes sense to talk of a particle corresponding to a soliton (that is, under conditions ensuring the validity of the semiclassical approximation), this ratio is always high. For example, in the model above this ratio equals $\frac{2}{3}a^2/\hbar$. Correspondingly, the semiclassical approximation is applicable if $a^2/\hbar \gg 1$, or $a^2 \gg 1$ if we take $\hbar = 1$.

10. A Two-Dimensional Model.
Abrikosov Vortices

We now describe more complicated examples of theories that have topological integrals of motion. We start with the analog of the action integral (9.3) for a complex scalar field Ψ in two dimensions:

$$(10.1) \qquad S = \int \mathcal{L}\, d^3x = \frac{1}{2} \int \partial_\mu \bar\Psi\, \partial^\mu \Psi\, d^3x - \frac{1}{8}\lambda \int (|\Psi|^2 - a^2)^2\, d^3x,$$

where $\mu = 0, 1, 2$, $x = (x^0, x^1, x^2) = (x^0, \mathbf{x}) \in \mathbf{R}^3$, $\partial_0 = \partial^0 = \partial/\partial x^0 = \partial/\partial t$, and $\partial_i = -\partial^i$ for $i = 1, 2$. We can also think of Ψ as a two-component real scalar field, instead of a complex field.

The energy functional can be expressed in terms of the generalized coordinates $\Psi(\mathbf{x})$ and the generalized momenta $\pi(\mathbf{x}) = \partial \mathcal{L}/\partial \dot\Psi(\mathbf{x})$ as follows:

$$(10.2) \quad E = T + V = \frac{1}{2} \int |\pi(x)|^2\, d^2x + \frac{1}{2} \int \left(\left| \frac{\partial \Psi}{\partial x^1} \right|^2 + \left| \frac{\partial \Psi}{\partial x^2} \right|^2 \right) d^2x$$

$$+ \frac{1}{8}\lambda \int (|\Psi|^2 - a^2)^2\, d^2x.$$

Again, the energy breaks down into two components: the kinetic energy $T = T[\pi]$ and the potential energy $V = V[\Psi]$; the configuration space consists of those fields $\Psi(x)$ for which $V[\Psi] < \infty$.

We first study the set of fields that satisfy

$$(10.3) \qquad \int (|\Psi|^2 - a^2)^2\, d^2x < \infty,$$

and behave nicely at infinity: more precisely, we assume that, in polar coordinates (r, φ), the limit

$$(10.4) \qquad \lim_{r \to \infty} \Psi(r, \varphi) = \Phi(\varphi)$$

exists and is finite for every φ, and that the convergence is uniform in φ. Since Ψ is continuous, so is $\Phi(\varphi)$. Furthermore, (10.3) implies that $|\Phi(\varphi)| = a$, so we can write $\Phi(\varphi) = ae^{-i\alpha(\varphi)}$, where $\alpha(\varphi)$ is a continuous, real-valued function on the interval $[0, 2\pi]$, and satisfies

$$(10.5) \qquad \alpha(2\pi) - \alpha(0) = 2\pi n,$$

because $\Phi(0) = \Phi(2\pi)$. The integer $n = n(\Phi)$ characterizes the topological type of Φ. Geometrically, n is the degree (T2.3) of Φ, seen as a map from the φ-circle into the circle $|z| = a$.

We assume that $\Phi(\varphi)$ varies continuously with the field Ψ. Since the integer $n(\Phi)$ does not change under a continuous variation of Φ, it does not change under a continuous variation of Ψ either. Thus the set of fields satisfying (10.3) splits into components characterized by the topological number n.

As an example of a field with $n = 1$ and satisfying (10.3), take $\Psi(\mathbf{x}) = a(r)e^{i\varphi}$, where $\lim_{r\to\infty} a(r) = a$. By the preceding discussion, it is not possible to connect this field with one that satisfies $\Psi(\mathbf{x}) = a$ as $|\mathbf{x}| \to \infty$ by means of a continuous family of fields such that (10.3) is finite.

Although the space of fields satisfying (10.3) is disconnected, it does not follow that the action integral (10.1) has topological integrals of motion, because in fact every field of finite energy is topologically trivial. To see this, we write in polar coordinates the second summand from the expression (10.2) for the energy:

$$\int \left(\left| \frac{\partial \Psi}{\partial x^1} \right|^2 + \left| \frac{\partial \Psi}{\partial x^2} \right|^2 \right)^2 d^2x = \int \left| \frac{\partial \Psi}{\partial r} \right|^2 r\, dr\, d\varphi + \int \left| \frac{\partial \Psi}{\partial \varphi} \right|^2 d\varphi \frac{dr}{r},$$

If this is finite, Φ has trivial asymptotic behavior, because the last term diverges logarithmically unless $\partial\Phi/\partial\varphi$ tends to 0 as $r \to \infty$.

We can modify the action integral (10.1) in such a way that topological integrals of motion do arise. To do this, we make Ψ interact with an electromagnetic field A_μ, obtaining the action integral

$$(10.6) \qquad S = \int (\tfrac{1}{2}\nabla_\mu \bar\Psi\, \nabla^\mu \Psi - \tfrac{1}{8}\lambda(|\Psi|^2 - a^2)^2)\, d^3x - \frac{1}{4e^2} \int F_{\mu\nu} F^{\mu\nu}\, d^3x,$$

where x is as in (10.1), $\nabla_\mu = \partial_\mu - iA_\mu$ for $\mu, \nu = 0, 1, 2$, and $F_{\mu\nu} = \partial_\mu A_\nu - \partial_\nu A_\mu$ is the electromagnetic field tensor, so that $H = \partial_1 A_2 - \partial_2 A_1$ represents the magnetic field and $(E_1, E_2) = (A_1 - \partial_1 A_0, A_2 - \partial_2 A_0)$ the electric field. (Usually the covariant derivative ∇_μ is defined as $\partial_\mu + ieA_\mu$, where e is the electric charge; for our purposes, however, it is more convenient to include e in the potential A_μ.) The action integral (10.6) is invariant under gauge transformations

$$(10.7) \qquad A_\mu \mapsto A_\mu - \partial_\mu\lambda, \qquad \Psi \mapsto \Psi e^{i\lambda}.$$

Using gauge invariance, we impose the gauge condition $A_0 = 0$. Viewing $\Psi(x)$, $A_1(x)$ and $A_2(x)$ as generalized coordinates, and $\pi(x) = \dot\Psi(x)$ and the electric field $E = (E_1, E_2) = \dot A$ as generalized momenta, we can write the energy functional corresponding to the action integral (10.6) as

$$(10.8) \quad E = \frac{1}{2} \int |\pi|^2 d^2x + \frac{1}{2e^2} \int (E_1^2 + E_2^2 + H^2)\, d^2x$$

$$+ \int (\tfrac{1}{2}(|\nabla_1\Psi|^2 + |\nabla_2\Psi|^2) + \tfrac{1}{8}\lambda(|\Psi|2 - a^2)^2)d^2x,$$

the second integral being the energy of the electromagnetic field.

Now we can construct fields (Ψ, A) of finite energy such that Ψ has any desired topological number n. For example, for $n = 1$ we can choose

$$\Psi(\mathbf{x}) = \alpha(r)\frac{x^1 + ix^2}{r},$$

with $\alpha(r) = a$ for $r \geq R$, and

$$A_0 = 0, \qquad A_i = \rho(r)\varepsilon_{ij}\frac{x^j}{r^2} \quad \text{for} \quad i, j = 1, 2,$$

with $\rho(r) = 1$ for $r \geq R$.

For n arbitrary we use polar coordinates, where the energy (10.8) takes the form

$$E = \frac{1}{2}\int\left(|\nabla_r\Psi|^2 + \frac{1}{r^2}|\nabla_\varphi\Psi|^2\right)r\,dr\,d\varphi$$

$$+ \tfrac{1}{8}\lambda\int(|\Psi|^2 - a^2)r\,dr\,d\varphi + \frac{1}{2e^2}\int\left(\dot{A}_r^2 + \frac{1}{r^2}\dot{A}_\varphi^2 + H^2\right)r\,dr\,d\varphi$$

(recall that we have set $\pi(\mathbf{x}) = 0$). The following field has finite energy:

(10.9) $\Psi(r, \varphi) = \alpha(r)e^{in\varphi}, \qquad A_0 = 0, \qquad A_r = 0, \qquad A_\varphi = \gamma(r),$

where $\alpha(r) = a + \beta(r)$, $\gamma(r) = -n + \lambda(r)$, and the functions $\beta(r)$ and $\lambda(r)$ decay rapidly as $r \to \infty$. The energy of this field is basically localized in a ball centered at the origin; the radius of the ball depends on $\alpha(r)$ and $\gamma(r)$.

Replacing d^3x by d^4x in the action integral (10.6), and letting μ range from 0 to 3, we obtain a theory of fields in three-dimensional space. This theory does not contain topologically nontrivial fields of finite energy. However, if we interpret (10.9) in cylindrical coordinates (r, φ, z), we get a field invariant under translations along the z-axis, and having a topologically nontrivial string. (For a precise definition, see Chapter 19). Such a field has energy density $e(r, \varphi, z)$ independent of z, so its total energy is infinite. It is an interesting field, nevertheless, because it has a finite linear energy density (the energy of the field within a region bounded by two horizontal planes is finite). To be more graphic, we can say that the energy of these fields is localized within a tubular region surrounding the z-axis.

We return to the two-dimensional case. For a field of finite energy, the topological number n equals $(2\varphi)^{-1}$ times the circulation of the field A_μ along an infinitely distant circle in the (x^1, x^2) plane. This, in turn, equals the integral of the magnetic field H with respect to x^1 and x^2:

(10.10) $$n = \frac{1}{2\pi}\oint A_\mu\,dx^\mu = \frac{1}{2\pi}\int H\,d^2x.$$

This is most easily verified in polar coordinates. The finiteness of the energy implies that $\nabla_\varphi\Psi = \partial_\varphi\Psi + iA_\varphi\Psi$ tends to 0 as r tends to ∞. Recalling that Ψ behaves as $ae^{-i\lambda(\varphi)}$ as $r \to \infty$, we see that

$$A_\varphi \approx i\Psi^{-1}\partial_\varphi \Psi = \partial_\varphi \lambda,$$

so that

$$n = \frac{1}{2\pi} \int_0^{2\pi} \partial_\varphi \lambda \, d\varphi = \frac{1}{2\pi} \int_0^{2\pi} A_\varphi \, d\varphi = \frac{1}{2\pi} \oint A_\mu \, dx^\mu.$$

To make this proof more rigorous, we must require that the covariant derivatives decay rapidly as $|\mathbf{x}| \to \infty$. It is sufficient to assume that $|\nabla_\mu \Psi| \leq \mathrm{const}\, |\mathbf{x}|^{-(1+\delta)}$, which in polar coordinates corresponds to $|\nabla_r \Psi| \leq \mathrm{const}\, r^{-(1+\delta)}$ and $|\nabla_\varphi \Psi| \leq \mathrm{const}\, r^{-\delta}$. If we impose the gauge condition $A_r = 0$, we conclude from the fact that $|\partial_r \Psi| = |\nabla_r \Psi| \leq \mathrm{const}\, r^{-(1+\delta)}$ that the limit $\Phi(\varphi) = \lim_{r \to \infty} \Psi(r, \varphi)$ exists, and that

$$|\Phi(\varphi) - \Psi(r, \varphi)| = \left| \int_r^\infty \partial_r \Psi(r, \varphi) \, dr \right| \leq \mathrm{const}\, r^{-\delta}.$$

Thus the above reasoning is justified in the gauge $A_r = 0$. Since both sides of (10.10) are gauge-invariant, the equality follows in all gauges.

To summarize, the phase space of the system with action integral (10.6) is disconnected: for every integer n there exist fields of finite energy with topological number n, and fields with different topological numbers are separated by an infinitely high energy barrier.

If we quantize this theory, we can again assign an integer to each quantum state of the system, and this topological quantum number is a topological integral of motion. If the number is non-zero, the state is topologically nontrivial. In particular, there may exist topologically nontrivial quantum particles.

Let $E_n(\lambda)$ be the infimum of the energy functional on the set of fields with topological number n. $E_n(\lambda)$ is a lower bound for the energy of quantum states with topological number n. It is easy to see that $E_{m+n}(\lambda) \leq E_m(\lambda) + E_n(\lambda)$. Indeed, if we have two fields with topological numbers m and n, we can assume (by translating one field if necessary) that their energies are localized in regions far apart. By adding the two fields, we get a field with topological number $m+n$ and an energy approximately equal to the sum of the two energies; hence the above inequality.

Later we will show that $E_n = \pi |n| a^2$ for $\lambda = e^2$. A heuristic reasoning suggests that $E_n(\lambda) < |n| E_1(\lambda)$ for $\lambda < e^2$ and $E_n(\lambda) = |n| E_1(\lambda)$ for $\lambda > e^2$. For $\lambda > e^2$ the infimum is not reached; the fields that best approximate it are superpositions of fields with topological numbers $n = \pm 1$.

As noted in Chapter 9, in order to study topologically nontrivial fields in the semiclassical approximation, we need to find a field that represents a local minimum for the energy function in the set of topologically nontrivial fields. In our two-dimensional model, it is easy to find extremal fields for the energy functional having arbitrary topological number (such a field is a time-invariant solution for the equations of motion). In fact, it is sufficient to look for such an extremum among fields of the form (10.9), for the following reason: if we write down Euler's variational equations for the action integral (10.6) and then plug in the field (10.9), the result is the same as if we reversed the order of the two operations, that is, if we first plugged (10.9) into (10.6) and then varied $\alpha(r)$ and $\gamma(r)$. This fact can easily be verified by direct calculation.

(That we can restrict ourselves to fields of the form (10.9) also follows from Chapter 17. In order to apply the results from that chapter, we note that the field (10.9) is symmetric with respect to rotations in the (x, y)-plane, because the change caused by a rotation can be compensated by a gauge transformation. More precisely, a rotation through an angle φ multiplies Ψ by $e^{in\varphi}$, and does not change the gauge field; this is equivalent to a gauge transformation with gauge function $n\varphi$ independent of position (a global gauge transformation). Conversely, if a field (Ψ, A) does not change after a rotation through φ and a gauge transformation with gauge function $n\varphi$, the field can be written in the form (10.9).)

Thus, to prove that the action integral (10.6) has an extremum with arbitrary topological number n, it is sufficient to show that this integral attains its minimum when considered only on fields of the form (10.9). But this follows from the fact that a continuous function on a compact space achieves its minimum: the action integral is weakly continuous in the appropriate Hilbert space, and a ball in a Hilbert space is compact in the weak topology.

Note that the field $(\Psi^{(n)}, A^{(n)})$ that realizes the minimum of (10.6) on fields of the form (10.9) does not necessarily correspond to a local minimum of (10.6) considered on all fields, even if $(\Psi^{(n)}, A^{(n)})$ is an extremum for (10.6). We will show that if $\lambda = e^2$, the field $(\Psi^{(n)}, A^{(n)})$ gives the minimum of (10.6) on the set of fields having topological number n. In other words, for $\lambda = e^2$, we have $E_n(\lambda) = E(\Psi^{(n)}, A^{(n)})$. A heuristic reasoning suggests that the same is true for $\lambda < e^2$, and also for $\lambda > e^2$ if $n = \pm 1$; but it is not true if $\lambda > e^2$ and $|n| > 1$. Thus, in the semiclassical approximation, the field $(\Psi^{(n)}, A^{(n)})$ with $\lambda > e^2$ corresponds to a topologically nontrivial stable particle only for $n = \pm 1$, while for $\lambda < e^2$ there exist stable charges with arbitrary topological charges.

To analyze the case $\lambda = e^2$ we integrate by parts the expression for the energy, reducing it to the form

$$(10.11) \quad E = \int \left(\tfrac{1}{2} a^2 H + \tfrac{1}{2}|\pi|^2 + \frac{1}{2e^2}(E_1^2 + E_2^2) + \tfrac{1}{2}|\nabla_1 \Psi - i\nabla_2 \Psi|^2 \right.$$
$$\left. + \frac{1}{2e^2}(H + \tfrac{1}{2}e^2(|\Psi|^2 - a^2))^2 \right) d^2x.$$

Here all terms on the right-hand side, except for the first, are non-negative; and, by (10.10), the first term coincides with the topological number n up to a factor. Therefore $E \geq \pi n a^2$, and $E = \pi n a^2$ if and only if the all but the first term of the integrand vanish. In other words, the energy is $\pi n a^2$ if $\pi(\mathbf{x}) = 0$, $\mathcal{E}(\mathbf{x}) = 0$,

$$(10.12) \qquad \nabla_1 \Psi - i\nabla_2 \Psi = 0, \quad \text{and} \quad H + \tfrac{1}{2}e^2(|\Psi|^2 - a^2) = 0.$$

This field clearly realizes the minimum of the energy functional among fields of the form (10.9), and therefore also the minimum of the energy functional among fields with a given topological number $n > 0$.

The case $n < 0$ is reduced to the case $n > 0$ by reflection in a plane; the energy functional is invariant under reflection, and the topological number changes sign. We thus get the bound $E \geq \pi |n| a^2$ and and an analog to (10.12) for fields such that $E = \pi |n| a^2$.

It can be proved that the system (10.12) and its analog for $n < 0$ has a $2|n|$-parameter family of solutions, where gauge-equivalent solutions are considered identical. Every field in this family realizes the minimum of the energy function on fields with topological number n, so that the second variation of the energy in each of these fields is non-negative definite. In can be proved that the second variation of the energy functional on $(\Psi^{(n)}, A^{(n)})$ is non-negative definite also if λ/e^2 is less than 1 but close enough to 1. This implies that the field $(\Psi^{(n)}, A^{(n)})$ corresponds to a minimum.

On the other hand, if λ/e^2 is greater than 1 and close enough to 1, the second variation of the energy ceases to have a constant sign, so that $(\Psi^{(n)}, A^{(n)})$ is not even a local minimum. The proof of this is based on the fact that the zero modes of the operator corresponding to the second variation of the energy functional at $(\Psi^{(n)}, A^{(n)})$ can be computed explicitly for $\lambda = e^2$; in addition to the zero modes corresponding to gauge degeneracy, there are $2|n|$ zero modes corresponding to the $2|n|$-parameter family of solutions of (10.12) that are gauge-inequivalent. This enables one to use perturbation methods to study the second variation for λ/e^2 close to one. For such values of λ one can show that $E(\Psi^{(n)}, A^{(n)}) > n(\Psi^{(1)}, A^{(1)})$ if $\lambda > e^2$; that is, fields having rotational symmetry turn out to be less energy-minimizing than superpositions of distant fields with $n = \pm 1$. The results obtained for λ/e^2 close to one agree with the assertions made earlier for the general case.

The mathematical problem discussed above arises also in statistical physics, and, in fact, it arose there before it did in quantum field theory. The free-energy functional in the Ginzburg–Landau model of superconductivity is given by

$$(10.13) \quad E = \frac{1}{2} \int |\operatorname{grad} \Psi + i\mathbf{A}\Psi|^2 \, d^3x$$
$$+ \tfrac{1}{8}\lambda \int (|\Psi|^2 - a^2)^2 \, d^3x + \frac{1}{2e^2} \int |\operatorname{rot} \mathcal{A}|^2 \, d^3x$$

where $\Psi(x)$ is the complex-valued order parameter and $\mathcal{A}$ is the electromagnetic field potential. The extrema of the functional (10.8) studied above are associated with topologically nontrivial strings of the functional (10.13), known as *Abrikosov vortices*. When $\lambda < e^2$ we get a type I superconductor, and when $\lambda > e^2$ a type II superconductor.

11. 't Hooft–Polyakov Monopoles

To construct a three-dimensional theory having topological integrals of motion, we start by considering the action functional

$$(11.1) \qquad S = \frac{1}{2} \int (\partial_\mu \varphi)^2 \, d^4x - \lambda \int (\varphi^2 - a^2)^2 \, d^4x,$$

where $\varphi = (\varphi^1, \varphi^2, \varphi^3)$ is a three-component scalar field. The energy functional then has the form

$$(11.2) \quad E = E_1 + E_2 = \frac{1}{2} \int \pi^2(\mathbf{x}) \, d^3x + \frac{1}{2} \int (\operatorname{grad} \varphi)^2 \, d^3x + \lambda \int (\varphi^2 - a^2)^2 \, d^3x,$$

where $\pi(\mathbf{x}) = \dot{\varphi}(\mathbf{x})$ is the generalized momentum corresponding to the generalized coordinate $\varphi(\mathbf{x})$.

We investigate the space of fields φ for which the integral

$$(11.3) \qquad \int (\varphi^2 - a^2)^2 \, d^3x$$

is finite. We assume that the fields considered are continuous, that the limit

$$(11.4) \qquad \lim_{\lambda \to \infty} \varphi(\lambda \mathbf{n}) = \boldsymbol{\Phi}(\mathbf{n})$$

exists for every vector $\mathbf{n}$, and that the convergence is uniform as $\mathbf{n}$ runs through the set of vectors of unit length. Under these conditions the vector $\boldsymbol{\Phi}(\mathbf{n})$ depends continuously on $\mathbf{n}$. From the finiteness of the integral it follows that $|\boldsymbol{\Phi}(\mathbf{n})| = a$; thus, if we look at $\boldsymbol{\Phi}(\mathbf{n})$ on the set of unit vectors $\mathbf{n}$, we obtain a continuous map from the two-sphere $|\mathbf{n}| = 1$ into the two-sphere $|\boldsymbol{\Phi}| = a$. The *topological number* of $\varphi(\mathbf{x})$ is the degree (T2.3) of this map $S^2 \to S^2$ describing the asymptotic behavior of $\boldsymbol{\Phi}(\mathbf{n})$. The topological number is an integer, and is therefore invariant under continuous changes in $\varphi(\mathbf{x})$, if we make the natural assumption that the asymptotic behavior of $\boldsymbol{\Phi}(\mathbf{n})$ varies continuously with $\varphi(\mathbf{x})$.

Thus, the set of fields for which (11.3) is finite is disconnected: it has one component for each value of the topological number. Just as for the model discussed in Chapter 10, this does not imply that the space of fields for which (11.2) is finite is disconnected, because, if $\int (\operatorname{grad} \varphi)^2 \, d^3x$ is finite, $\varphi(\mathbf{x})$ tends to the same value as one goes away to infinity in any direction. This means that the topological number is zero for any field for which (11.2) is finite.

To obtain a theory that has topological integrals of motion, we incorporate into from (11.1) the *Yang–Mills* or *gauge field* $\mathbf{A}_\mu(x) = (A^1_\mu, A^2_\mu, A^3_\mu)$, which is a generalization of the electromagnetic field. Namely, we consider the action integral

$$(11.5) \qquad S = \frac{1}{2} \int (\nabla_\mu \varphi)^2 \, d^4x - \lambda \int (\varphi^2 - a^2)^2 \, d^4x - \frac{1}{4g^2} \int \langle \mathcal{F}_{\mu\nu}, \mathcal{F}^{\mu\nu} \rangle d^4x,$$

where $\nabla_\mu \varphi(x) = \partial_\mu \varphi(x) + [\mathbf{A}_\mu(x), \varphi(x)]$ and $\mathcal{F}_{\mu\nu} = \partial_\mu \mathbf{A}_\nu - \partial_\nu \mathbf{A}_\mu + [\mathbf{A}_\mu, \mathbf{A}_\nu]$. Here $\mu = 0, 1, 2, 3$, the square brackets stand for the cross product, and the angle brackets for the scalar (dot) product.

The passage from (11.1) to (11.5) is a particular case of the procedure presented in Chapter 4, whereby one localizes the internal symmetry group on an action functional. Here, (11.1) has internal symmetry group SO(3), since S is invariant under transformations $\varphi(x) \mapsto V\varphi(x)$, for V an orthogonal matrix. After localization, S acquires local gauge symmetry: the functional (11.5) is invariant under the transformation $\varphi(x) \mapsto V(x)\varphi(x)$, for $V(x)$ an orthogonal matrix varying smoothly with x, provided the gauge field $\mathbf{A}_\mu(x)$ is also transformed appropriately.

Generally speaking, a gauge field takes its values in the Lie algebra of the internal symmetry group. In the present case, we think of $\mathbf{A}_\mu(x)$ as a three-dimensional vector, using the fact that the Lie algebra of SO(3) is isomorphic to $\mathbf{R}^3$, with multiplication given by the cross product.

The action functional (11.5) gives the boson part of the model suggested by H. Georgi and S. Glashow to describe weak and electromagnetic interactions; the model also contains a three-component fermion field that transforms via the vector representation of SO(3). Currently, after the discovery of neutral currents, this model cannot be considered realistic. Nonetheless, its study is important to the understanding of the physics of grand unification, because, like the grand unification models, it is based on a simple gauge group.

The energy functional derived from (11.5) is

$$(11.6) \qquad E = \frac{1}{2} \int \pi^2(\mathbf{x}) \, d^3x + \frac{1}{2} \int (\nabla_i \varphi)^2 \, d^3x + \lambda \int (\varphi^2 - a^2)^2 \, d^3x + E_{\mathrm{YM}},$$

where

$$E_{\mathrm{YM}} = \frac{1}{4g^2} \int \left(2\langle \mathcal{F}_{0i}, \mathcal{F}_{0i} \rangle + \langle \mathcal{F}_{ij}, \mathcal{F}_{ij} \rangle \right) d^3x$$

is the energy functional for the gauge field. Here we have imposed the gauge condition $\mathbf{A}_0(\mathbf{x}) = 0$, taking advantage of local gauge invariance.

The space of fields of finite energy is disconnected. Indeed, if (11.6) is finite, so is (11.3), and thus we can define the topological number n of the field, as in Chapter 10. Conversely, for every integer n, we can construct a field of finite energy having that topological number. For example, for $n = 1$ we can set

$$(11.7) \qquad A^i_j(\mathbf{x}) = \varepsilon_{kij} x^k \frac{\alpha(r) + 1}{r^2}, \qquad A^i_0(\mathbf{x}) = 0, \qquad \varphi(\mathbf{x}) = (\beta(r) + a)\mathbf{n},$$

where $k, i, j = 1, 2, 3$, $r = |\mathbf{x}|$, $n = \mathbf{x}/r$, and the functions $\alpha(r)$ and $\beta(r)$ decay rapidly and satisfy $\alpha(0) = -1$ and $\beta(0) = -a$. From now on we assume $\boldsymbol{\pi}(\mathbf{x}) \equiv 0$ and $\mathbf{E}(\mathbf{x}) \equiv 0$. This corresponds to studying the configuration space instead on the phase space, and obviously has no effect on the topological properties of the space.

To construct a field of finite energy with n arbitrary, we consider an arbitrary field $\boldsymbol{\Phi}(\mathbf{n})$ defined on the unit sphere $|\mathbf{n}| = 1$ and satisfying $|\boldsymbol{\Phi}(\mathbf{n})| = a$. We extend this field to the exterior of the unit ball, by setting $\varphi(\mathbf{x}) = \boldsymbol{\Phi}(\mathbf{x}/|\mathbf{x}|)$. We choose the gauge field $A_\mu(\mathbf{x})$ in such a way that the covariant derivative $\nabla_\mu \varphi(\mathbf{x}) = \partial_\mu \varphi(\mathbf{x}) + [A_\mu(\mathbf{x}), \varphi(\mathbf{x})]$ vanishes. This is possible because $|\boldsymbol{\Phi}(\mathbf{n})| = a$ implies that $\varphi(\mathbf{x})$ is orthogonal to the vectors $\partial_1 \varphi(\mathbf{x})$, $\partial_2 \varphi(\mathbf{x})$ and $\partial_3 \varphi(\mathbf{x})$, and therefore the field

$$A_\mu(\mathbf{x}) = \frac{1}{|\varphi(\mathbf{x})|^2}[\varphi(\mathbf{x}), \partial_\mu \varphi(\mathbf{x})] + \lambda_\mu(\mathbf{x})\varphi(\mathbf{x}),$$

where the $\lambda_\mu(\mathbf{x})$ are arbitrary functions, satisfies the desired condition. Inside the unit ball we define $\varphi(\mathbf{x})$ and $A_\mu(\mathbf{x})$ arbitrarily, the only requirement being that they be smooth functions of $\mathbf{x}$. It is easy to see that the energy of the field thus constructed is finite, at least in the case $\lambda_\mu(\mathbf{x}) \equiv 0$. Indeed, the energy density outside the unit ball reduces to the energy density of the gauge field $A_\mu(\mathbf{x})$, and the energy of the gauge field is finite because we can write $A_\mu(\mathbf{x}) = |x|^{-1} A_\mu(\mathbf{x}/|\mathbf{x}|)$, so that the energy density decreases as $|\mathbf{x}|^{-4}$.

We now discuss the physical meaning of the topological number of a field. We first observe that every vector φ can be transformed, by means of a rotation, into a vector with $\varphi^1 = \varphi^2 = 0$ and $\varphi^3 = |\varphi|$. In view of this, it is natural to impose on $\varphi(\mathbf{x})$ the gauge condition $\varphi^1(\mathbf{x}) = \varphi^2(\mathbf{x}) = 0$. It turns out, however, that this gauge condition cannot be imposed if the topological number of the field is non-zero, because a field satisfying this gauge condition has trivial asymptotic behavior at infinity: $\boldsymbol{\Phi}(\mathbf{n}) = V(\mathbf{n})\boldsymbol{\Phi}_0$, where $\boldsymbol{\Phi}_0 = (0, 0, a)$ and $V(\mathbf{n})$ is a continuous function defined on the unit two-sphere $|\mathbf{n}| = 1$ and taking values in $SO(3)$. But every mapping $S^2 \to SO(3)$ is null-homotopic, because $\pi_2(SO(3)) = 0$ (T10.2). This implies that $\boldsymbol{\Phi}(\mathbf{n}) = V(\mathbf{n})\boldsymbol{\Phi}_0$ is a null-homotopic map from the two-sphere S^2 into itself, and hence that $\varphi(\mathbf{x})$ has topological number zero. Therefore, if a field is gauge-equivalent to a field satisfying the condition $\varphi^1 = \varphi^2 = 0$, it is topologically trivial. The converse is also true.

The field $\varphi(\mathbf{x}) \equiv \boldsymbol{\Phi}_0 = (0, 0, a)$ has energy zero, and thus minimizes the functional (11.6), which is clearly non-negative. As usual, the energy of the field considered here is the semiclassical approximation to the ground-state energy (the energy of the physical vacuum). It is natural, then, to call this field the *classical vacuum*. To find the energies of weakly excited states, one must expand the action integral in a power series in the deviations from the ground-state field, and retain only the quadratic terms. Doing this to (11.5), we arrive at the following expression for the quadratic part of the action integral:

(11.8) $S_{\text{qu}} = -\dfrac{1}{4g^2} \displaystyle\int \left((\partial_\mu \mathbf{A}_\nu - \partial_\nu \mathbf{A}_\mu)^2 - 2a^2 g^2 ((A_\mu^1)^2 + (A_\mu^2)^2) \right) d^4x$

$$+ \frac{1}{2} \int \left((\partial_\mu \eta)^2 - 8\lambda a^2 \eta^2 \right) d^4x,$$

where $\eta(x) = \varphi^3(x) - a$. Quantization gives a theory that contains a scalar particle of mass $2a\sqrt{2\lambda}$, two vector particles of mass ag, and one massless vector particle. The massless particle corresponds to the filed A_μ^3 and must be identified with the photon, and the field A_μ^3 with the electromagnetic field.

Notice that the classical vacuum $\boldsymbol{\Phi}_0$ has SO(2)-symmetry, instead of the SO(3)-symmetry of the original Lagrangian. The vector fields corresponding to the broken symmetries acquire mass. This phenomenon is a manifestation of the *Higgs effect*.

In the gauge $\varphi^1 = \varphi^2 = 0$, the field $A_\mu^3(x)$ can be seen as the electromagnetic field, so $F_{\mu\nu} = \partial_\mu A_\nu^3 - \partial_\nu A_\mu^3$ can be seen as the electromagnetic field tensor. The electromagnetic field tensor can be expressed in a form independent of the choice of gauge as

(11.9) $F_{\mu\nu} = \left\langle \mathcal{F}_{\mu\nu}, \dfrac{\varphi}{|\varphi|} \right\rangle - \dfrac{1}{|\varphi|^3} \varepsilon_{klm} \varphi^k (\nabla_\mu \varphi)^l (\nabla_\nu \varphi)^m ;$

this is the correct expression, because it is gauge-invariant, and reduces to the standard expression in the gauge $\varphi^1 = \varphi^2 = 0$. This tensor satisfies Maxwell's equation

$$\partial_\sigma F_{\mu\nu} + \partial_\mu F_{\nu\sigma} + \partial_\nu F_{\sigma\mu} = 0;$$

this can be seen immediately in the gauge $\varphi^1 = \varphi^2 = 0$, and is true in any gauge because of gauge invariance. In particular, the magnetic field $\mathbf{H} = (F_{23}, F_{31}, F_{12})$ has divergence zero.

The expression (11.9) for the electromagnetic field tensor retains its meaning for topologically nontrivial fields at any point where $\varphi(\mathbf{x}) \neq 0$. The justification given above, involving the imposition of the global gauge condition $\varphi^1 = \varphi^2 = 0$, cannot be used non-trivial fields; however, the electromagnetic field is determined locally, and in any small region where $\varphi(\mathbf{x})$ does not vanish the gauge condition $\varphi^1 = \varphi^2 = 0$ can be imposed.

Using (11.9), we can link the topological number to the magnetic charge of the field. The magnetic charge is defined as

(11.10) $\mathfrak{m} = \dfrac{1}{4\pi} \displaystyle\oint \mathbf{H} \, d\mathbf{S},$

where the integral stands for the flux of $\mathbf{H}$ through an infinitely distant sphere. In classical electrodynamics the equation $\operatorname{div} \mathbf{H} = 0$ holds true at all points, so that (11.10) is always zero. In the present model, on the other hand, the magnetic field is only defined at points where $\varphi(\mathbf{x}) \neq 0$, so that $\operatorname{div} \mathbf{H} = 0$ only at those points. This leaves open the possibility of the existence of fields with a non-zero magnetic charge. In particular, it is easy to verify that

$$(11.11) \qquad\qquad \mathbf{H} = \frac{\mathbf{x}}{|\mathbf{x}|^3},$$

for the field (11.7), so this field has magnetic charge 1. (The definitions of A_μ and F_μ adopted here differ from the standard definitions by a factor of g^{-1}. If we use the standard notation, the magnetic charge is g^{-1}.) We see that for the field (11.7) the topological number coincides with the magnetic charge. This is not accidental, and it holds true in general.

We now analyze in greater detail the concepts of topological number and magnetic charge, and in particular prove that the two coincide. First we note that the topological number of a field $\varphi(\mathbf{x})$ can be defined in much greater generality. It is enough to assume that there is some ball D^3 outside of which $\varphi(\mathbf{x})$ does not vanish. Then on the sphere S^2 bounding this ball, $\varphi(\mathbf{x})$ can be thought of a map from S^2 to $\mathbf{R}^3 \setminus \{0\}$. Such a mapping can be assigned a degree (really the degree of the mapping $S^2 \to S^2$ obtained by composing with the projection $\mathbf{x} \mapsto \mathbf{x}/|\mathbf{x}|$); this integer is then, by definition, the *topological number* of $\varphi(\mathbf{x})$. The topological number is obviously independent of the choice of S^2, and when the field has asymptotic behavior $\boldsymbol{\Phi}(\mathbf{n})$, this number coincides with the topological number defined earlier. The topological number (under the new definition) has the analytic expression

$$(11.12) \qquad n = -\frac{1}{4\pi} \sum_{\mu<\nu} \oint \frac{\varepsilon_{abc}\varphi^a\, \partial_\mu\varphi^b\, \partial_\nu\varphi^c}{|\varphi|^3}\, dx^\mu\, dx^\nu,$$

where the integral is over any sphere outside of which $\varphi(\mathbf{x})$ does not vanish. This expression is derived in T2.3, and generalized in T2.1.

Formula (11.9) for the electromagnetic field tensor is meaningful whenever $\varphi(\mathbf{x})$ does not vanish anywhere outside some ball D^3—the same circumstances under which we can talk of the topological number of $\varphi(\mathbf{x})$. When calculating the magnetic charge, we can use any sphere S^3 bounding such a ball; by Gauss's divergence theorem, the flux of $\mathbf{H}$ is the same for all such spheres.

The magnetic charge does not change under continuous variations of φ and $\mathbf{A}_\mu$, the only requirement being that in the process φ does not vanish outside a certain ball. Indeed, suppose that $(\varphi(\mathbf{x}, \tau), \mathbf{A}_\mu(\mathbf{x}, \tau))$ is a family of fields that depends continuously on the parameter τ, for $0 \leq \tau \leq 1$. Consider the field $(\varphi(\mathbf{x}, \tau(\mathbf{x})), \mathbf{A}_\mu(\mathbf{x}, \tau(\mathbf{x})))$, where $\tau(\mathbf{x})$ equals 0 for $|\mathbf{x}| < 2L$ and 1 for $|\mathbf{x}| > 3L$. Expressing the magnetic charge of this field first in terms of the flux of $\mathbf{H}$ through a sphere of radius less than $2L$ and then in terms of the flux of $\mathbf{H}$ through a sphere of radius greater than $3L$, we find that it coincides with the magnetic charge of the field $(\varphi(\mathbf{x}, \tau), \mathbf{A}_\mu(\mathbf{x}, \tau))$ at $\tau = 0$ and $\tau = 1$. On the other hand, the flux does not depend on the choice of sphere, so the magnetic charges of $(\varphi(\mathbf{x}, \tau), \mathbf{A}_\mu(\mathbf{x}, \tau))$ for $\tau = 0$ and $\tau = 1$ coincide.

We are now ready to prove that the topological number given by formula (11.12) coincides with the magnetic charge. Indeed, consider the family of fields $(\varphi(\mathbf{x}), \tau\mathbf{A}_\mu(\mathbf{x}))$, which depends continuously on the parameter τ. For $\tau = 0$ the magnetic charge of this field equals the expression (11.12) for the topological number of φ. Since the magnetic charge does not depend on τ, the equality remains true for $\tau = 1$, that is, for an arbitrary field $(\varphi(\mathbf{x}), \mathbf{A}_\mu(\mathbf{x}))$. We remark that $(\varphi(\mathbf{x}), \tau\mathbf{A}_\mu(\mathbf{x}))$ can have infinite energy for certain values of τ, even if the energy of $(\varphi(\mathbf{x}), \mathbf{A}_\mu(\mathbf{x}))$ is finite. This, however, is not a problem, because the magnetic charge still has meaning.

12. Topological Integrals of Motion in Gauge Theory

The existing models combining strong, weak and electromagnetic interactions (grand unification models) are built using the same principles as the theory of electroweak interaction. We take the Lagrangian

$$(12.1) \qquad \mathcal{L} = \mathcal{L}_0 - \Gamma \bar{\psi} \psi \varphi - U(\varphi),$$

where $\psi = (\psi^1, \ldots, \psi^m)$ is a multicomponent fermion field, $\varphi = (\varphi^1, \ldots, \varphi^m)$ a multicomponent scalar field, and $\mathcal{L}_0$ the free Lagrangian describing the interaction of these fields. Assume that (12.1) is invariant under an internal symmetry group G. This means that ψ and φ transform in a certain way under transformations in G (they take values in some representation space of G) and that the Lagrangian is a scalar with respect to this group; in particular, the polynomial $U(\varphi)$ and the expression $\Gamma \bar{\psi} \psi \varphi = G_{ijk} \bar{\psi}^i \psi^j \varphi$ are G-invariant.

As explained in Chapter 4, we construct from $\mathcal{L}$ a Lagrangian $\hat{\mathcal{L}}$, invariant under local gauge transformations corresponding to G-valued functions $g(x)$, and involving the gauge fields $A_\mu(x)$, which take values in the Lie algebra $\mathcal{G}$ of G. We do this by replacing all derivatives ∂_μ by covariant derivatives ∇_μ, and adding the term

$$\mathcal{L}_{\text{YM}} = -\frac{1}{4e^2} \langle \mathcal{F}_{\mu\nu}, \mathcal{F}^{\mu\nu} \rangle,$$

where $\mathcal{F}_{\mu\nu} = \partial_\mu A_\nu - \partial_\nu A_\mu + [A_\mu, A_\nu]$ is the gauge field strength, and the angle brackets stand for the invariant scalar (dot) product in the Lie algebra of G.

Thus, a "field" in the theory described by $\hat{\mathcal{L}}$ comprises a fermion field $\psi(x)$, a scalar field $\varphi(x)$ and a gauge field $A_\mu(x)$. Fields differing by a gauge transformation are physically equivalent. This allows us to restrict the class of fields that must be considered by the imposition of a *gauge condition*, chosen in such a way that any field is equivalent, under a gauge transformation, to a field satisfying the gauge condition. We usually impose the gauge condition $A_0 = 0$.

What topological integrals of motion exist in the theory described by the Lagrangian $\hat{\mathcal{L}}$ obtained from (12.1)? To answer this question, we note first that fermions are inessential to the problem, so we can keep just the boson part of the Lagrangian:

$$(12.2) \qquad \hat{\mathcal{L}} = \tfrac{1}{2} \langle \nabla_\mu \varphi, \nabla^\mu \varphi \rangle - U(\varphi) + \mathcal{L}_{\text{YM}}.$$

Here $\varphi(x) = (\varphi^1(x), \ldots, \varphi^n(x))$ is an n-component (that is, $\mathbf{R}^n$-valued) scalar field, which transforms according to some representation T of the compact Lie group G; the gauge field A_μ takes values in the Lie algebra $\mathcal{G}$ of G, and $\nabla_\mu = \partial_\mu + t(A_\mu)\varphi$, where t is the representation of $\mathcal{G}$ corresponding to the representation T of G. The subset of $\mathbf{R}^n$ on which the function $U(\varphi) = U(\varphi^1, \ldots, \varphi^n)$ achieves its minimum is called the set of *classical vacuums*, and is denoted by R. Assume that R is a submanifold of $\mathbf{R}^n$. Without loss of generality we can assume that the value of $U(\varphi)$ on R is zero.

What are the topological properties of the phase space in this theory? We impose the gauge condition $A_0 = 0$. Then the energy functional can be written as

$$(12.3) \qquad E = \int (\tfrac{1}{2}\pi^2 + \tfrac{1}{2}\langle \nabla_i\varphi, \nabla_i\varphi \rangle + U(\varphi))\, d^3x + E_{\mathrm{YM}},$$

where

$$E_{\mathrm{YM}} = \int \left(\frac{1}{2e^2}\langle E_i, E_i \rangle + \frac{1}{4e^2}\langle \mathcal{F}_{ij}, \mathcal{F}_{ij} \rangle \right) d^3x$$

with $i, j = 1, 2, 3$, $\pi(x) = \dot{\varphi}(x)$ is the generalized momentum associated with the generalized coordinate $\varphi(x)$, and $E_i(x) = -\dot{A}_i(x) = \mathcal{F}_{i0}(x)$ is proportional to the generalized momentum associated with $A_i(x)$. The phase space P consists of fields $(\pi(\mathbf{x}), \varphi(\mathbf{x}), E_i(\mathbf{x}), A_i(\mathbf{x}))$ for which (12.3) is finite.

Obviously, P has one connected component for each component of the space P_0 of fields $(\varphi(\mathbf{x}), A_i(\mathbf{x}))$ for which the functional

$$(12.4) \qquad E[\varphi, A] = -\int \mathcal{L}\, d^3x$$

$$= \int \left(\tfrac{1}{2}\langle \nabla_i\varphi, \nabla_i\varphi \rangle + U(\varphi) + \frac{1}{4e^2}\langle \mathcal{F}_{ij}, \mathcal{F}_{ij} \rangle \right) d^3x$$

is finite, where $i, j = 1, 2, 3$. For this reason we restrict our attention to P_0 from now on.

Consider the mapping σ that assigns to each point $(\varphi^1, \ldots, \varphi^n) \in \mathbf{R}^n$ the nearest point of R. This map is well defined, one-to-one and continuous in a small enough neighborhood $\mathcal{R}$ of R (T4.4). We fix such a neighborhood.

The finiteness of (12.4) requires that, as $|\mathbf{x}| \to \infty$, the field $\varphi(\mathbf{x})$ approach the manifold R of classical vacuums. We therefore insist that the values of φ outside a ball D^3 must belong to the neighborhood $\mathcal{R}$ of R. In other words, we require that the set of points for which $\varphi(\mathbf{x}) \notin \mathcal{R}$ be bounded.

To each field $\varphi(\mathbf{x})$ we assign the homotopy class $\zeta(\varphi)$ of the mapping $\sigma\varphi$ from a large enough sphere into R. This is well-defined because, by assumption, φ takes large spheres into $\mathcal{R}$, and σ takes $\mathcal{R}$ into R. We call $\zeta(\varphi)$ the *topological type* or *topological charge* of φ.

From now on we assume that fields with finite energy behave well at infinity, that is, that the covariant derivative $\nabla_i\varphi(x)$ decays fast enough as $|\mathbf{x}| \to \infty$:

$$(12.5) \qquad |\nabla_i\varphi(\mathbf{x})| \le \mathrm{const}\, |\mathbf{x}|^{-(1+\delta)}, \qquad \text{with } d > 0.$$

The function $\nabla_i\varphi(\mathbf{x})$ is square-integrable because the energy functional is assumed finite. If $\nabla_i\varphi(\mathbf{x}) \sim |\mathbf{x}|^{-\alpha}$ as $|\mathbf{x}| \to \infty$, square integrability implies that $\alpha > \frac{3}{2}$. This means that the condition we have imposed will certainly be met if $\nabla_i\varphi(\mathbf{x})$ tends to zero is more or less at the same rate in all directions.

We also assume that along each ray $\lambda\mathbf{n}$, with $|\mathbf{n}| = 1$ and $0 < \lambda < \infty$, the field φ has a finite limit $\Phi(\mathbf{n}) = \lim_{n\to\infty}\varphi(\lambda\mathbf{n})$, and that it approaches that limit fast enough: $|\Phi(\mathbf{n}) - \varphi(\lambda\mathbf{n})| < \mathrm{const}\,\lambda^{-\delta}$ for $\delta > 0$. Note that this is true if the gauge field satisfies (12.5) and the gauge condition $x^i A_i = 0$, for then

$$|x^i \partial_i \varphi(\mathbf{x})| = |x^i \nabla_i \varphi(\mathbf{x})| \leq \mathrm{const}\,|\mathbf{x}|^{-\delta},$$

so that

$$\left|\frac{\partial \varphi(\lambda\mathbf{n})}{\partial \lambda}\right| \leq \mathrm{const}\,\alpha^{-(1+\delta)}.$$

Under the conditions above, the finiteness of (12.4) implies that $\Phi(\mathbf{n}) \in R$ for every direction $\mathbf{n}$. The topological charge of φ can be defined as the homotopy type of Φ, considered as a map from the unit sphere $|\mathbf{n}| = 1$ into R. This definition coincides with the one given earlier because $\varphi(\mathbf{n})$ and $\sigma\varphi(\lambda\mathbf{n})$ differ little for large $|\mathbf{n}|$, so the homotopy types of Φ and $\sigma\varphi$ coincide, the latter being seen as a map on a sphere of large radius.

If the field (φ, A_μ) depends continuously on a parameter τ and has finite energy for every τ, its magnetic charge is independent of τ. More precisely, we must require that when τ varies within a bounded interval, there exists a ball D^3 outside of which φ_τ takes values in the neighborhood $\mathcal{R}$ of R. Then the topological charge of φ_τ can be calculated using the map defined by φ_τ on the boundary of this ball.

In particular, since the field changes continuously with time, the topological charge is an integral of motion. In other words, we can talk of the topological charge of a time-dependent field $\varphi(x) = \varphi(t, \mathbf{x})$, defining it as the topological charge of $\varphi(t_0, \mathbf{x})$, for a fixed value of t_0. It suffices to assume that for every bounded interval $[t_0, t_1]$ there is a ball $D^3 \subset \mathbf{R}^3$ such that $\varphi(t, \mathbf{x}) \in \mathcal{R}$ for all $t_0 \leq t \leq t_1$ and all $x \notin D^3$.

The group G acts on the manifold R of classical vacuums: since $U(\varphi)$ is invariant under G, an element of G takes a point that minimizes U to another such point. We will always assume that G acts on R transitively, that is, that any classical vacuum can be obtained from any other by the action of a transformation in G. Physically, this means that the degeneracy of the classical vacuum is due solely to invariance under G. Then R can be thought of as the right coset space G/H, where H, the *group of unbroken symmetries*, is the subgroup of G that fixes a given classical vacuum (Chapter 40 and T9.3).

Under the assumptions above, the set of homotopy classes $\{S^2, R\}$ can easily be calculated. We assume that G is simply connected; as explained in Chapter 8 (see also T14.1), this can be done without loss of generality, by replacing G with its universal cover. Then the homotopy group $\pi_2(R) = \pi_2(G/H)$ is isomorphic to the group $\pi_1(H)$ (T14.1). If H is connected, $R = G/H$ is simply connected and the set of homotopy classes $\{S^2, R\}$ can be identified with the homotopy group $\pi_2(R) = \pi_1(H)$, so that the topological type of the field is determined by an element of $\pi_1(H)$.

If H is isomorphic to $\mathrm{SU}(3) \times U(1)$, we have $\pi_1(H) = \pi_1(U(1)) = \mathbf{Z}$, that is, the topological type is an integer. (Later we will see how this number can be identified with the magnetic charge.) In the $\mathrm{SU}(5)$ grand unification model, and in all grand unification models encountered in the literature, this is still true, although H and $\mathrm{SU}(3) \times U(1)$ are only locally isomorphic (the symmetries in the color group $\mathrm{SU}(3)$ and in the electromagnetic group are not broken). To compute $\pi_1(H)$ in the case of local isomorphism $H \approx \mathrm{SU}(3) \times U(1)$, we notice that the universal cover of $\mathrm{SU}(3) \times U(1)$ is $\mathrm{SU}(3) \times \mathbf{R}$ (since $\mathrm{SU}(3)$ and $\mathbf{R}$ are simply connected, and the map $\lambda \mapsto e^{i\lambda}$ is a covering map $\mathbf{R} \to U(1)$). Therefore there is a covering homomorphism $\mathrm{SU}(3) \times \mathbf{R}_+ \to H$, and $\pi_1(H)$ is isomorphic to the kernel D of this homomorphism, since $\mathrm{SU}(3) \times \mathbf{R}_+/D = H$ (T14.1).

For the $\mathrm{SU}(5)$ model, the covering homomorphism takes $(u, \lambda) \in \mathrm{SU}(3) \times \mathbf{R}$ to the matrix (7.2), with $K = ue^{-i\lambda/3}$ and $l = e^{i\lambda}$. Thus the kernel D is the set of pairs $(e^{i2\pi k/3}, e^{2\pi k})$, with $k \in \mathbf{Z}$, and in particular is isomorphic to $\mathbf{Z}$. A loop in H that begins and ends at the identity can be lifted to a path in the universal cover $\mathrm{SU}(3) \times \mathbf{R}$ connecting the identity with a point of D. Therefore it has the form $(K, l) = (u(t)e^{-i\lambda(t)/3}, e^{i\lambda(t)})$, where $u : [0,1] \to \mathrm{SU}(3)$ and $\lambda : [0,1] \to \mathbf{R}$ are continuous. Its homotopy class is given by the integer $(2\pi)^{-1}(\lambda(1) - \lambda(0))$, and depends only on $l(t)$, not on $K(t)$—that is, it is determined solely by its projection in $U(1)$. In other words, the projection map $H \to U(1)$ taking the matrix (7.2) to $l \in U(1)$ gives rise to an isomorphism between $\pi_1(H)$ and $\pi_1(U(1)) = \mathbf{Z}$.

Generally, D is a discrete subgroup of the center of $\mathrm{SU}(3) \times \mathbf{R}$. This center is isomorphic to $\mathbf{Z}_3 \dotplus \mathbf{R}$, where $\mathbf{Z}_3$ is the group of order three. Every discrete subgroup of $\mathbf{Z}_3 \dotplus \mathbf{R}$ is isomorphic to either $\mathbf{Z}$ or $\mathbf{Z}_3 \dotplus \mathbf{Z}$. If $D = \mathbf{Z}_3 \dotplus \mathbf{Z}$, we have $H = (\mathrm{SU}(3)/\mathbf{Z}_3) \times U(1)$. This is possible only if all scalar fields transform according to representations of $\mathrm{SU}(3)$ in which the center $\mathbf{Z}_3$ of this group acts trivially. In the existing models this is not the case, so $\pi_1(H) = \mathbf{Z}$.

Under the assumption that G is simply connected, $\pi_1(R)$ is isomorphic to $\pi_0(H)$, which in turn equals H/H_{con} (we recall from Chapter 8 that H_{con} is the group of *continuous unbroken symmetries*, the maximal connected subgroup of H). If there are discrete unbroken symmetries, that is, if H is not connected, it follows that R is not simply connected. Furthermore, $\{S^2, R\}$ can be thought of as the set of orbits of $\pi_1(R) = \pi_0(H)$ acting on $\pi_2(R) = \pi_1(H)$ (see T8.2). In this chapter and in Chapter 15 we assume that H is connected, and defer to Chapters 19 and 20 the study of effects that arise when there are discrete unbroken symmetries.

The topological charge of $\varphi(x)$ does not change under gauge transformations. This is because, for a given gauge transformation specified by a function $g(\mathbf{x})$, one can find a continuous family of gauge transformations $g_\tau(\mathbf{x})$ joining $g(\mathbf{x})$ to the identity transformation.

If $g(\mathbf{x})$ is defined in all of $\mathbf{R}^3$, this follows immediately from the contractibility of $\mathbf{R}^3$. It remains true when $g(\mathbf{x})$ is defined only outside a ball, for in this case the domain of definition of $g(\mathbf{x})$ is homotopically equivalent to the sphere S^2. The analysis of a map form this domain into G reduces to the the analysis of a map $S^2 \to G$, and this implies the assertion in general because $\pi_2(G) = 0$ for any Lie group G.

We now show that the space P_0 of fields (φ, A) with finite energy (12.4) contains fields of arbitrary topological type. (Naturally, we mean the topological type of φ.) We must construct, for each homotopy class of maps $S^2 \to R$, a field φ whose asymptotic behavior determines the prescribed homotopy class, and then find for φ a field A such that $E(\varphi, A)$ is finite. Consider an arbitrary smooth map α from the unit sphere S^2 into R, and define the scalar field for $|x| \geq 1$ by the formula $\varphi(\mathbf{x}) = \alpha(\mathbf{x}/|\mathbf{x}|)$. To define the gauge field, set

$$(12.6) \qquad\qquad A_\mu(\mathbf{x}) = \frac{1}{|\mathbf{x}|} a_\mu \left(\frac{\mathbf{x}}{|\mathbf{x}|} \right)$$

for $|\mathbf{x}| \geq 1$, where the function $a_\mu(\mathbf{n})$ is chose so that $\nabla_\mu \varphi = 0$ for $|\mathbf{x}| \geq 1$ (we will see in the next paragraph that such a function must exist). Then extend $\varphi(\mathbf{x})$ and $A_\mu(\mathbf{x})$ to the interior of the unit ball arbitrarily, the only condition being that they be smooth. It is clear that the energy of the resulting field (φ, A) is finite: outside the unit ball all terms in (12.4) vanish, except for the gauge field, and the energy of the gauge field is finite because the field strength decreases as $|\mathbf{x}|^{-2}$.

We show now that we can choose a_μ so that $\nabla_\mu \varphi = 0$. The desired equation can also be written as

$$(12.7) \qquad\qquad \alpha(\mathbf{n} + d\mathbf{n}) = \alpha(\mathbf{n}) + t(a_\mu(\mathbf{n}))\alpha(\mathbf{n}) \, dn^\mu,$$

where $|\mathbf{n}| = 1$ and $(\mathbf{n}, d\mathbf{n}) = 0$. Now, for every $\mathbf{n}$ we can find $a_\mu(\mathbf{n})$ satisfying (12.7); this is because G acts transitively on R, so a point $\alpha(\mathbf{n} + d\mathbf{n})$ lying infinitely close to $\alpha(\mathbf{n})$ can be obtained from $\alpha(\mathbf{n})$ by an infinitesimal transformation in G, that is, by an element of the Lie algebra of G. This element depends linearly on $d\mathbf{n}$, and so can be represented as $a_\mu(\mathbf{n})dn^\mu$.

The choice of $a_\mu(\mathbf{n})$ is not unique; for example, we could add to $a_\mu(\mathbf{n})$ any element in the Lie algebra of the stabilizer of $\alpha(\mathbf{n})$, that is, any $b_\mu(\mathbf{n})$ such that $t(b_\mu(\mathbf{n}))\alpha(\mathbf{n}) = 0$. Therefore, we still have to check whether the choice of $a_\mu(\mathbf{n})$ can be made to depend continuously in $\mathbf{n}$. To do this, we use a topological argument: for each $\mathbf{n}$ the vectors $a_\mu(\mathbf{n})$ that obey (12.7) form a linear space $Q(\mathbf{n})$. The collection of spaces $Q(\mathbf{n})$ forms a fiber bundle over the sphere S^2; a continuous choice of $a_\mu(\mathbf{n})$ means the choice of a continuous section of this bundle. Such a section exists because the bundle has a contractible fiber.

We have established that the space of fields of finite energy has exactly one connected component for each element of the group $\pi_1(H) = \pi_2(R)$. The group $\pi_1(H)$ is isomorphic to the direct sum of a finite abelian group with r copies of $\mathbf{Z}$, where r is the dimension of the center of H. This follows from the fact that a connected compact Lie group H is locally isomorphic to the product of r copies of $U(1)$ and a compact, simply connected Lie group (T14.1). We thus see that the topological type of a field is characterized by the choice of r integers and an element of a finite abelian group. Since a finite abelian group can be written as a direct sum $\mathbf{Z}_{m_1} + \cdots + \mathbf{Z}_{m_n}$, an element of it can be thought of as an n-tuple $(k_1, \ldots, k_n)$, where k_i is a residue modulo m_i. These integers and residues will be called the *topological numbers* or *topological charges* of the field.

The topological charge can be defined even if the domain of definition V of the field is not all of $\mathbf{R}^3$, provided that, on the boundary Γ of V, the field takes values near a classical vacuum (say in the neighborhood $\mathcal{R}$ of R.) In this case, we consider the topological charge of the field φ defined on V as the homotopy class ζ_Γ of the map $\sigma : \Gamma \to R$. We assume here that the boundary Γ of V is homeomorphic to S^2, so that $V \cup \Gamma$ is homeomorphic to the ball D^3. Then ζ_Γ can be thought of as an element of $\pi_2(R) = \pi_1(H)$.

We denote by K_Γ the set of points $\mathbf{x} \in \Gamma$ such that $\varphi(\mathbf{x}) \notin \mathcal{R}$. If K_Γ is empty, the mapping $\sigma : \Gamma \to R$ can be continued into V, and must be null-homotopic, so that $\zeta_\Gamma = 0$. Furthermore, $\zeta_{\Gamma_0} = \zeta_{\Gamma_1}$ if $K_{\Gamma_0} = K_{\Gamma_1}$, that is, the topological type of the field inside Γ is determined solely by K_Γ. Indeed, if $K_{\Gamma_0} = K_{\Gamma_1}$, we can find a continuous family of surfaces Γ_t joining Γ_0 and Γ_1, all homeomorphic to the sphere and satisfying $\varphi(\mathbf{x}) \in \mathcal{R}$ at every point $\mathbf{x} \in \Gamma_t$. Then $\sigma\varphi$ is defined on Γ_t, so the homotopy type of $\sigma\varphi : \Gamma_t \to R$ does not depend on t.

The topological charge of φ is additive: if K_Γ is the union of disjoint sets K_{Γ_1} and K_{Γ_2}, we have $\zeta_\Gamma = \zeta_{\Gamma_1} + \zeta_{\Gamma_2}$. This follows from the results in T7.1: we assume, without loss of generality, that Γ_1 and Γ_2 are disjoint and lie inside a ball bounded by Γ, and we look at the region bounded outside by Γ and inside by Γ_1 and Γ_2.

To interpret physically the additivity of the topological charge, we consider a field φ whose energy is localized in a domain D equal to the union of domains D_1 and D_2 that are far apart. Outside $D = D_1 \cup D_2$ the field must be close to a vacuum field—that is, have values in $\mathcal{R}$—so we can talk of the topological charge of φ in D, D_1 and D_2. Additivity means that the topological charge in D equals the sum of the topological charges in D_1 and in D_2.

Up to this point our reasoning has been classical. However, as we noted in Chapter 9, the existence of topological integrals of motion in the classical theory (corresponding to the fact that the phase space is disconnected) usually implies their existence in the corresponding quantum theory as well. The current situation is no exception.

Mathematically, the concept of the topological type of a field inside a surface Γ is similar to the concept of the topological type of a defect of local equilibrium, as defined in Chapter 8. The points at which the field $\varphi(x)$ takes values in $\mathcal{R}$ are to be seen as analogous to the points where local equilibrium is present. The set K_Γ is the analog of the set of defects in local equilibrium (points in Γ at which local equilibrium is violated). The proof that the topological charge is additive is similar to the proof of a similar assertion in Chapter 8.

13. Particles in Gauge Theories

We now analyze the particles that appear in the theory described by the Lagrangian (12.2). As usual, we must expand the Lagrangian in the neighborhood of a classical vacuum φ_0, having lifted the degeneracy of the classical vacuum by imposing a gauge condition. A natural gauge condition is

$$(13.1) \qquad \sigma\varphi(x) = \varphi_0,$$

where σ, as before, is the map taking a point in $\mathbf{R}^n$ to the nearest point in R. We recall that, in general, σ is one-to-one and continuous only in a neighborhood of R, say $\mathcal{R}$, so that the gauge condition (13.1) is only meaningful for fields that are near the vacuum manifold. Generally, the fields considered here satisfy this condition only for $|\mathbf{x}|$ large enough, so we will consider gauge transformations defined outside a certain ball, rather than on the whole of $\mathbf{R}^3$. (Also, the gauge transformation can depend on time, but this is not important to the current discussion.)

The map σ commutes with transformations in G, since such transformations preserve distances and map R into itself:

$$(13.2) \qquad \sigma T(g) = T(g)\sigma.$$

In view of this and of the assumption that G acts transitively on R, we can find for every point $\varphi \in \mathcal{R}$ a transformation $g \in G$ such that $T(g)\varphi$ satisfies the gauge condition (13.1): we just choose the element that takes $\sigma\varphi$ to φ_0, so that

$$\sigma T(g)\varphi = T(g)\sigma\varphi = \varphi_0.$$

This argument implies that every field $\varphi(\mathbf{x})$ can be transformed into a field satisfying (13.1) by means of a gauge transformation. If φ is topologically nontrivial, the gauge transformation must be discontinuous, because any field satisfying (13.1) is topologically trivial, and continuous gauge transformations, even when defined only in the complement of a ball, preserve topological type, as we have seen. Since we are only considering continuous (and in fact smooth) gauge transformations, it is not the case that an arbitrary field $\varphi(\mathbf{x})$ can be gauge-transformed to one satisfying (13.1). However, it is always possible to effect such a gauge transformation inside a contractible domain $V \subset \mathbf{R}^3$ (provided, of course, that the values of φ in V belong to the set $\mathcal{R}$ where the mapping σ is one-to-one and continuous).

To verify this, consider for each point $\mathbf{x}$ the set $H_{\mathbf{x}}$ of elements $g \in G$ satisfying $T(g)\sigma\varphi = \varphi_0$; we already know that this set is non-empty. In fact, $H_{\mathbf{x}}$ is homeomorphic to the stabilizer $H = H(\varphi_0)$: if $T(g)\sigma\varphi(x) = T(g')\sigma\varphi(x)\varphi_0$, we have $(T(g'))^{-1}\varphi_0 = (T(g))^{-1}\varphi_0$, so that $g'g^{-1} \in H$. The union of all the sets $H_{\mathbf{x}}$ forms a fiber bundle over the domain of definition of $\varphi(\mathbf{x})$, the fibers being the $H_{\mathbf{x}}$. Finding the required gauge transformation means choosing for each fiber $H_{\mathbf{x}}$ an element $g(\mathbf{x})$ that depends continuously on $\mathbf{x}$, or, equivalently, constructing a section of the bundle. But a bundle over a contractible base space always has a section (T11.2), so the desired gauge transformation exists.

Consider the Lagrangian (12.2) under the gauge condition (13.1). The expansion of this Lagrangian in terms of the gauge field, in the neighborhood of the vacuum φ_0, contains a quadratic term

$$(13.3) \qquad \tfrac{1}{2}\langle t(A_\mu)\varphi_0, t(A^\mu)\varphi_0\rangle,$$

which comes from the term $\tfrac{1}{2}\langle \nabla_\mu\varphi, \nabla^\mu\varphi\rangle$ in (12.2). This term gives rise to the mass terms in the gauge fields. The Lie algebra $\mathcal{G}$ of G is the direct sum of subspaces $\mathcal{H}$ and $\mathcal{H}^\perp$, where $\mathcal{H}$ is the Lie algebra of the subgroup H: that is, any $A \in \mathcal{G}$ can be represented uniquely as $A' + A''$, with $A' \in \mathcal{H}$ and A'' orthogonal to $\mathcal{H}$ (in the sense of the invariant scalar product). Clearly, $t(A')\varphi_0 = 0$ for $A' \in \mathcal{H}$. Writing $A_\mu(x) = A'_\mu(x) + A''_\mu(x)$, where $A'_\mu(x) \in \mathcal{H}$ and $A''_\mu(x) \in \mathcal{H}^\perp$, we see that (13.3) equals

$$(13.4) \qquad \tfrac{1}{2}\langle t(A''_\mu)\varphi_0, t(A''^\mu)\varphi_0\rangle.$$

This means that the gauge fields corresponding to the generators of the subgroup H remain massless after spontaneous symmetry breaking.

If we select an orthonormal basis $\{e_1, \ldots, e_n\}$ for the Lie algebra $\mathcal{G}$, such that the first r vectors $e_1, \ldots, e_r$ lie in $\mathcal{H}$, the gauge field A_μ can be written in this basis as $A_\mu = A_\mu^k e_k$. The mass term for A_μ^k can be written as $m_{ij}A_\mu^i A^{j\mu}$, with $m_{ij} = \tfrac{1}{2}\langle t(e_i)\varphi_0, t(e_j)\varphi_0\rangle$. In particular, the fields $A_\mu^1, \ldots, A_\mu^r$ correspond to massless vector particles, since $m_{ij} = 0$ if $i \le r$ or $j \le r$. The masses of the other vector particles are defined as the eigenvalues of the matrix (m_{ij}), for $r + 1 \le i, j \le n$: see Chapter 2.

Thus, as a result of symmetry breaking, $n - r$ vector fields acquire mass, where $n = \dim G$ and $r = \dim H$, and $n - r$ scalar fields disappear as a result of the gauge condition (13.1). The number of degrees of freedom does not change in the process, because a massive vector particle has three spin states for a given momentum, while a massless vector particle has two.

The particles studied so far do not exhaust the particle spectrum of the Lagrangian (12.2). They can be called *elementary particles* or *elementary excitations* of the physical vacuum. Of course, bound states of elementary particles also exist. In quantum field theory, elementary particles correspond to the poles of two-point Green's functions, and their bound states to the poles in multipoint Green's functions.

Moreover, in the theories considered here there must be *topologically nontrivial* particles, that is, particles with non-zero topological charge. (It is easy to see that elementary particles and their bound states have zero topological charge.) For, as we have said, the concept of a topological charge is defined for quantum states of finite energy, as well as for classical fields. The topological charge is an element $\pi_2(R) = \pi_1(H)$ that does not change as the state evolves with time. On the other hand, it is assumed in quantum field theory that every state with a finite energy splits, as $t \to \infty$, into stable particles positioned far apart. In view of additivity, the sum of the topological charges of these stable particles must be equal to the topological charge of the initial state. Therefore, when there are topologically nontrivial states (that is, when H is not simply connected), there are also topologically nontrivial stable particles.

As we have said, in grand unification theories the topological charge can be identified with the magnetic charge, so topologically nontrivial particles carry a magnetic charge: we say they are *magnetic monopoles*.

If $\pi_1(H) \neq \mathbf{Z}$, we can not only state that there are topologically nontrivial particles, but also compute the number of topologically nontrivial stable particles. Indeed, since there are quantum states with arbitrary topological charge, any element of $\pi_2(R) = \pi_1(H)$ can be represented as the sum of topological charges of stable particles. Each particle has an antiparticle with the opposite topological charge (just as for all other charges). Hence, each element of $\pi_1(H)$ is a linear combination, with integral coefficients, of the topological charges $\zeta_1, \ldots, \zeta_s$ of stable particles, that is, $\{\zeta_1, \ldots, \zeta_s\}$ generates the group $\pi_1(H)$. (Of course, the classification in particles and antiparticles is purely conventional; but once a convention has been agreed upon, we can represent an element of $\pi_1(H)$ in terms of particles and not of antiparticles). The number s of generators cannot be smaller than the dimension r of the center of the Lie group H. We conclude that there must be at least r types of topologically nontrivial stable particles.

Actually, the reasoning in the previous paragraph needs a bit more care, because, as it is, it assumes implicitly that the number of particles obtained as a result of the decay of a state of finite energy is finite. If the theory contains massless particles, this assumption is unjustified. However, in Chapter 18 we will find a lower bound for the mass of topologically nontrivial particles, and the existence of this bound implies that all massless particles in this theory are topologically trivial, and so irrelevant to the argument.

The particles considered above correspond to the Lagrangian (12.2), the boson part of Lagrangian (12.1). As mentioned earlier, the presence of fermions does not alter our results. The fermion spectrum can be found by standard methods from the quadratic part of the Lagrangian, expanded in powers of deviations from the classical vacuum.

14. The Magnetic Charge

Suppose that the unbroken symmetry group H is a connected one-parameter group. Then $H = U(1)$, since that is the only connected one-dimensional compact Lie group. (H is compact because it is a closed subspace of the compact topological space G.)

We saw in Chapter 13 that the number of gauge fields that remain massless after symmetry breaking is equal to the dimension of H. Thus, for $H = U(1)$ there remains only one massless field, which we call the electromagnetic field. In the gauge (13.1), the potential of the electromagnetic field is the component $A_\mu^1 = \langle A_\mu(x), e_1 \rangle$ of $A_\mu(x)$ associated with the generator e_1 of $H = U(1)$. The electromagnetic field tensor is expressed in terms of the potential $A_\mu^1(x)$ by the usual formula

$$(14.1) \qquad F_{\mu\nu}(x) = \partial_\mu A_\nu^1(x) - \partial_\nu A_\mu^1(x).$$

Now consider a field $(\varphi(x), A_\mu(x))$, and assume that, within a certain domain, $\varphi(x)$ takes on values in the vacuum manifold R and satisfies $\nabla_\mu \varphi(x) = 0$. Then, in the gauge (13.1), the field $A_\mu(x)$ belongs to the Lie algebra $\mathcal{H}$ of $H = U(1)$, that is, $A_\mu(x) = A_\mu^1(x) e_1$. (In this gauge, we clearly have $\varphi(x) \equiv \varphi_0$, so $\nabla_\mu \varphi(x) = t(A_\mu)\varphi_0 = 0$.) In the same gauge, therefore,

$$(14.2) \qquad \mathcal{F}_{\mu\nu}(x) = F_{\mu\nu}(x) e_1,$$

or, equivalently,

$$(14.3) \qquad F_{\mu\nu}(x) = \langle \mathcal{F}_{\mu\nu}(x), e_1 \rangle,$$

since e_1 was assumed to have unit norm. If the energy of the field is finite, $\varphi(x)$ approaches R and $\nabla_\mu \varphi(x)$ approaches zero as $|\mathbf{x}|$ tends to infinity. This means we can apply equation (14.2) for the electromagnetic field strength (or the strength of the gauge field) as $|\mathbf{x}|$ tends to infinity. Thus, at great distances, only the electromagnetic part $F_{\mu\nu}(x)$ of the tensor $\mathcal{F}_{\mu\nu}(x)$ remains; this is related to the fact that the remaining gauge fields acquires mass.

Since $H = U(1)$, we have

$$\pi_2(R) = \pi_2(G/H) = \pi_1(H) = \pi_1(U(1)) = \mathbf{Z},$$

which means that the topological type of the field is characterized by an integer, the topological charge. We will show that this integer coincides, to within a factor, with the magnetic charge.

First we explain what we mean by the magnetic charge of a field $(\varphi(x), A_\mu(x))$. As noted earlier, in the gauge (13.1) the electromagnetic field tensor $F_{\mu\nu}(x)$ has the standard form (14.1). To find this tensor for an arbitrary field $(\varphi(x), A_\mu(x))$ at point x, we must first gauge-transform $(\varphi(x), A_\mu(x))$ so it satisfies (13.1), and then apply (14.1). The required gauge transformation must exist if the value of φ at x belongs to the neighborhood $\mathcal{R}$ of the vacuum manifold introduced in Chapter 13. The gauge transformation is not unique, but if two gauge fields obtained from φ by applying gauge transformations both satisfy (13.1), they can be transformed into one another by a gauge transformation with function $g(x) \in H = U(1)$. Indeed, if $\sigma\varphi(x) = \varphi_0$ and $\sigma T(g(x))\varphi(x) = \varphi_0$, equation (14.2) implies that

$$T(g(x))\sigma\varphi(x) = T(g(x))\varphi_0 = \varphi_0,$$

so that $g(x) \in H = H_{\varphi_0}$. This means that A_μ^1 is defined to within the gauge equivalence usual in electrodynamics, and that the tensor $F_{\mu\nu}$ is defined uniquely. Thus, we can talk about the electromagnetic field strength $F_{\mu\nu}(x)$ of a field $(\varphi(x), A_\mu(x))$ at any point where the value of φ belongs to $\mathcal{R}$. Later we will show an explicit formula to compute $F_{\mu\nu}(x)$ from $(\varphi(x), A_\mu(x))$. Here we note only that, if $\varphi(x)$ belongs to the vacuum manifold R and $\nabla_\mu\varphi(x)$ vanishes in the neighborhood of x, we have

(14.4) $$F_{\mu\nu}(x) = \langle \mathcal{F}_{\mu\nu}(x), h(\varphi(x)) \rangle,$$

where $h(\varphi)$ is a generator for the stabilizer H_φ of $\varphi \in R$, satisfying the normalization condition $\|h(\varphi)\| = 1$. Equation (14.4) follows from (14.3) in the gauge (13.1), and therefore in all gauges because it is gauge-invariant.

Formula (14.4) is also true for any field of finite energy $\varphi(x)$ as $|x| \to \infty$, because in this case $\varphi(x)$ approaches R and $\nabla_\mu\varphi(x)$ approaches zero as $|\mathbf{x}|$ tends to ∞. More precisely, what we have is the equation

(14.5) $$F_{\mu\nu}(x) = \langle \mathcal{F}_{\mu\nu}(x), h(\sigma\varphi(x)) \rangle,$$

which differs from (14.4) by the introduction of σ, necessary because the value of $\varphi(x)$ does not belong to the vacuum manifold R, but is merely close to it.

The electromagnetic field tensor satisfies Maxwell's equation

(14.6) $$\partial_\mu F_{\rho\sigma} + \partial_\rho F_{\sigma\mu} + \partial_\sigma F_{\mu\rho} = 0;$$

this is clear in the gauge (13.1), and is true in general by the gauge invariance of $F_{\mu\nu}$. In particular, the magnetic field $\mathbf{H} = (F_{23}, F_{31}, F_{12})$ satisfies $\operatorname{div}\mathbf{H} = 0$. Thus we can define the magnetic charge of any field $(\varphi(x), A_\mu(x))$ for which the values of $\varphi(x)$ outside a three-dimensional ball lie in $\mathcal{R}$. Namely, we set

(14.7) $$\mathfrak{m} = \frac{1}{4\pi} \oint \mathbf{H}\, d\mathbf{S},$$

where the integral is over any two-dimensional sphere enclosing all points at which $\varphi(x) \notin \mathcal{R}$. Since $\operatorname{div}\mathbf{H} = 0$, this integral does not depend on the choice of a sphere. The integral may be non-zero if $\mathbf{H}$ is not defined everywhere.

The magnetic charge of a field of finite energy also satisfies

(14.8)
$$\mathfrak{m} = \frac{1}{4\pi} \oint \langle \mathcal{H}(\mathbf{x}), h(\sigma\varphi(\mathbf{x})) \rangle \, d\mathbf{S},$$

where $\mathcal{H} = (\mathcal{F}_{23}, \mathcal{F}_{31}, \mathcal{F}_{12})$ and the integral is taken over a sphere infinitely far away. This is obtained by applying (14.5) to a sphere at infinity.

It is easy to see that the magnetic charge does not vary when the field $(\varphi(\mathbf{x}), A_\mu(\mathbf{x}))$ varies continuously, so long as we assume, as in Chapter 12, that there is some ball outside of which φ takes values in $\mathcal{R}$ at all times. In other words, the magnetic charge is the topological number of the field $(\varphi(\mathbf{x}), A_\mu(\mathbf{x}))$. The proof is the same as the one given in Chapter 11 for the Georgi–Glashow model.

As in the case of the topological charge, in the definition of the magnetic charge we look at the field (φ, A_μ) at a fixed time t, but the charge does not depend on t, because it is invariant under continuous changes in φ. Therefore, in what follows we assume that φ and A_μ depend only on $\mathbf{x}$.

Also as before, if Γ is a surface on which $\varphi(\mathbf{x})$ takes values in $\mathcal{R}$, we can define the magnetic charge of $(\varphi(\mathbf{x}), A_\mu(\mathbf{x}))$ enclosed by Γ as $\mathfrak{m}_\Gamma = (4\pi)^{-1} \int_\Gamma \mathbf{H} \, d\mathbf{S}$. Then $\mathfrak{m}_\Gamma$ is invariant under continuous changes in the field, so long as the value of the field on Γ remains in $\mathcal{R}$ at all times.

The magnetic charge of $(\varphi(\mathbf{x}), A_\mu(\mathbf{x}))$ depends only on $\varphi(\mathbf{x})$, that is, $(\varphi(\mathbf{x}), A_\mu(\mathbf{x}))$ and $(\varphi(\mathbf{x}), \tilde{A}_\mu(\mathbf{x}))$ have the same magnetic charge, because they can be joined by a continuous family of fields

$$(\varphi(\mathbf{x}), t\tilde{A}_\mu(\mathbf{x}) + (1-t)A_\mu(\mathbf{x})).$$

Moreover, the magnetic charge $\mathfrak{m}_\Gamma$ of $(\varphi(\mathbf{x}), A_\mu(\mathbf{x}))$ in the region enclosed by a surface Γ depends only on the topological charge of $\varphi(\mathbf{x})$: if $\varphi_1(\mathbf{x})$ and $\varphi_2(\mathbf{x})$ have the same topological charge, the corresponding mappings $\Gamma \to \mathcal{R}$ are in the same homotopy class, and so can be connected by a continuous transformation, so the magnetic charges also coincide. The magnetic charge is, in fact, a linear function of the topological number,

(14.9)
$$\mathfrak{m}_\Gamma(\varphi) = C\zeta_\Gamma(\varphi),$$

because both charges are additive. We will prove that $C = (4\pi)^{-1}\nu$, where ν is the smallest positive number such that $\exp(\nu e_1) = 1$. Geometrically, ν is the length of the circle $H = U(1)$.

In view of (14.9), it is sufficient to check that

(14.10)
$$\mathfrak{m}_\Gamma(\varphi, A) = \frac{\nu}{4\pi}\zeta_\Gamma(\varphi, A)$$

for at least one topologically nontrivial field $(\varphi(x), A_\mu(x))$. We will check this when $\varphi(x)$ has values in R and $\partial_\mu\varphi(x) = 0$. (Alternatively, we can derive (14.10) without using additivity, since any field can be transformed into a field for which (14.10) can be verified directly.)

We assume that Γ is homeomorphic to a sphere S^2, and consider on S^2 the coordinate system (ρ, α), where $\alpha \in [0, 2\pi)$ is the longitude and $\rho \in [0, 1)$ is the distance to the south pole. We can use either the chordal distance or the intrinsic distance along the surface of S^2; in either case we normalize the distance between the poles to one, by dividing by the diameter or by the half-circumference. We have $\rho = 0$ for the south pole and $\rho = 1$ for the north pole, and in both of these cases α is undefined.

We transfer this coordinate system to Γ by a fixed homeomorphism $S^2 \to \Gamma$, still denoting the coordinates by ρ and α. We then construct a map $f : D^2 \to \Gamma$ by taking the point with polar coordinates (ρ, α) on the disk D^2 to the point on Γ with the same coordinates; thus f is a homeomorphism on the interior of the disk, and maps the whole boundary of the disk (the circle $\rho = 1$) to the north pole on Γ. By T14.1, there is a standard isomorphism between $\pi_2(R)$ and $\pi_1(H)$ obtained as follows: given a map $\varphi : \Gamma \to R$, we find a lift $\beta : D^2 \to G$, that is, a map such that

$$(14.11) \qquad\qquad \varphi(\rho, \alpha) = T(\beta^{-1}(\rho, \alpha))\varphi_0,$$

where the exponent -1 denotes inversion in G (rather than the inverse map), and φ_0 is the value of φ at the north pole of Γ. We have $T(\beta^{-1}(1, \alpha))\varphi_0 = \varphi_0$ by (14.11), so we can consider the restriction of β^{-1} to the boundary of D^2 as a map from S^1 into H, the stabilizer of φ_0. The isomorphism is defined by taking the element of $\pi_2(R)$ defined by φ to the element of $\pi_1(H)$ defined by β.

To verify that (14.11) is equivalent to the relations used in T14.1, notice that the projection map on the bundle $(G, G/H, H, \pi)$ assigns to each element $g \in G$ the point $\pi(g) = T(g^{-1})\varphi_0 \in R = G/H$.

It follows that, for $H = U(1)$, the topological number of φ can be defined as

$$(14.12) \qquad\qquad \zeta_\Gamma = \frac{1}{\nu} \int_0^{2\pi} \frac{\partial\lambda}{\partial\alpha}\, d\alpha = \frac{1}{\nu}(\lambda(2\pi) - \lambda(0)),$$

where $\lambda(\alpha)$ is the continuous function specified by the formula $\beta(1, \alpha) = \exp(-\lambda(\alpha)e_1)$. This formula follows by taking $2\pi\lambda/\nu$ as the angular coordinate on $H = U(1)$; we recall that every element of $H = U(1)$ can be uniquely written in the form $\exp(\lambda e_1)$, where $0 \le \lambda < \nu$.

To show that ζ_Γ coincides with the magnetic charge, we note that the gauge transformation with function $\beta(\rho, \alpha)$ maps φ into a field having value φ_0 everywhere except perhaps at the north pole. The resulting field, therefore, satisfies the gauge condition (13.1). In view of the condition $\nabla_\mu \varphi = 0$ imposed before, our gauge transformation takes A_μ to a field A'_μ with values in the Lie algebra of the subgroup H, that is, $A'_\mu = a_\mu e_1$. We can think of a_μ as the electromagnetic potential in gauge (13.1); the electromagnetic field tensor is expressed in terms of a_μ by means of the standard formula. The magnetic charge equals $(4\pi)^{-1}$ times the magnetic flux through Γ. Now the electromagnetic field has no singularities; this means that we can delete a small neighborhood of the north pole without affecting the flux considerably. More precisely, let Γ_ε denote the subset of Γ where $\rho < 1 - \varepsilon$; then, as $\varepsilon \to 0$, the flux through Γ_ε approaches the flux through Γ. The boundary of Γ_ε is a small circle L_ε; it easy to see that the magnetic flux through Γ_ε equals the circulation of a_μ along this circle. Thus,

$$(14.13) \qquad \mathfrak{m} = \lim_{\varepsilon \to 0} \frac{1}{4\pi} \int \mathbf{H} \, d\mathbf{S}$$
$$= \frac{1}{4\pi} \lim_{\varepsilon \to 0} \oint a_\mu \, dx^\mu = \frac{1}{4\pi} \lim_{\varepsilon \to 0} \int_0^{2\pi} a_\alpha(1 - \varepsilon, \alpha) \, d\alpha.$$

Here $a_\rho(\rho, \alpha)$ and $a_\alpha(\rho, \alpha)$ are the components of the vector potential a_μ in terms of the coordinates ρ and α on Γ. We assume, for simplicity, that the gauge field A_μ vanishes near the north pole. Then a gauge transformation transforms this field into

$$A'_\alpha(\rho, \alpha) = a_\alpha(\rho, \alpha)e_1 = -\frac{\partial \beta^{-1}(\rho, \alpha)}{\partial \alpha} \beta(\rho, \alpha)$$

for $\rho \geq 1 - \varepsilon$. Since $\beta(\rho, \alpha)$ approaches $\beta(1, \alpha) = \exp(-\lambda(\alpha)e_1)$ as $\rho \to 1$, we see that $a_\alpha(\rho, \alpha)$ approaches $\partial \lambda / \partial \alpha$, which implies that

$$(14.14) \qquad \mathfrak{m}_\Gamma = \frac{1}{4\pi}(\lambda(2\pi) - \lambda(0))$$

by (14.13). Combining this with (14.12), we get the desired equation (14.10).

We now show that in the theories considered here the electric charge is quantized, and establish the link between the quantization of the electric and magnetic charges. The electric charges of particles corresponding to the field φ are defined as $-i\lambda_k$, where the λ_k are the eigenvalues of the operator $t(e_1)$, which are purely imaginary, since $t(e_1)$ is anti-Hermitian. Indeed, if $\varphi = \sum_k \varphi^k f_k$, where $f_1, \ldots, f_n$ are the eigenvectors of $t(e_1)$, and if $A_\mu = a_\mu e_1$, that is, only the electromagnetic part of the gauge field A_μ is non-zero, we get

$$\nabla_\mu \varphi = \partial_\mu \varphi + t(a_\mu e_1)\varphi = \sum_k (\partial_\mu \varphi^k + \lambda_k a_\mu \varphi^k) f_k.$$

Since $\exp(\nu e_1) = 1$, we have $T(\exp(\nu e_1)) = \exp(\nu T(e_1)) = 1$. Thus, every eigenvalue λ_k of $t(e_1)$ satisfies $\exp(\nu \lambda_k) = 1$. It follows that $\nu \lambda_k = 2\pi n i$, where n is an integer, so the electric charges are multiples of $2\pi/\nu$. Since the magnetic charges are multiples of $\nu/4\pi$, we see that the product of the electric charge of a particle by the magnetic charge of another particle is a half-integer:

$$(14.15) \qquad e\mathfrak{m} = \tfrac{1}{2}n.$$

Note that we used a non-standard normalization condition for the electro-magnetic potential a_μ. If we employ the standard normalization, in which the coefficient of $F_{\mu\nu}^2$ in the Lagrangian is $-\frac{1}{4}$ instead of $-\frac{1}{4}g^2$ as it is here, the electric and magnetic charges will be equal to $i\lambda_k g = (2\pi n/\nu)g$ and $(\nu/4\pi g)\zeta$, respectively, where n and ζ are integers, and g is the coupling constant. This clearly shows that (14.15) remains valid under the standard normalization.

Until now we have assumed that the unbroken symmetry group H is isomorphic to $U(1)$. The transition to an arbitrary connected group H requires only small changes to the reasoning. It is always the case that H, as a compact Lie group, is locally isomorphic to the product of r copies of $U(1)$ with a simply connected group K; r is the dimension of the center of Lie algebra of H. In the gauge (13.1), the generator e_i of the i-th copy of $U(1)$ has associated to it an "electromagnetic" potential and an "electromagnetic" field tensor; the potentials are defined by $A_\mu^k = \langle A_\mu, e_k \rangle$ and the field tensors are expressed in terms of the potentials by the standard formulas. In grand unification theories we have $r = 1$, and we can talk of *the* electromagnetic potential and field tensor without quotation marks.

We can say that each element e of the center Z of the Lie algebra $\mathcal{H}$ corresponds to an electromagnetic potential $a_\mu = \langle A_\mu, e \rangle$; among these potentials we can choose r linearly independent ones, corresponding to the generators $e_1, \dots, e_r$. The field-strength tensor for any potential can be calculated by passing to gauge (13.1) via a gauge transformation, and using the formula $F_{\mu\nu} = \partial_\mu a_\nu - \partial_\nu a_\mu$. It is easy to check that the result is independent of the choice of gauge transformation. Indeed, as we noted earlier, two gauge-equivalent potentials A_μ and A'_μ satisfying (13.1) are, in fact, gauge-related by a transformation whose corresponding function has values in H. This implies that the fields $a_\mu = \langle A_\mu, e \rangle$ and $a'_\mu = \langle A'_\mu, e \rangle$ are gauge-related by a transformation of the type used in electrodynamics, and therefore have the same strength. For example, in the case of an infinitesimal gauge transformation, $A'_\mu = A_\mu - \partial_\mu \lambda - [A_\mu, \lambda]$, and our assertion follows from the fact that the scalar product of the commutator $[A_\mu, \lambda]$ with any element in the center of $\mathcal{H}$ is zero. The case of a finite gauge transformation with H connected can be reduced to that of an infinitesimal transformation.

Magnetic charges associated with the electromagnetic field tensors just introduced can be defined in a standard manner. Working as in Chapter 11, we can prove that these magnetic charges are invariant under continuous variations of the fields, that is, they are topological invariants. Moreover, we defined the topological type of a field by looking at the asymptotic behavior of the scalar field φ. Thus the topological type of a field is an element of $\{S^2, R\}$, that is, a homotopy class of mappings from S^2 into the vacuum manifold. As we mentioned in Chapter 12, when H is connected, the set $\{S^2, R\}$ can be identified with the group $\pi_2(R) = \pi_1(H)$, which is a direct sum of r copies of $\mathbf{Z}$ with a finite group. Thus there are r topological charges, as many as there are magnetic charges. Clearly, the magnetic charges completely determine the integral values of the topological charges of a field. In addition to integral topological charges, there may be charges taking values in a finite group $\mathbf{Z}_m$; these do not correspond to magnetic charges.

15. Electromagnetic Field Strength and Magnetic Charge in Gauge Theories

Formula (14.5) gives the electromagnetic field strength $F_{\mu\nu}(x)$, as $|\mathbf{x}| \to \infty$, for a field of finite energy, so long as the stabilizer H is one-dimensional. This formula leads to the expression (14.8) for the magnetic charge of such a field. However, to obtain an explicit expression for the magnetic charge $\mathfrak{m}_\Gamma(\varphi, A)$ of the field in the region bounded by a surface Γ, we need a way to compute the electromagnetic field strength of (φ, A_μ) directly, without using the gauge (13.1). To this end it suffices to find a gauge-invariant expression that coincides with (14.1) in gauge (13.1). It is natural to look for an expression of the form

$$(15.1) \qquad F_{\mu\nu}(x) = \langle \mathcal{F}_{\mu\nu}(x), h(\sigma\varphi(x)) \rangle + \omega_{ab}(\varphi(x))(\nabla_\mu\varphi)^a(\nabla_\nu\varphi)^b,$$

where $\omega_{ab}(\varphi) = -\omega_{ba}(\varphi)$. The first term in this expression coincides with (14.5), and the second generalizes the corresponding term in the expression for the field strength in the Georgi–Glashow model (Chapter 11).

Setting $\omega_\varphi(x, y) = \omega_{ab}(\varphi)x^a y^b$, we can rewrite (15.1) as

$$(15.2) \qquad F_{\mu\nu}(x) = \langle \mathcal{F}_{\mu\nu}(x), h(\sigma\varphi(x)) \rangle + \omega_\varphi(\nabla_\mu\varphi, \nabla_\nu\varphi).$$

The function $\omega_{ab}(\varphi)$ can be found from the requirements

$$(15.3) \qquad\qquad\qquad \omega_\varphi(x, y) = \omega_{\sigma\varphi}(\sigma_* x, \sigma_* y),$$
$$(15.4) \qquad \omega_{\varphi_0}(t(A)\varphi_0, t(B)\varphi_0) = -\langle [A, B], h(\varphi_0) \rangle,$$

where σ_* denotes the differential of the map σ, that is, $(\sigma_* x)^a = (\partial\sigma^a/\partial\varphi^b)x^b$.

The vectors $\sigma_* x$ and $\sigma_* y$ are tangent to R, so (15.3) reduces the problem of finding $\omega_\varphi(x, y)$ to the case where $\varphi \in R$ and x, y are tangent to R. Now, every vector tangent to R at the point $\varphi_0 \in R$ can be written in the form $t(A)\varphi_0$, where A is an element of the Lie algebra $\mathcal{G}$ of G; this is because G acts transitively on R, so a point in R infinitesimally close to φ_0 can be obtained from φ_0 by the action of an infinitesimal transformation in G, that is, an element of $\mathcal{G}$. Then (15.4) determines $\omega_\varphi(x, y)$ in the particular case that $\varphi \in R$ and x, y are tangent to R, and therefore $\omega_\varphi(x, y)$ is determined everywhere.

Take, for example, the Georgi–Glashow model. Here σ assigns to each vector $\varphi \in \mathbf{R}^3$ the point $a\varphi/|\varphi|$, and the stabilizer of φ consists of all rotations around an axis parallel to φ. From this we see that $h(\varphi) = \varphi/|\varphi|$; this also follows from the relation

$$t(h(\varphi))\varphi = [\varphi/|\varphi|, \varphi] = 0.$$

The differential of $\sigma(\varphi)$ is given by

$$\sigma_*(\varphi)^i_j = \frac{a}{|\varphi|}\left(\delta^i_j - \frac{1}{|\varphi|^2}\varphi^i\varphi_j\right),$$

so that

$$(\sigma_*\nabla_\mu\varphi)^i = \left(\frac{-\varphi^i\partial_\mu\varphi_j + \varphi_j\partial_\mu\varphi^i}{|\varphi|^2} + t(A_\mu)^i_j\right)a\frac{\varphi^j}{|\varphi|}.$$

Combining this with (15.4), we find

$$w_\varphi(x, y) = \frac{1}{|\varphi|^3}\varepsilon_{ijk}\varphi^i x^j y^k.$$

This shows that (15.1) agrees with formula (11.9) for the electromagnetic field strength in the Georgi–Glashow model.

We now check that, by selecting $w_{ab}(\varphi)$ using (15.3) and (15.4), we do get a gauge-invariant expression for $F_{\mu\nu}(x)$. We show that

(15.5) $$h(T(g)\sigma\varphi) = \tau_g h(\sigma\varphi),$$

(15.6) $$w_{T(g)\varphi}(T(g)x, T(g)y) = w_\varphi(x, y).$$

Formula (15.5) expresses the gauge invariance of the first term in the right-hand side of (15.2), since both $F_{\mu\nu}(x)$ and $h(\sigma\varphi(x))$ transform by the adjoint representation under a gauge transformation, and the scalar product $\langle,\rangle$ is invariant under the adjoint representation. Similarly, (15.6) implies the gauge invariance of the second term in (15.2).

To prove (15.5), note that the stabilizer $H_{T(g)\varphi}$ of $T(g)\varphi$ can be obtained from H_φ by an inner automorphism: if $T(h)\varphi = \varphi$, we have

$$T(ghg^{-1})T(g)\varphi = T(g)T(h)T^{-1}(g)T(g)\varphi = T(g)\varphi.$$

This implies that a generator of $H_{T(g)\varphi}$ can be obtained from a generator of H_φ by an inner automorphism. Recalling that the adjoint representation acts by means of inner automorphisms, we see that (15.5) holds if $\varphi \in R$. For arbitrary φ, the formula holds because

$$h(\sigma(T(g)\varphi)) = h(T(g)\sigma\varphi) = \tau_g h(\sigma\varphi),$$

where we have used (15.2).

To verify (15.6), we start with the case where $\varphi = \varphi_0 \in R$ and $x = t(A)\varphi_0$, $y = t(B)\varphi_0$ are tangent to R; then $w_\varphi(x, y)$ is defined by (15.4). Notice first that

(15.7) $$T(g)t(A)\varphi = t(\tau_g(A))T(g)\varphi;$$

this follows by direct computation:

$$T(g)\varphi + T(g)t(A)\varphi = T(g)T(1+A)\varphi = T(g)T(1+A)T^{-1}(g)T(g)\varphi$$
$$= T(g(1+A)g^{-1})T(g)\varphi = T(g)(1+\tau_g(A))T(g)\varphi$$
$$= (1+t(\tau_g(A)))T(g)\varphi = T(g)\varphi + t(\tau_g(A))T(g)\varphi,$$

where we have used the equalities $T(1+A)\varphi = (1+t(A))\varphi$ and $g(1+A)g^{-1} = 1 + \tau_g(A)$, which hold for A infinitesimal. Combining (15.7), (15.5) and the invariance of the scalar product under the adjoint representation τ_g, we see that

$$\omega_{T(g)\varphi_0}(T(g)t(A)\varphi_0, T(g)t(B)\varphi_0)$$
$$= \omega_{T(g)\varphi_0}(t(\tau_g(A))T(g)\varphi_0, t(\tau_g(B))T(g)\varphi_0)$$
$$= -\langle [\tau_g(A), \tau_g(B)], h(T(g)\varphi_0) \rangle = -\langle \tau_g[A,B], \tau_g h(\varphi_0) \rangle$$
$$= -\langle [A,B], h(\varphi_0) \rangle = \omega_{\varphi_0}(t(A)\varphi_0, t(B)\varphi_0).$$

The verification of (15.6) in general can be reduced to the case just considered by the use of (13.2) and (15.3).

Thus, we have proved that the right-hand side of (15.2) is gauge-invariant. To check that (15.2) gives the right expression for the electromagnetic field strength, we need only show that, if the gauge condition (13.1) is satisfied, the right-hand side of (15.2) reduces to the standard expression $\partial_\mu a_\nu(x) - \partial_\nu a_\mu(x)$, where $a_\mu(x) = \langle A_\mu(x), h(\varphi_0) \rangle$. By differentiation, we see that if (13.1) is satisfied we have

$$\frac{\partial \sigma^a}{\partial \varphi^b}(\partial_\mu \varphi)^b = 0;$$

in other words, $\sigma_* \partial_\mu \varphi = 0$. Thus, in gauge (13.1),

$$F_{\mu\nu}(x) = \langle \mathcal{F}_{\mu\nu}(x), h(\sigma\varphi(x)) \rangle + \omega_{\sigma\varphi}(\sigma_* \nabla_\mu \varphi, \sigma_* \nabla_\nu \varphi)$$
$$= \langle \mathcal{F}_{\mu\nu}(x), h(\varphi_0) \rangle + \omega_{\varphi_0}(\sigma_* t(A_\mu)\varphi, \sigma_* t(A_\nu)\varphi)$$
$$= \langle \mathcal{F}_{\mu\nu}(x), h(\varphi_0) \rangle + \omega_{\varphi_0}(t(A_\mu)\varphi_0, t(A_\nu)\varphi_0)$$
$$= \langle \partial_\mu A_\nu - \partial_\nu A_\mu + [A_\mu, A_\nu], h(\varphi_0) \rangle - \langle [A_\mu, A_\nu], h(\varphi_0) \rangle$$
$$= \langle \partial_\mu A_\nu - \partial_\nu A_\mu, h(\varphi_0) \rangle = \partial_\mu a_\nu - \partial_\nu a_\mu,$$

where we have used the equality $\sigma_* t(A)\varphi = t(A)\sigma\varphi$, which follows from applying (13.2) to $g = 1 + A$, where A is an infinitesimal transformation. This concludes the proof that formulas (15.2), (15.3) and (15.4) determine the electromagnetic field strength for $H = U(1)$.

The same formula (15.2) can be used when H is an arbitrary disconnected one-dimensional group. However, in this case $F_{\mu\nu}(x)$ may be a two-valued function, and it may not be possible to make it single-valued by choosing a continuous branch for it. For more details, see Chapter 19.

As observed before, the generator $h(\varphi)$ is defined only up to sign, so, strictly speaking, the value of $F_{\mu\nu}$ in (15.2) is also defined only up to sign. Fixing $h(\varphi)$ for a given φ and extending by continuity, we can choose a well-defined value for $h(\varphi)$ for every $\varphi \in \mathcal{R}$. For $H = U(1)$ this extension can be carried out consistently, and we get a continuous single-valued function $h(\varphi)$. But if H is disconnected, one may run into inconsistencies, so $h(\varphi)$ cannot be made single-valued.

It is convenient to express the electromagnetic field strength using differential forms. We can associate to the antisymmetric tensor $\omega_{ab}(\varphi)$ the two-form $\omega = \frac{1}{2}\omega_{ab}(\varphi)\,d\varphi^a \wedge d\varphi^b$. Equality (15.3) says that this form can be written as $\omega = \sigma^*\rho$, where ρ is a G-invariant closed two-form on the manifold R. In the principal fiber bundle $(G, G/H, H) = (G, R, U(1))$ we can construct a G-invariant connection, that is, a gauge field on H, whose change under a transformation in G can be compensated for by a gauge transformation. The form ρ can be interpreted as the strength of this gauge field (Chapter 17).

We now turn to the case where the stabilizer $H = H_0$ of $\varphi_0 \in R$ is not one-dimensional. We choose an element $h_0 \in \mathcal{H}$ invariant under the adjoint representation of H, that is, such that

$$(15.8) \qquad\qquad \tau_g h_0 = h_0$$

for every $g \in H$. If H is connected, this is equivalent to saying that h_0 is invariant under the adjoint representation of the Lie algebra $\mathcal{H}$. Now the adjoint representation σ of a Lie algebra is given by $\sigma_A h = [A, h]$, so saying that h_0 is invariant under σ is the same as saying that it commutes with every element of $\mathcal{H}$, or that it lies in the center $\mathcal{Z}$ of $\mathcal{H}$.

If $h_0 \in \mathcal{H}$ is invariant under σ we can construct a function $h(\varphi)$ on R satisfying

$$(15.9) \qquad\qquad h(T(g)\varphi) = \tau_g h(\varphi),$$
$$(15.10) \qquad\qquad h(\varphi_0) = h_0;$$

we just use (15.9) and the transitivity of the action of G on R to extend h from φ_0 to other points on R, and (15.8) guarantees that this extension is consistent. From $h(\varphi)$, we can build ω using (15.3) and (15.4), and then define the electromagnetic field strength by (15.2). From (15.9), (15.10) and (15.2) it follows that conditions (15.5) and (15.6) are met, and hence that the field strength is gauge invariant. This reasoning shows that in gauge (13.1) the field strength is given by the standard expression, with $a_\mu = \langle A_\mu, h_0 \rangle$ playing the role of the electromagnetic potential:

$$F_{\mu\nu} = \partial_\mu \langle A_\nu, h_0 \rangle - \partial_\nu \langle A_\mu, h_0 \rangle.$$

Therefore the electromagnetic field strength satisfies Maxwell's equation

$$\partial_\sigma F_{\mu\nu} + \partial_\mu F_{\nu\sigma} + \partial_\nu F_{\sigma\mu} = 0.$$

Because of gauge invariance, Maxwell's equation is also satisfied in gauges other than (13.1).

Thus, if H is connected, every element h_0 of the center $\mathcal{Z}$ of $\mathcal{H}$ can be used to construct the electromagnetic field strength and the corresponding magnetic charge (this was proved in a somewhat different way in Chapter 14). If the center is an r-dimensional Lie algebra, there are r independent magnetic charges.

What happens if H is disconnected? In general, an element $h_0 \in \mathcal{Z}$ does not satisfy (15.10). However, the set of elements of the form $\tau_g h_0$, with $g \in H$, is finite, because $\tau_g h_0 = h_0$ if g is in the connected component of the identity, and there are finitely many connected components. We denote the number of distinct elements $\tau_g h_0$ by s. Defining $h(\varphi)$ on R by means of (15.9) and (15.10), we obtain an s-valued function. Indeed, if $\varphi \in R$ can be written both as $T(g)\varphi_0$ and as $T(\tilde{g})\varphi_0$, we have $\tau_g h_0 = h(\varphi) = \tau_{\tilde{g}} h_0$, by (15.9). It may happen that $\tau_g h_0$ and $\tau_{\tilde{g}} h_0 = \tau_g \tau_{g^{-1}\tilde{g}} h_0$ do not coincide, so that $h(\varphi)$ is not single-valued, but since $\tau_{g^{-1}\tilde{g}} h_0$ can take on at most s different values, we conclude that $h(\varphi)$ is at most s-valued. Thus, for H disconnected, $F_{\mu\nu}$ is in general a multivalued function. For more details, see Chapter 19.

16. Extrema of Symmetric Functionals

Consider a linear representation of a compact Lie group G on a vector space M, and a differentiable function $f(x)$ on M invariant under this representation. We will show that the G-invariance of $f(x)$ can be used to determine its *critical points* (those where the gradient vanishes).

Let N be the set of points of V invariant under all transformations in G; since G acts linearly, N is a vector subspace of V. We will show that a critical point of the restriction of f to N is also a critical point of f considered as a function on all of M.

This statement remains true even when M is infinite-dimensional, if we define differentiability appropriately. (In this case we use the term "functional" instead of "function," and sometimes also "extremal point" instead of "critical point".) A functional f is differentiable if

$$(16.1) \qquad f(x+h) - f(x) = \langle A, h \rangle + \alpha(x, h),$$

where the angle brackets denote a scalar product invariant under G (it is well-known that such a product always exists: see Chapter 39), and $\alpha(x, h)$ vanishes at a higher order than h, that is, $\lim_{\varepsilon \to 0} \varepsilon^{-1} \alpha(x, \varepsilon h) = 0$ for any h. Notice that differentiability, then, depends on the choice of a scalar product. The expression $\delta f = \langle A, h \rangle$ is called the *variation* or *differential* of f, and the vector $A = \text{grad} f$ is the gradient of f at the point x. We say that x is a critical point if the gradient of f at x is 0.

Here is an example to illustrate the application of the symmetry properties of a functional to the search for extremals. Take the functional

$$(16.2) \qquad F[\varphi] = \frac{1}{2} \int (\text{grad}\, \varphi(\mathbf{x}))^2\, d^3x + \int U(\varphi(\mathbf{x}))\, d^3x$$

defined on fields $\varphi(x)$ that vanish on the sphere $|\mathbf{x}| = R$. On the space M of such fields we consider the following scalar product:

$$(16.3) \qquad \langle \varphi, \psi \rangle = \int \varphi(\mathbf{x}) \psi(\mathbf{x})\, d^3x.$$

The increment of functional (16.2) can be written in the form (16.1) if we set $A = -\nabla^2 \varphi + U'(\varphi)$. Accordingly, the extremality condition (Euler's equation) takes the form $-\nabla^2 \varphi + U'(\varphi) = 0$. Clearly, (16.2) and (16.3) are invariant under rotations; more precisely, the group $G = \text{SO}(3)$ of rotations of $\mathbf{R}^3$ can also be considered to act on the space M, and it preserves the functional (16.2) and the scalar product (16.3). The set N of G-invariant points consists of all spherically symmetric fields φ, that is, fields that depend only on $|\mathbf{x}|$. We can say, then, that an extremal of (16.1) among spherically symmetric fields is also an extremal among all fields in M.

This fact simplifies drastically the task of finding extremals, because it allows one to reduce Euler's equation to a set of ordinary differential equations. Of course, there is no guarantee that all extremals are spherically symmetric; all we are saying is that we can find *some* solutions to Euler's equation by applying a spherically symmetric substitution.

Note that, in searching for extremals, we need not require that the points be invariant with respect to the complete symmetry group of the functional. For example, one can examine the behavior of (16.1) on fields invariant under the group $SO(2)$ of rotations about some fixed axis. An extremal of the functional considered on this set is also an extremal of the functional on M.

We now prove the statement made at the beginning of this chapter. The gradient of f at a point $x_0 \in N$ is invariant under the action of G, because of the invariance of f and of the scalar product. Thus $\operatorname{grad}_V f(x_0) \in N$, where the subscript indicates the space on which f is being considered. By the defining equation (16.1), this implies that $\operatorname{grad}_V f(x_0) = \operatorname{grad}_N f(x_0)$, so if x_0 is a critical point for f restricted to N it is also a critical point for f on V.

This result can be generalized for functions defined on a (possibly infinite-dimensional) Riemannian manifold. Suppose G is a compact Lie group acting on a smooth manifold M, and let $f(x)$ be a differentiable function on M invariant under this action. Let $N \subset M$ be the set of fixed points under G. Under an additional condition, which is usually met, we can say that the critical points of f restricted to N are also critical points of f on M.

To formulate the condition, take an arbitrary $x_0 \in N$. By definition, any element of G leaves x_0 invariant, and so takes a tangent vector at x_0 to another such vector. Therefore G acts by a linear representation T on the tangent space at x_0. For points close to x_0 the map φ_g corresponding to a group element g can be written as

$$(4) \qquad \varphi_g(x_0 + h) = x_0 + T(g)h + \cdots ,$$

where the dots indicate terms of order greater than one in h. From this it follows that for a curve in N the tangent vector at x_0 must satisfy $T(g)h = h$ for all $g \in G$, and so must belong to the vector space $\mathcal{N}$ of G-invariant vectors at x_0 (where G-invariant means invariant under the representation T).

The condition we want is that $x_0 \in N$ be *non-degenerate*, which means, by definition, that N is a manifold in a neighborhood of x_0 and $\mathcal{N}$ is the tangent space to N at x_0 (that is, every vector in $\mathcal{N}$ at x_0 is tangent to N). We show that, in this case, if x_0 is a critical point for f on N, it is also one for f on M. Since G is compact we can give tangent spaces to M a G-invariant scalar product (in other words, M has a G-invariant Riemannian metric). Equation (16.1) is still meaningful. The vector $\operatorname{grad}_M f(x_0)$ lies in $\mathcal{N}$, by the invariance of the function and of the scalar product. We conclude the argument exactly as before, bearing in mind that $\mathcal{N}$ coincides with the tangent space to N at x_0 by the non-degeneracy assumption. This reason is still valid if M is infinite-dimensional (in which case differentiability is defined by means of (16.1), as in the case of an infinite-dimensional vector space V).

Note that this result is still interesting when N is zero-dimensional, that is, consists of isolated points. In this case a G-invariant point x_0 is a critical point of f if there are no G-invariant tangent vectors at x_0.

17. Symmetric Gauge Fields

Consider n scalar fields $\varphi(x) = (\varphi^1(x), \ldots, \varphi^n(x))$, interacting with $\mathcal{K}$-valued gauge fields $A_\mu(x)$, where $\mathcal{K}$ is the Lie algebra of the gauge group K. As we have seen, the extremals of the energy functional are connected with quantum particles in the semiclassical approximation. How does one find such extremals? Using the results in the preceding chapter, one can take advantage of the invariance of the energy functional under gauge transformations and (usually) under spatial transformations such as rotations. More precisely, let L be the group generated by the group K^∞ of local gauge transformations together with the group O of spatial symmetries. If $G \subset L$ is a subgroup, an extremal of the functional restricted to the space of G-invariant fields is also an extremal for the unrestricted functional. (We will see below that the non-degeneracy condition is met.)

The simplest way to choose a subgroup $G \subset L$ is the following: Let λ be a homomorphism from the rotation group SO(3) into the gauge group K. To each element $g \in$ SO(3) assign the transformation $\rho(g) \in L$ obtained by applying the rotation $g \in$ SO(3) followed by the global gauge transformation corresponding to $\lambda(g) \in K$. Clearly, $\rho : $ SO(3) $\rightarrow L$ is a homomorphism, and its image G_λ is a subgroup of L.

It is useful to generalize this construction slightly by assuming that λ is a two-valued homomorphism from SO(3) into K, which we can lift to a single-valued homomorphism $\tilde\lambda : $ SU(2) $\rightarrow K$; we recall that SU(2) is the universal cover of SO(3), with covering map $p : $ SU(2) $\rightarrow$ SO(3). From λ we get the subgroup $G_\lambda \subset L$ consisting of products of a rotation $p(g)$, for $g \in$ SU(2), with the global gauge transformation corresponding to $\tilde\lambda(g)$.

A field invariant under G_λ can be seen as spherically symmetric in a certain sense, for the change it undergoes under a spatial rotation can be compensated for by a global gauge transformation. The compensation is specified by the homomorphism $\lambda : $ SO(3) $\rightarrow K$; we can say that the type of spherical symmetry is determined by λ, up to conjugacy. (If λ and λ' are conjugate, a field (φ, A) with symmetry type λ can be transformed into a field (φ', A') with symmetry type λ' by the action of the element $k \in K$ such that $\lambda'(g) = k\lambda(g)k^{-1}$.) For example, if the gauge group is SO(3), the identity map SO(3) $\rightarrow$ SO(3) $= K$ is essentially distinct from the trivial homomorphism that takes every element to the identity element of K.

We now describe all G_λ-invariant fields $(\varphi(\mathbf{x}), A_i(\mathbf{x}))$, with $i = 1, 2, 3$, also called fields with symmetry of type λ. First we note that they satisfy the conditions

$$(17.1) \qquad\qquad T(\lambda(g))\varphi(\mathbf{x}) = \varphi(g\mathbf{x}),$$
$$(17.2) \qquad\qquad g_i^j \tau(\lambda(g))A_j(\mathbf{x}) = A_i(g\mathbf{x}),$$

where g_i^j is a rotation matrix $g \in SO(3)$, T is the representation of K by which the multicomponent scalar field φ transforms, and $\lambda(g) \in K$, as usual, acts on the gauge field via the adjoint representation τ. Equations (17.1) and (17.2) imply that $(\varphi(\mathbf{x}), A_i(\mathbf{x}))$ will be known everywhere if we know its value on the positive x^3-semiaxis (or on any other ray starting at the origin); this is because any point of $\mathbf{R}^3$ can be taken to the semiaxis by a rotation. Also, $(\varphi(\mathbf{x}), A_i(\mathbf{x}))$ does not change under a transformation in G_λ that corresponds to a rotation about the x^3-axis. In other words, we have

$$(17.3) \qquad\qquad T(\lambda(g))\varphi(r) = \varphi(r),$$
$$(17.4) \qquad\qquad \tau(\lambda(g))g_i^j A_j(r) = A_i(r),$$

where $\varphi(r)$ and $A_i(r)$ are the values of $\varphi(\mathbf{x})$ and $A_i(\mathbf{x})$ for $\mathbf{x} = (0, 0, r)$, with $r \geq 0$, and g_i^j is the rotation matrix corresponding to g, a rotation about the x^3-axis. In infinitesimal terms, if t denotes the representation of $\mathcal{K}$ corresponding to the representation T of K, and $I_i \in \mathcal{K}$, for $i = 1, 2, 3$, is the infinitesimal rotation around the x^i-axis, (17.3) and (17.4) become

$$(17.5) \qquad t(I_3)\varphi(r) = 0$$
$$(17.6) \qquad [I_3, A_3(r)] = 0, \quad [I_3, A_1(r)] = A_2(r), \quad [I_3, A_2(r)] = -A_1(r).$$

For example, if $K = SO(3)$ and λ is the identity map, (17.6) gives

$$A_1(r) = r^{-1}(\beta(r)I_1 + \gamma(r)I_2),$$
$$A_2(r) = r^{-1}(\beta(r)I_2 - \gamma(r)I_1),$$
$$A_3(r) = r^{-1}\alpha(r)I_3,$$

where $\alpha(r)$, $\beta(r)$ and $\gamma(r)$ are arbitrary functions. If $\varphi(\mathbf{x})$ transforms by the vector representation of $K = SO(3)$, it follows from (17.5) that $\varphi(r) = (0, 0, r^{-1}\xi(r))$, where $\xi(r)$ is likewise arbitrary.

To summarize, if we prescribe values for (φ, A) on the positive x^3-semiaxis, satisfying (17.5) and (17.6), we can reconstruct the spherically symmetric field on the whole of $\mathbf{R}^3$ by using (17.1) and (17.2). In particular, if $K = SO(3)$ and λ is the identity, we obtain

$$(17.7) \qquad \varphi^i(\mathbf{x}) = \xi(r)\frac{x^i}{r^2},$$

$$(17.8) \qquad A^i(\mathbf{x}) = \frac{x^k x^i}{r^3}(\alpha(r) - \beta(r))I_k + \frac{\beta(r)}{r}I_i - \gamma(r)\varepsilon_{ijk}\frac{x^j}{r^2}I_k.$$

For $\alpha \equiv \beta \equiv 0$ we obtain the fields used to analyze magnetic monopoles in the Georgi–Glashow model.

Note that all quantities associated with a spherically symmetric field can be expressed in terms of the field values along the positive x^3-semiaxis. For example, let us compute the energy of a spherically symmetric field $(\varphi(\mathbf{x}), A_i(\mathbf{x}))$. The energy density $\mathcal{E}(\mathbf{x})$ depends only on $r = |\mathbf{x}|$, because it is gauge-invariant, and a rotation takes the field to a gauge-equivalent one. In terms of the energy density $\mathcal{E}(\mathbf{x})$ on the positive x^3-semiaxis, the total energy $\mathcal{E}$ can be found by integrating

$$(17.9) \qquad \mathcal{E} = 4\pi \int_0^\infty \mathcal{E}(r) r^2 \, dr.$$

To find $\mathcal{E}(r) = \mathcal{E}(0,0,r)$, we need the derivatives of $\varphi(\mathbf{x})$ and $A_i(\mathbf{x})$, evaluated along the x^3-axis. To obtain them, we apply (17.1) and (17.2) to the case where $\mathbf{x} = (0,0,r)$ and g is an infinitesimal rotation $g = I_1$ or $g = I_2$. We get

$$(17.10) \qquad \begin{aligned} -\frac{\partial\varphi(\mathbf{x})}{\partial x^2}\bigg|_{\mathbf{x}=(0,0,r)} &= \frac{1}{r} t(I_1)\varphi(r), \\ \frac{\partial\varphi(\mathbf{x})}{\partial x^1}\bigg|_{\mathbf{x}=(0,0,r)} &= \frac{1}{r} t(I_2)\varphi(r) \end{aligned}$$

and

$$(17.11) \qquad \begin{aligned} -\frac{\partial A_i(\mathbf{x})}{\partial x^2}\bigg|_{\mathbf{x}=(0,0,r)} &= \frac{1}{r}((I_1)_i^j A_j(r) + [I_1, A_i(r)]), \\ \frac{\partial A_i(\mathbf{x})}{\partial x^1}\bigg|_{\mathbf{x}=(0,0,r)} &= \frac{1}{r}((I_2)_i^j A_j(r) + [I_2, A_i(r)]). \end{aligned}$$

Using (17.10) and (17.11), we conclude that
(17.12)

$$\begin{aligned} \mathcal{E}(r) = \frac{1}{2r^4}\bigg(&([I_1, W_1(r)] + [I_2, W_2(r)] + [W_1(r), W_2(r)])^2 \\ &+ \left([I_2, W_3(r)] + [W_1(r), W_3(r)] - r\frac{\partial W_1(r)}{\partial r}\right)^2 \\ &+ \left([I_1, W_3(r)] - [W_2(r), W_3(r)] + r\frac{\partial W_2(r)}{\partial r}\right)^2\bigg) \\ &+ \frac{1}{2r^2}\left(r^2\left(\frac{\partial\varphi(r)}{\partial r}\right)^2 + (t(W_1 + I_2)\varphi(r))^2 + (t(W_2 - I_1)\varphi(r))^2\right) \\ &+ V(\varphi), \end{aligned}$$

where $W_i(r) = r A_i(r)$ and the charge $e = 1$.

In order to use the results of Chapter 16 to find the extremum of the energy functional on the space of fields $(\varphi(\mathbf{x}), A_i(\mathbf{x}))$, we define a scalar product on this space as follows:

(17.13)
$$\langle(\varphi(\mathbf{x}), A_i(\mathbf{x})), (\varphi'(\mathbf{x}), A_i'(\mathbf{x}))\rangle = \int \langle\varphi(\mathbf{x}), \varphi'(\mathbf{x})\rangle\, dx + \int \langle A_i(\mathbf{x}), A_i'(\mathbf{x})\rangle\, dx,$$

where the angle brackets stand, in the first integrand, for the invariant scalar product on the space of representation T, and in the second term for the invariant scalar product on the Lie algebra $\mathcal{K}$. To guarantee that the integrals in (17.12) are finite, we assume that the fields fall off fast enough at infinity. We denote by $\mathcal{V}$ the space of such fields, endowed with the scalar product (17.13).

The elements of G are linear transformations on $\mathcal{V}$ and preserve the scalar product. Local gauge transformations, in contrast with global transformations, are not, in general, linear transformations on $\mathcal{V}$; but they are isometries, that is, they preserve distance between fields, where distance is defined by the scalar product (17.13).

One can easily check that the energy functional is a differentiable functional on $\mathcal{V}$, with respect to the scalar product (17.13). In view of the preceding discussion, the extremals of this functional considered only on G_λ-invariant fields (fields with symmetry of type λ) are extremals of the functional on the whole of $\mathcal{V}$. Recall that on G_λ-invariant fields the energy functional has the form (17.9), where $\mathcal{E}(r)$ is given by (17.12).

In particular, if the gauge group is $K = \mathrm{SO}(3)$ and λ is the identity map from the group $\mathrm{SO}(3)$ of spatial rotations into K, and if φ transforms by the vector representation, the energy functional (17.9) on spherically symmetric fields of the form (17.7) and (17.8) is given by

$$(17.14) \quad \mathcal{E} = 4\pi \int \frac{dr}{r^2}\left((\tfrac{1}{2}(2\gamma + \beta^2 + \gamma^2)^2 + (\alpha + \alpha\gamma - r\beta')^2 + (r\gamma' + \alpha\beta)^2) \right.$$
$$\left. + \frac{1}{2}\left(r\frac{\partial\xi}{\partial r} - \xi\right)^2 + ((1+\gamma)^2 + \beta^2)^2\xi^2 + V(\xi)r^4 \right).$$

Thus, in this case the conditions stated in the previous chapter are met, and any extremal of the energy functional considered on symmetric fields is also an extremal of the functional on all of $\mathcal{V}$. In particular, the extremals of functional (17.14) are extremals of the energy functional in the Georgi–Glashow model. They may be thought of as spherically symmetric monopoles in this model (or, more precisely, as classical fields—solitons—corresponding in the semiclassical limit to particles carrying magnetic charge). Standard mathematical methods allow one to prove the existence of extremals for functional (17.12).

It turns out that in other models, too, spherically symmetric extremals can be thought of as magnetic monopoles. This follows from the fact, to be proved momentarily, that the topological type (and hence the magnetic charge) of a spherically symmetric field is completely determined by the type of symmetry. Recall that, by the definition given in Chapter 12, the topological type depends on the behavior of the scalar field $\varphi(x)$ at infinity. For simplicity, we assume that the asymptotics of the scalar field is $\varphi(\mathbf{x}) \sim \Phi(\mathbf{x}/|\mathbf{x}|)$, where $\Phi(\mathbf{n})$ is a map from the sphere $|\mathbf{n}| = 1$ into the vacuum manifold $R = K/H$, where H is the group of unbroken symmetries. The symmetry properties of $\varphi(\mathbf{x})$, expressed by (17.1), imply that its asymptotic behavior has similar properties:

$$(17.15) \qquad \Phi(g\mathbf{n}) = T(\lambda(g))\Phi(\mathbf{n})$$

for every rotation $g \in \mathrm{SO}(3)$. In particular, if $\mathbf{n} = \mathbf{k} = (0,0,1)$ is the unit vector in the positive z-direction, (17.15) implies that $\lambda(g)\Phi_0 = \Phi_0$, where $\Phi_0 = \Phi(\mathbf{k})$. In other words, λ can be interpreted as a map from the group $\mathrm{SO}(2)$ of rotations around the z-axis into the group H of elements of K that fix the classical vacuum Φ_0. Topologically, $\mathrm{SO}(2)$ is just a circle, so the homotopy class $[\lambda]$ of λ is an element of $\pi_1(H)$. On the other hand, the topological type of the scalar field φ is determined by the homotopy class of the map Φ, that is, by an element of the group $\pi_2(R) = \pi_2(K/H)$, which, as we know, is isomorphic to $\pi_1(H)$ when K is simply connected.

We now show that, for a spherically symmetric field, the topological type is determined by the element $2[\lambda] \in \pi_1(H) = \pi_2(K/H)$. (We denote the group law in $\pi_1(H)$ additively.) In fact, we prove a slightly more general assertion, valid also for the case where the symmetry type is determined by a two-valued homomorphism $\lambda : \mathrm{SO}(3) \to K$, or, which is the same thing, a homomorphism $\tilde{\lambda} : \mathrm{SU}(2) \to K$. Then the symmetry condition (17.1) for the scalar field $\varphi(\mathbf{x})$ is replaced with

$$(17.16) \qquad \varphi(p(g)\mathbf{x}) = T(\tilde{\lambda}(g))\varphi(\mathbf{x}),$$

where $g \in \mathrm{SU}(2)$ and p is the covering homomorphism $\mathrm{SU}(2) \to \mathrm{SO}(3)$. Likewise, (17.15) is replaced by

$$(17.17) \qquad \Phi(p(g)\mathbf{n}) = T(\tilde{\lambda}(g))\Phi(\mathbf{n}),$$

which implies that

$$(17.18) \qquad \tilde{\lambda}(g)\Phi(\mathbf{k}) = \Phi(\mathbf{k}) \qquad \text{if } p(g)\mathbf{k} = \mathbf{k},$$

where $\mathbf{k} = (0,0,1)$.

The set of elements of $\mathrm{SU}(2)$ satisfying $p(g)\mathbf{k} = \mathbf{k}$ is a topological circle, and is a double cover of the subgroup $\mathrm{SO}(2) \subset \mathrm{SO}(3)$ of rotations about the z-axis, the covering map being p. We denote this circle by $\widetilde{\mathrm{SO}}(2)$. From (17.18) it follows that $\tilde{\lambda}$ maps $\widetilde{\mathrm{SO}}(2)$ into H, so the homotopy class $[\tilde{\lambda}]$ of the map $\tilde{\lambda}$ considered on $\widetilde{\mathrm{SO}}(2)$ defines an element of $\pi_1(H) = \pi_2(K/H)$. We wish to show that the topological class of a field whose symmetry is determined by $\tilde{\lambda}$ is given by $[\tilde{\lambda}] \in \pi_1(H) = \pi_2(K/H)$. This will imply the assertion in the previous paragraph about the topological type of a field with symmetry of type λ, because $[\tilde{\lambda}] = 2[\lambda]$ for $\tilde{\lambda} = \lambda p$ (since p is a double cover).

To find the element of $\pi_1(H)$ corresponding to the homotopy class of Φ we must find a map $\tilde{\Phi}$ from the disk D^2 into K that is a lifting of Φ. Then the desired element of $\pi_1(H)$ is the homotopy class of $\tilde{\Phi}$ restricted to the boundary of D^2 (T14.1). More precisely, given $\Phi : S^2 \to R = K/H$, we look at $\Phi\rho : D^2 \to R$, where $\rho : D^2 \to S^2$ is a homeomorphism in the interior of D^2 and maps the whole boundary of D^2 to the north pole $\mathbf{k}$ of S^2. The map $\tilde{\Phi}$ is a lifting of Φ if the element $\tilde{\Phi}(x) \in K$ takes the point $\Phi_0 \in R$ to $\Phi(\rho(x))$; in symbols,

$$(17.19) \qquad\qquad T(\tilde{\Phi}(x))\Phi_0 = \Phi(\rho(x))$$

for all $x \in D^2$.

We first find a map $\Psi : D^2 \to \mathrm{SU}(2)$ satisfying

$$(17.20) \qquad\qquad p(\Psi(x))\mathbf{k} = \rho(x)$$

for all $x \in D^2$. Then we get $\tilde{\Phi}$ by setting $\tilde{\Phi}(x) = \tilde{\lambda}(\Psi(x))$. Indeed, using (17.17), (17.19) and (17.20), we see that

$$T(\tilde{\Phi}(x))\Phi_0 = T(\tilde{\lambda}(\Psi(x)))\Phi(\mathbf{k}) = \Phi(p(\Psi(x))\mathbf{k}) = \Phi(\rho(x)).$$

For points x on the boundary S^1 of D^2 we have $\rho(x) = \mathbf{k}$, so the rotation $p(\Psi(x))$ takes $\mathbf{k}$ to itself, that is, $\Psi(x)$ lies in $\mathrm{SO}(2)$.

We will shortly give an explicit procedure for finding Ψ, which implies that Ψ, considered as a map from $S^1 \subset D^2$ to $S^1 = \widetilde{\mathrm{SO}}(2)$, has degree one, and is in fact a homeomorphism. Recalling that the topological type of the field $\varphi(x)$ is given by the homotopy class $[\tilde{\Phi}]$ of the map $\tilde{\Phi} = \tilde{\lambda}\Psi : S^1 \to H$, we conclude that $[\tilde{\Phi}] = [\tilde{\lambda}]$.

We now proceed to construct Ψ. This can be done, for example, by thinking of D^2 as the closed unit disk $|w| \leq 1$ in the complex plane, and by setting

$$\Psi(w) = \begin{pmatrix} w & \sqrt{1 - |w|^2} \\ -\sqrt{1 - |w|^2} & \bar{w} \end{pmatrix}.$$

To check that the map $\rho : w \mapsto p(\Psi(w))\mathbf{k}$ is a homeomorphism in the interior of the disk, it is enough to show that $\rho(w) \neq \rho(w')$ if w and w' are distinct and have absolute value less than 1. Now $p(g)\mathbf{k} = p(g')\mathbf{k}$ if and only if $g' = gh$ with $h \in \mathrm{SO}(2)$, so the assertion follows because $\Psi(w')\Psi^{-1}(w)$ does not belong to $\widetilde{\mathrm{SO}}(2)$ for such values of w and w'.

We thus arrive at the following result. Let λ be a homomorphism from $\mathrm{SO}(3)$ into the gauge group K, and let I_3 be the element of the Lie algebra $\mathcal{K}$ of K corresponding under λ to an infinitesimal rotation about the z-axis. If I_3 belongs to the Lie algebra of the stabilizer H of the classical vacuum Φ_0, there exists an extremal of the energy functional having spherical symmetry of type λ. The magnetic charge of the extremal is determined by the homotopy class of the map that takes $\alpha \in S^1$ (where α is the angular coordinate) to $e^{2\alpha I_3} \in H$.

The first assertion follows from the fact that there are fields of finite energy with symmetry type λ, which in turn follows from the explicit formula (17.12) for the energy functional. An energy-minimizing field (among those of symmetry type λ) is the desired extremal. However, one must still prove that the minimum is attained; this is done by standard mathematical methods, which lie outside the scope of this book. The second assertion of the preceding paragraph follows from the fact that the map from the circle $SO(2) \subset SO(3)$ into K induced by λ has the form $\alpha \mapsto e^{\alpha I_3}$.

We now use these results to study spherically symmetric magnetic monopoles in the $SU(5)$ grand unification model. Given what we have proved above, we must consider the various homomorphisms $\lambda : SO(3) \rightarrow SU(5)$ that map an infinitesimal rotation about the z-axis to an element I_3 of the Lie algebra of the stabilizer H. The latter consists of block matrices of the form (7.2). Without loss of generality we can assume that I_3 is a diagonal matrix with entries $(im_1, im_2, im_3, im_4, im_5)$. Then the conditions above imply that $m_5 = 0$. The magnetic charge of a field with symmetry type λ can be computed as the homotopy class of the map $S^1 \rightarrow H$ that takes a point with angular coordinate α to the matrix $e^{2\alpha I_3}$; this matrix is diagonal, with elements $e^{2im_k\alpha}$ along the diagonal. From Chapter 12 it follows that the desired homotopy class is determined by the dependence of the fourth diagonal entry on α, because the map $H \rightarrow U(1)$ taking $m \in U(1)$ to the matrix (7.2) generates the group isomorphism $\pi_1(H) \simeq \pi_1(U(1))$. Thus, the magnetic charge is determined by the homotopy class of the map $\alpha \mapsto e^{2im_4\alpha}$ from S^1 into itself; it is therefore equal to $2m_4$. This is true also when λ is a two-valued homomorphism and the m_i are half-integers.

There is a certain arbitrariness in the determination of the magnetic charge, namely, the choice of a generator for the electromagnetic group $U(1)$. A different choice of generator results in the charge having the opposite sign.

We can easily enumerate all homomorphisms from $SO(3)$ into $SU(5)$, that is, all five-dimensional representations of $SO(3)$. The simplest possibilities are $I_3 = \mathrm{diag}(0, 0, -i/2, i/2, 0)$ and $I_3 = \mathrm{diag}(0, 0, i/2, -i/2, 0)$; each corresponding representation splits into the direct sum of a two-dimensional and a three-dimensional representation. The magnetic charge in either case is ± 1.

The possible values of the magnetic charge of a spherically symmetric field in the $SU(5)$-model are 0, ± 1, ± 2, ± 3 and ± 4. Indeed, every five-dimensional representation of $SO(3)$ is a direct sum of irreducible representations. If the representation is reducible, the dimensions of the irreducible components correspond to the possible partitions of the number 5:

$$4 + 1, \quad 3 + 1 + 1, \quad 3 + 2, \quad 2 + 2 + 1, \quad 2 + 1 + 1 + 1, \quad 1 + 1 + 1 + 1 + 1.$$

Since the matrix entries of I_3 in an irreducible representation are half-integers in the interval $(-l, l)$, where l is related to the dimension of the representation by the formula $d = 2l + 1$, we see that the magnetic charge of a spherically symmetric field is an integer ranging from -4 to 4. Moreover, all such integers can be realized as magnetic charges of spherically symmetric fields: for example, the value 4 for the magnetic charge corresponds to the irreducible representation given by $I_3 = \mathrm{diag}(-i, i, -2i, 2i, 0)$.

Using (17.12), one can easily write down the expression for the energy of a spherically symmetric field in the $SU(5)$-model.

18. Estimates of the Energy of a Magnetic Monopole

As already mentioned, in order to find the mass of magnetically charged particles in the semiclassical approximation, we must calculate the minimum of the energy functional on the set of fields with non-zero magnetic charge. We first do this for models where the scalar field transforms according to the adjoint representation of the gauge group G. These include the Georgi–Glashow model, where $G = \mathrm{SO}(3)$, and the SU(5) grand unification model, if we restrict ourselves to the first symmetry breaking.

In this case, the action integral is given by

$$(18.1) \quad S = -\frac{1}{4e^2} \int \langle \mathcal{F}_{\mu\nu}, \mathcal{F}^{\mu\nu} \rangle \, d^4x$$

$$+ \frac{1}{2} \int \langle \partial_\mu \varphi + [A_\mu, \varphi], \partial^\mu \varphi + [A^\mu, \varphi] \rangle \, d^4x - \lambda \int U(\varphi) \, d^4x,$$

where φ and A_μ take values in the Lie algebra $\mathcal{G}$ of G, and $U(\varphi)$ is an invariant polynomial on $\mathcal{G}$ that assumes its minimum on the manifold R of classical vacuums.

The group of unbroken symmetries H corresponding to a classical vacuum φ_0 consists of the elements $g \in G$ that commute with φ_0. We are assuming that G, and consequently its Lie algebra $\mathcal{G}$, are realized by means of matrices; then the adjoint representation is given by $T_g(\varphi) = g\varphi g^{-1}$. We let $H^{(0)} \subset H$ be the subgroup generated by φ_0, that is, the one-parameter subgroup whose elements are of the form $e^{\alpha \varphi_0}$. This clearly belongs to the center of H. We will consider the electromagnetic field strength $F_{\mu\nu}$ and the magnetic charge $\mathfrak{m}$ corresponding to $H^{(0)}$. (In the Georgi–Glashow model and the SU(5)-model the center of H is one-dimensional, so there is essentially no other possibility for the definition of the electromagnetic field strength.)

At great distances the electromagnetic field strength can be written as

$$F_{\mu\nu} = \left\langle \mathcal{F}_{\mu\nu}, \frac{\varphi}{\|\varphi\|} \right\rangle,$$

because $h(\varphi) = \varphi/\|\varphi\|$ is the normalized generator of the "electromagnetic subgroup" of the unbroken symmetry group of the vacuum $\varphi \in R$. We can therefore write the magnetic charge as $(4\pi)^{-1}$ times the flux of the magnetic field strength $\mathcal{H}$ through a sphere at infinity, where $\mathcal{H}$ has components $H^a = \frac{1}{2}\varepsilon^{ajk}F_{jk} = \frac{1}{2}\varepsilon^{ajk}\langle\mathcal{F}_{jk},\varphi\rangle/\|\varphi\|$. Note that the norm of any classical vacuum is the same, since any two classical vacuums can be taken to one another by an element of G, which preserves the norm. This allows us to set $a = \|\varphi\|$ for any $\varphi \in R$.

Using Stokes' theorem to replace the flux by a volume integral, we get

$$(18.2) \qquad \mathfrak{m} = \frac{1}{4\pi}\int \mathcal{H}\, dS = \frac{1}{8\pi a}\int \partial_i \varepsilon^{ijk}\langle\mathcal{F}_{jk},\varphi\rangle\, d^3x.$$

We now note that

$$\varepsilon^{ijk}\partial_i\langle\mathcal{F}_{jk},\varphi\rangle = \varepsilon^{ijk}\nabla_i\langle\mathcal{F}_{jk},\varphi\rangle$$
$$= \varepsilon^{ijk}(\langle\nabla_i\mathcal{F}_{jk},\varphi\rangle + \langle\mathcal{F}_{jk},\nabla_i\varphi\rangle) = \varepsilon^{ijk}\langle\mathcal{F}_{jk},\nabla_i\varphi\rangle,$$

and that for the gauge-invariant quantity $\langle\mathcal{F}_{jk},\varphi\rangle$, the partial derivative ∂_i coincides with the covariant derivative ∇_i. Using Bianchi's identity $\varepsilon^{ijk}\nabla_i\mathcal{F}_{jk} = 0$ (T15.2), we obtain

$$(18.3) \qquad \mathfrak{m} = \frac{1}{8\pi a}\int \varepsilon^{ijk}\langle\mathcal{F}_{jk},\nabla_i\varphi\rangle\, d^3x.$$

We now notice that, if $b(x)$ and $c(x)$ are vector-valued functions, we have $\int\langle c-b, c-b\rangle\, dx \geq 0$, and therefore

$$(18.4) \qquad \int\langle c,b\rangle\, dx \leq \frac{1}{2}\int(\langle c,c\rangle + \langle b,b\rangle)\, dx,$$

and that equality holds if and only if b and c coincide identically. Applying this inequality to the vectors $(2e)^{-1}\varepsilon^{ijk}\mathcal{F}_{jk}$ and $\nabla_i\varphi$, we get

$$(18.5) \qquad \int \frac{1}{2e}\varepsilon^{ijk}\langle\mathcal{F}_{jk},\nabla_i\varphi\rangle\, d^3x$$
$$\leq \frac{1}{2}\int\left(\frac{1}{4e^2}\langle\mathcal{F}_{jk},\mathcal{F}_{jk}\rangle + \langle\nabla_i\varphi,\nabla_i\varphi\rangle\right) d^3x.$$

The integral on the right differs from the energy E of the field (φ, A_μ) only by the absence of the term $E_1 = \lambda\int U(\varphi)\, d^3x$. On the other hand, the left side of the equality equals the magnetic charge up to a factor, so we get $\mathfrak{m} \leq e(E-E_1)/(4\pi a)$, or

$$(18.6) \qquad E \geq 4\pi\mathfrak{m}\frac{a}{e} + E_1,$$

with equality only if

$$(18.7) \qquad \frac{1}{2e}\varepsilon^{ijk}\mathcal{F}_{jk} = \nabla^i\varphi.$$

Since E_1 is non-negative and the magnetic charge is integer-valued (under our normalization), the energy of a magnetically charged field is bounded below:

(18.8)
$$E \geq 4\pi \frac{a}{e}$$

We now show that this bound becomes sharp as $\lambda \to 0$, if the parameters a and e remain constant. This is known as the *Bogomolnyi–Prasad–Sommerfeld limit*. In this limit we have $E_1 = 0$, that is,

(18.9)
$$E = \frac{1}{4e^2} \int \langle \mathcal{F}_{jk}, \mathcal{F}_{jk} \rangle \, d^3x + \frac{1}{2} \int \langle \nabla_i \varphi, \nabla_i \varphi \rangle \, d^3x.$$

Note that, although the term E_1 disappears from the energy functional, it leaves a trace in that, for finite λ, only fields for which $\varphi(\mathbf{x}) \to R$ as $|\mathbf{x}| \to \infty$ can have finite energy, and this condition must be preserved for $\lambda = 0$.

The reasoning above shows that fields satisfying (18.7) are extremals for the functional (18.9), since they realize the minimum

$$E_{\min}(\mathfrak{m}) = 4\pi \mathfrak{m} \frac{e}{a}$$

(among fields with $\|\varphi(\mathbf{x})\| \to a$ at infinity and having fixed magnetic charge $\mathfrak{m}$). It is easy to verify that such fields exist. For example, in the Georgi–Glashow model with $\mathfrak{m} = 1$ we can look for a solution of the form (11.7): substituting (11.7) in (18.7) and solving the resulting first-order equations in $\alpha(r)$ and $\beta(r)$, we obtain

(18.10)
$$A_0^a(\mathbf{x}) = 0,$$
$$A_i^a(\mathbf{x}) = \frac{\varepsilon_{aik} x^k}{|\mathbf{x}|^2} \left(1 - \frac{ae\,|\mathbf{x}|}{\sinh ae\,|\mathbf{x}|} \right),$$
$$\varphi^a(\mathbf{x}) = \frac{x^a}{e\,|\mathbf{x}|^2} (ae\,|\mathbf{x}| \coth ae\,|\mathbf{x}| - 1).$$

For an arbitrary positive integer $\mathfrak{m}$ it can be shown that (18.7) has a $(2\mathfrak{m} + 2)$-parameter family of gauge-inequivalent solutions.

For $\lambda > 0$ the lower bound (18.8) is not sharp. But, in addition to the lower bound provided by (18.6) for the minimum energy of a field with magnetic charge $\mathfrak{m}$, it is easy to obtain an upper bound by selecting a trial function that satisfies (18.7). We get

(18.11)
$$4\pi \mathfrak{m} \frac{a}{e} \leq E_{\min} \leq 4\pi \mathfrak{m} \frac{a}{e} + \lambda C,$$

where C does not depend on λ.

In the semiclassical approximation, ae coincides with the mass M_W of a charged vector particle, the intermediate vector boson (see Chapter 11). The bounds in (18.11) imply that the mass of a magnetic monopole is of the same order of magnitude as M_W/e^2.

By a slight modification of the reasoning above, we can obtain a lower bound for the energy of a magnetically charged field in any grand unification model.

Recall from (14.8) that in such a model the magnetic charge of a field of finite energy is the flux of the vector field

$$\frac{1}{4\pi} H^i = \frac{1}{8\pi} \varepsilon^{ijk} \langle \mathcal{F}_{jk}(\mathbf{x}), h(\varphi(\mathbf{x})) \rangle$$

through a sphere at infinity. Take a G-invariant smooth function $\alpha(\varphi)$ equal to one in a neighborhood of the manifold R of classical vacuums, and equal to zero outside a larger neighborhood of R.

To construct this function we can start by making $\rho(\varphi)$ equal to the distance from φ to R. We then set $a(\varphi) = \beta(\rho(\varphi))$, where $\beta : \mathbf{R}_+ \to \mathbf{R}_+$ is a smooth function equal to 1 for values of the argument less than $\varepsilon/3$ and equal to 0 for values greater than $2\varepsilon/3$. One must select ε so that in an ε-neighborhood of R the function $\rho(\varphi)$ is continuous and single-valued.

Obviously, the magnetic charge also equals the flux of

$$A^i = \frac{1}{8\pi} \varepsilon^{ijk} \langle \mathcal{F}_{jk}, \alpha(\varphi)h(\varphi) \rangle$$

through a sphere at infinity, since far from the origin the function $\varphi(\mathbf{x})$ is close to R, and therefore $\alpha(\varphi(\mathbf{x})) = 1$. Notice that, unlike H^i, the vector A^i is defined for all $\mathbf{x}$. Indeed, the element $h(\varphi(x)) \in \mathcal{G}$ is defined only when $\varphi(x)$ is close enough to R; but when this element is not defined, we have $\alpha(\varphi(\mathbf{x})) = 0$. Thus the product $h(\varphi(\mathbf{x}))\alpha(\varphi(\mathbf{x}))$ is always defined.

Expressing the flux of $A^i(x)$ in terms of a volume integral, we have

$$(18.12) \qquad \mathfrak{m} = \frac{1}{8\pi} \int \partial_i \varepsilon^{ijk} \langle \mathcal{F}_{jk}, a(\varphi)h(\varphi) \rangle \, d^3x.$$

Using the gauge invariance of $\langle \mathcal{F}_{jk}, \alpha(\varphi)h(\varphi) \rangle$ and repeating the argument that led to (18.3), we get

$$(18.13) \qquad \mathfrak{m} = \frac{1}{8\pi} \int \varepsilon^{ijk} \langle \mathcal{F}_{jk}, \nabla_i(\alpha(\varphi)h(\varphi)) \rangle \, d^3x.$$

Applying (18.4) to the functions $d^{1/2}\varepsilon^{ijk}\mathcal{F}_{jk}$ and $d^{-1/2}\nabla_i(\alpha(\varphi)h(\varphi))$, we get
(18.14)

$$\mathfrak{m} \le \frac{1}{16\pi} \left(2d \int \langle \mathcal{F}_{jk}, \mathcal{F}_{jk} \rangle \, d^3x + \frac{1}{d} \int \langle \nabla_i(\alpha(\varphi)h(\varphi)), \nabla_i(\alpha(\varphi)h(\varphi)) \rangle \, d^3x \right),$$

where d is an arbitrary number. The G-invariance of α and the fact that $h(\tau_g(\varphi)) = \tau_g h(\varphi)$ imply that

$$(18.15) \qquad \nabla_i(\alpha(\varphi)h(\varphi)) = \frac{\partial(\alpha(\varphi)h(\varphi))}{\partial\varphi^a}(\nabla_i\varphi)^a.$$

We see that

$$(18.16) \qquad \mathfrak{m} \le \frac{1}{16\pi} \int \left(2d\langle \mathcal{F}_{jk}, \mathcal{F}_{jk} \rangle + \frac{1}{d}g_{ab}(\varphi)(\nabla_i\varphi)^a(\nabla_i\varphi)^b \right) d^3x,$$

where

$$g_{ab}(\varphi) = \left\langle \frac{\partial(\alpha(\varphi)h(\varphi))}{\partial\varphi^a}, \frac{\partial(\alpha(\varphi)h(\varphi))}{\partial\varphi^b} \right\rangle.$$

We now choose a number η such that $g_{ab}(\varphi)\xi^a\xi^b \leq \eta^2 |\xi|^2$ for all ξ; this number can be chosen independently of φ, since the tensor $g_{ab}(\varphi)$ is non-zero only on a bounded set. Setting $d = \eta/(2e)$, we get from (18.16) the estimate

$$\mathfrak{m} \leq \frac{\eta e}{4\pi} \int \left(\frac{1}{4e^2} \langle \mathcal{F}_{jk}, \mathcal{F}_{jk} \rangle + \tfrac{1}{2}\langle \nabla_i\varphi, \nabla_i\varphi \rangle \right) d^3x \leq \frac{\eta e}{4\pi} E(\varphi, A).$$

We see that the energy $E(\varphi, A)$ of the field (φ, A_μ) with magnetic charge $\mathfrak{m}$ satisfies

$$E(\varphi, A) \geq 4\pi\frac{\mathfrak{m}}{\eta e}.$$

This implies that the mass of a magnetic monopole has the same order of magnitude as M_W/e^2, where M_W is the mass of the heaviest vector particle. Indeed, M_W has same order of magnitude as $e|a|$, where $|a|$ is the maximum of the expectation values of components of a field φ corresponding to a classical vacuum. (Instead of $|a|$, one could also take the length of φ.) On the other hand, η has same order of magnitude as $|a|^{-1}$; this gives the desired bound for the mass of a magnetic monopole.

19. Topologically Non-Trivial Strings

We have seen that in grand unification theories there exist topologically nontrivial fields, that is, fields that cannot be reduced to a vacuum field by a continuous deformation in the class of fields of finite energy. Roughly speaking, such fields are objects whose energy is located within a bounded region of space and whose stability is guaranteed on topological grounds. In a quantum theory they represent stable particles carrying a topological charge; this charge, as we say, can be identified with magnetic charge, and topologically charged particles with magnetic monopoles.

We now show that in grand unification theories there can also exist topologically nontrivial strings, which are objects whose energy is located within a tubular region, and whose stability is also topologically guaranteed. We saw some examples of strings in Chapter 10.

A string can be defined as a field (φ, A_μ) that differs significantly from a vacuum field only in the neighborhood of a curve Γ. We will require that outside this neighborhood φ has values in a neighborhood $\mathcal{R}$ of the vacuum manifold R (see Chapter 12). Consider a circle S^1 surrounding Γ, and lying in the region where φ takes values in $\mathcal{R}$. For example, if Γ is a vertical line and φ is close to the vacuum except in a cylinder of radius δ, we can take for S^1 a horizontal circle of radius greater than δ and center along the vertical line. The restriction of φ to S^1 determines a map $S^1 \to \mathcal{R}$, and we will say that the topological type of the string is the homotopy class of this map, or the homotopy class of the map $\sigma\varphi : S^1 \to R$, where σ takes a point in $\mathcal{R}$ to the nearest point in R.

The two definitions in the last sentence are equivalent, because σ is a homotopy equivalence (if $\mathcal{R}$ is a small enough neighborhood of R), and thus establishes a one-to-one correspondence between $\{S^1, \mathcal{R}\}$ and $\{S^1, R\}$.

A string is *topologically nontrivial* if its topological type is nontrivial, that is, if the map $\sigma\varphi : S^1 \to R$ is not null-homotopic.

As noted in Chapter 10, strings occur in the model described by the Lagrangian (10.6). We now study another simple model admitting strings. Let $G = \mathrm{SO}(3)$ be the gauge group, and assume that the scalar field transforms by the five-dimensional irreducible representation of $\mathrm{SO}(3)$. In other words, we think of the scalar field as a symmetric traceless tensor Φ^{ij}, for $i, j = 1, 2, 3$. We choose a polynomial $V(\Phi)$ such that the manifold R consists of matrices Φ^{ij} with two equal eigenvalues λ, for λ fixed. For example, we can take

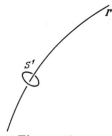

Figure 13

$$V(\Phi) = a(\operatorname{tr}\Phi^2 - b^2)^2 + \varepsilon\operatorname{tr}\Phi^3,$$

where ε is a small enough positive number.

To each matrix with two equal eigenvalues λ we assign the straight line going through the origin and parallel to the eigenvector with eigenvalue -2λ. This gives a one-to-one correspondence between R and lines through the origin. Thus, R is homeomorphic to the space of lines through the origin, which is the projective plane RP^2. We see that $\pi_1(R) = \pi_1(RP^2) = \mathbf{Z}_2$, so that $\{S^1, R\}$ has exactly two elements, as does $\pi_1(R)$ (T3.2).

An example of a topologically nontrivial string that occurs in this model is

(19.1)
$$\Phi(\rho, \theta) = \alpha(\rho)V(-\tfrac{1}{2}\theta)\Phi_0 V(\tfrac{1}{2}\theta),$$
$$A_\theta(\rho, \theta) = \tfrac{1}{2}L_z\beta(\rho), \qquad A_\rho = A_z = 0,$$

where ρ, θ and z are cylindrical coordinates, L_z is the element of the Lie algebra of SO(3) corresponding to an infinitesimal rotation around the z-axis, $V(\theta) = \exp(L_z\theta)$ is the matrix expressing a rotation by θ around the z-axis, Φ_0 is a diagonal matrix with eigenvalues λ, λ and -2λ, and $\alpha(\rho)$ and $\beta(\rho)$ are numerical functions decaying rapidly to zero as $\rho \to \infty$ and vanishing at 0. It is possible to select α and β so that the string (19.1) satisfies the equations of motion. (We will prove more general results below, valid in arbitrary gauge models.)

We always assume that the degeneracy of the classical vacuum is due solely to symmetry. This allows us to identify R with G/H, where H is the stabilizer of a point $\Phi_0 \in R$. If G is simply connected, $\pi_1(R)$ is isomorphic to the quotient $\pi_0(H) = H/H_{\mathrm{con}}$. Thus, to every discrete unbroken symmetry $h \in H$ we assign an element of $\pi_1(R)$ (T14.1). This correspondence is defined as follows. To each path $g(t)$, with $0 \le t \le 1$, in G we assign a path

(19.2)
$$r(t) = T(g(t))\varphi_0$$

in R; every path in R can be written in the form (19.2). If $g(0)$ and $g(1)$, the endpoints of g, lie in H, the path $r(t)$ is closed. Since G is simply connected, the element $[r] \in \pi_1(R)$ defined by the path $r(t)$ depends only on the endpoints of $g(t)$. Moreover, since $g(t)$ and $g(t)h$ give the same path $r(t) \in R$, for any $h \in H$, the element $[r] \in \pi_1(R)$ depends only on $g(0)g^{-1}(t)$. If $g(0)g^{-1}(t)$ changes continuously, $[r]$ does not change. This means there is a correspondence between H/H_{con} and $\pi_1(R)$, which is easily seen to be an isomorphism (T14.1).

Two elements of $\pi_1(R)$ give the same homotopy class of mappings $S^1 \to R$ if they are conjugate; in other words, $\{S^1, R\}$ can be identified with the set of conjugacy classes in $\pi_0(H) = H/H_{\mathrm{con}}$ (T3.2). Thus, the topological type of a string is determined by an element of $\pi_0(H)$, but conjugate elements give the same type. Likewise, an element of H can be seen as an element of $\pi_0(H)$, if we keep in mind that two elements of H joined by a continuous curve are the same element of $\pi_0(H)$.

We now consider in greater detail linear strings, that is, fields localized in the neighborhood of a straight line. If a field is localized in the neighborhood of an arbitrary curve Γ, it can still be considered linear if the width of the neighborhood is much smaller than the radius of curvature of Γ. A typical example of a linear string is a field (φ, A_μ) invariant under translations along the z-axis and having a finite linear energy density; this latter condition means that the energy in the region bounded by two horizontal planes is finite. Because the field is translation-invariant along the z-axis, the total energy is infinite.

In order for the linear energy density to be finite, the covariant derivative $\nabla_\mu(\varphi)$ must tend to zero as we move away from the z-axis. We assume that

$$(19.3) \qquad \nabla_\mu \varphi \leq \mathrm{const} \, \rho^{-(1+\delta)},$$

where $\rho = \sqrt{x^2 + y^2}$ is the distance from the z-axis. We also assume that in cylindrical coordinates (ρ, θ, z) the field $\varphi(\rho, \theta, z) = \varphi(\rho, \theta)$ has a limit as $\rho \to \infty$:

$$(19.4) \qquad \Phi(\theta) = \lim_{\rho \to \infty} \varphi(\rho, \theta).$$

In the gauge $xA_x + yA_y = 0$ the existence of this limit follows from (19.3). Even more: one can show that $|\Phi(\theta) - \varphi(\rho, \theta)| < \mathrm{const} \, \rho^{-\delta}$. This is shown in the same way as the corresponding assertions in Chapters 10 and 12.

The function $\Phi(\theta)$ can be thought of as a map $S^1 \to R$; the topological type of (φ, A_μ) obviously coincides with the homotopy class of Φ.

For a linear string (φ, A_μ) satisfying (19.3), the topological type can be expressed in terms of the gauge field A_μ. More precisely, the element of $\pi_0(H)$ corresponding to the topological type can be represented as

$$(19.5) \qquad \alpha = \left(\mathrm{P} \exp\left(-\oint A_\mu \, dx^\mu \right) \right)^{-1} = \left(\mathrm{P} \exp\left(-\int_0^{2\pi} A_\theta \, d\theta \right) \right)^{-1},$$

where the integral is taken along a circle at infinity going around the string, $\mathrm{P}\exp$ stands for a path-ordered exponential, and $A_\theta = A_\mu(\partial x^\mu/\partial\theta)$ is the θ-component of the gauge field in cylindrical coordinates (ρ, θ, z). We will see below that (19.5) defines an element of the stabilizer H of $\Phi(0) \in R$, and consequently also an element of $\pi_0(H)$.

We can also define α as $\alpha = \lim_{\rho \to \infty} \alpha(\rho, 2\pi)$, where $\alpha(\rho, \theta)$ is a solution to the differential equation

$$\partial_\theta \alpha(\rho, \theta) - \alpha(\rho, \theta) A_\theta(\rho, \theta) = 0$$

with initial condition $\alpha(\rho, 0) = 1$. In other words,

$$\alpha^{-1}(\rho, \theta) = \mathrm{P} \exp\left(-\int_0^{2\pi} A_\theta(\rho, \psi)\, d\psi\right).$$

In order to prove (19.5), we take the limit $\rho \to \infty$ in the equation

$$|\nabla_\theta \varphi| = |(\partial_\theta + t(A_\theta(\rho, \theta)))\varphi| \le \mathrm{const}\, \rho^{-\delta},$$

which follows from (19.3). Here t is the representation of the Lie algebra corresponding to the representation T of the gauge group G according to which φ transforms. We get

(19.6) $(\partial_\theta + t(A_\theta(\theta)))\Phi(\theta) = 0,$

with $A_\theta(\theta) = \lim_{\rho \to \infty} A_\theta(\rho, \theta)$. This implies that

(19.7) $\Phi(\theta) = T(\alpha^{-1}(\theta))\Phi(0),$

where $\alpha^{-1}(\theta) = \mathrm{P} \exp\left(-\int_0^\theta A_\theta\, d\theta\right)$. Since $\Phi(2\pi) = \Phi(0)$, we see that $\alpha = \alpha(2\pi)$ belongs to stabilizer H of Φ_0. The path $g(\theta) = \alpha^{-1}(\theta)$, for $0 \le \theta \le 2\pi$, in G corresponds to the path $\Phi(\theta)$ in R. As explained above, the homotopy class of $\Phi(\theta)$ corresponds to the element of H/H_{con} defined by $g(0)g^{-1}(2\pi) = \alpha^{-1}(0)\alpha(2\pi) = \alpha \in H$. This proves (19.5).

The reasoning above needs the assumption that $A_\theta(\rho, \theta)$ has a limit as $\rho \to \infty$. This restriction can be removed at the cost of complicating the proof somewhat.

To analyze the quantum analog of strings in the semiclassical approximation, we must find strings that satisfy the classical equations of motion. We do so now, finding for any topological class linear strings with finite linear energy density and satisfying the equations of motion. It is natural to try out axially symmetric fields, that is, fields where the change caused by a rotation can be compensated for by a global gauge transformation corresponding to an element $g_\theta = \exp(M\theta) \in G$, where M is an element of the Lie algebra of G. An axially symmetric field (φ, A_μ) that is also invariant under z-translations has the following expression in cylindrical coordinates:

(19.8)
$$\varphi(\rho, \theta, z) = T(\exp(M\theta))\varphi(\rho),$$
$$A_\theta(\rho, \theta, z) = \exp(M\theta)A_\theta(\rho)\exp(M\theta),$$
$$A_\rho(\rho, \theta, z) = \exp(M\theta)A_\rho(\rho)\exp(-M\theta),$$
$$A_z(\rho, \theta, z) = \exp(M\theta)A_z(\rho)\exp(-M\theta),$$

For these expressions to have the same value at $\theta = 2\pi$ and $\theta = 0$, we must have $T(\alpha)\varphi(\rho) = \varphi(\rho)$, where $\alpha = \exp(-2\pi M)$, and α must commute with $A_\theta(\rho)$, $A_\rho(\rho)$ and $A_z(\rho)$. For the field not to be singular along the z-axis, we must have $A_\rho(0) = A_z(0) = 0$. We assume that $\varphi(\rho) = \Phi_0 + \sigma(\rho)$ and $A_\theta(\rho) = M + \beta(\rho)$, with the functions $\alpha(\rho)$, $\beta(\rho)$, $A_\rho(\rho)$ and $A_z(\rho)$ decaying fast enough as $\rho \to \infty$.

Then the linear energy density ε of (19.8) is finite, as can be seen from its explicit expression:

(19.9)

$$\varepsilon = \frac{1}{2}\int\left(\left|\frac{1}{\rho}t(M+A_\theta(\rho))\varphi(\rho)\right|^2 + \left|\frac{\partial\varphi(\rho)}{\partial\rho} + tA_\rho(\rho)\varphi(\rho)\right|^2 + |t(A_z(\rho))\varphi(\rho)|^2\right)$$
$$\cdot\,\rho\,d\rho\,d\theta$$
$$+ \int V(\varphi(\rho))\rho\,d\rho\,d\theta + \varepsilon_{\mathrm{YM}}.$$

The extremals of (19.9) satisfy the classical equations of motion. This can be checked directly, or derived from the general results of Chapter 16. From this it is easy to deduce that every class of axial symmetry contains a string that satisfies the classical equations. Such a string can also be found in the narrower class of fields that satisfy (19.8) and the additional conditions $A_\rho = A_z = 0$ and $A_\theta(\rho) = M\gamma(\rho)$, where $\gamma(\rho)$ is a real-valued function.

Note that the topological type of an axisymmetric string is completely determined by the class of axial symmetry of the string, that is, by the element $M \in \mathcal{G}$. Indeed, the topological type of a string is the homotopy class of the map $\Phi(\theta) = T(\exp(M\theta))\Phi_0$; the element $\alpha = \exp(-2\pi M) \in H$ corresponds to this homotopy class. Since every element $\alpha \in H \subset G$ is of the form $\exp(-2\pi M)$, we see that every topological class contains linear strings of finite energy density that are solutions to the equations of motion. Using these solutions, one can construct curved closed strings that satisfy the equations of motion approximately. In the semiclassical approach one can construct, using these approximate solutions, quantum objects that we will call *closed strings*.

The preceding discussion allows one to predict the existence of strings in particular grand unification models. We mention the following cases:

If the gauge group G is simply connected, topologically nontrivial strings exist if and only if there are discrete unbroken symmetries.

If $G = \mathrm{SO}(n)$, we replace G by its universal cover $\mathrm{Spin}(n)$, which is simply connected. It $H \subset \mathrm{SO}(n)$ is the group of unbroken symmetries, we consider the inverse image $\tilde{H}$ of H under the covering map. The vacuum manifold $R = \mathrm{SO}(n)/H$ is isomorphic to $\mathrm{Spin}(n)/\tilde{H}$, and by the preceding paragraph topologically nontrivial strings exist if and only if $\tilde{H}$ is disconnected. This is certainly the case if H is disconnected; but it may also happen even if H is connected. Indeed, if $\tilde{H}$ is connected, it contains a path connecting the two inverse images of the identity in $\mathrm{Spin}(n)$ (recall that $\mathrm{Spin}(n)$ is a double cover of $\mathrm{SO}(n)$). The image of this path under the covering map is a loop in H that is not null-homotopic. We conclude that $\tilde{H}$ is necessarily disconnected if H is contained in a simply connected subset of $\mathrm{SO}(n)$, for in this case all loops in H are null-homotopic. We therefore have the following criterion: if $G = \mathrm{SO}(n)$, a sufficient condition for the existence of topologically nontrivial strings is that the scalar field transforms by a single-valued representation of $\mathrm{SO}(n)$ and there exists a simply connected subset of $\mathrm{SO}(n)$ containing the group H of unbroken symmetries.

In the $SO(10)$ grand unification model the group H is connected and contained in the simply connected group $SU(5) \subset SO(10)$. This means that $\tilde{H}$ is disconnected and hence that there are topologically nontrivial strings in the model. In particular, the axial symmetry associated with the generator M of $SO(2) \subset SO(10)$, normalized so that $\exp(2\pi M) = 1$ and $\exp(\theta M) \neq 1$ for $0 \leq \theta \leq 2\pi$, yields such strings. (If M is regarded as a generator of $Spin(10)$, it does not satisfy $\exp(2\pi M) = 1$.)

Topologically nontrivial strings may exist not only in grand unification models but also in models describing electroweak interactions. There are none in the standard Weinberg–Salam model, but it can be proved that they exist in the Lee–Weinberg model based on $SU(3) \times U(1)$. (This group is not simply connected; if we want to apply the results of this chapter, we must replace it with a simply connected non-compact gauge group. It is easier, however, to compute $\pi_1(R)$ using an exact homotopy sequence.)

20. Particles in the Presence of Strings

We now show that, when a particle goes around a string, it may change its quantum numbers, and in particular it may change its electric charge in some models. These models are not realistic, but it is possible to construct realistic grand unification models that contain "mirror" particles that interact weakly with ordinary ones, and in which a particle is turned into its mirror image when going around a string. Such strings can, in principle, be discovered through astronomical observations.

For simplicity we start with models where the unbroken symmetry group H is one-dimensional. In this case, as shown in Chapter 15, the electromagnetic field strength $F_{\mu\nu}(x)$ is of the form (15.1). This formula contains the function $h(\sigma\varphi(x))$, where $h(\varphi)$ is the generator of the one-dimensional group H_φ of transformations that fix a given classical vacuum φ, and σ is the map taking a point in the set $\mathcal{R}$ to the nearest point in R. In particular, for a finite-energy field, $F_{\mu\nu}(x)$ tends to $\langle F_{\mu\nu}(x), h(\sigma\varphi(x)) \rangle$ as $|x| \to \infty$: see (14.5). (In the discussion below we do not use the explicit form (15.1) for $F_{\mu\nu}$.) As noted in Chapter 14, after the gauge condition has been imposed, the generator $h(\varphi)$ is defined up to sign. We now study when this ambiguity can be eliminated.

First we note that both $h(\sigma\varphi(x))$ and $F_{\mu\nu}(x)$ are defined only on the set V of points x where $\varphi(x)$ is close enough to a vacuum, that is, $\varphi(x) \in \mathcal{R}$. If this set is simply connected, we can take a continuous single-valued branch from the function $h(\sigma\varphi(x))$; this is a general property of simply connected spaces.

If a function is defined on an interval, we can construct a single-valued branch of it by partitioning the interval into small subintervals and extending the chosen branch from one subinterval to the next. For a general simply connected space, we fix the value of the function at a point x_0, connect this point with other points x_1 by paths, and choose a continuous branch along each of these paths. The value of the function at x_1 defined in this way does not change if the path is deformed continuously. Since in a simply connected space any two paths between x_0 and x_1 can be continuously deformed into one another, this construction gives a well defined value for the function everywhere.

Next, if R is simply connected, that is, if H is connected, the same reasoning allows us to choose a continuous, single-valued branch of $h(\sigma\varphi)$ on $\mathcal{R}$, and hence a single-valued branch of $h(\sigma\varphi(x))$ defined on V. The most interesting case is when R is not simply connected, so topologically nontrivial strings exist. Then it may happen that the ambiguity in $h(\sigma\varphi(x))$, and consequently in the magnetic field strength, cannot be eliminated. For example, in the model of Chapter 19, for the field of (19.1), we cannot get a continuous single-valued branch for $h(\sigma\varphi(x))$. This is clear from the expression

$$(20.1) \qquad h(\sigma\varphi(\rho, \theta)) = \pm V(-\tfrac{1}{2}\theta)h_0 V(\tfrac{1}{2}\theta),$$

where h_0 is the generator of the group $H = H_{\varphi(\infty,0)}$.
    For the field (19.8),

$$h(\sigma\varphi(\rho, \theta)) = \pm \exp(M\theta)h(\sigma\varphi(\rho)) \exp(-M\theta),$$

which shows that the ambiguity in $h(\sigma\varphi(x))$ cannot be eliminated if

$$(20.2) \qquad \alpha h_0 \alpha^{-1} = -h_0,$$

where $\alpha = \exp(-2\pi M)$ and h_0 is the generator of $U(1) \subset H = H_{\varphi(\infty,0)}$.
    This situation is general: a necessary and sufficient condition for the existence of a field that cannot be assigned an unambiguous electromagnetic field strength is that the unbroken symmetry group H contains an element α satisfying (20.2), where h_0 is a generator of $U(1)$. (In this case α conjugates *any* element of $U(1)$ to its inverse). For, if such an α exists, $h(\sigma\varphi(x))$ has no continuous single-valued branch for any string (φ, A_μ) of topological type α, and therefore the ambiguity cannot be eliminated unless the electromagnetic field strength vanishes identically. (If it does vanish, as in the case of the field (19.8), there is no ambiguity to eliminate.)
    We now consider the electric and magnetic charges of particles, assuming that far from any particles the fields can be considered classical. By definition, the electric charge of a particle is $(4\pi)^{-1}$ times the flux of the electric field strength through a sphere surrounding the particle. Similarly, the magnetic charge of a particle is defined in terms of the magnetic flux. However, if a particle is in the field of a string whose topological type α satisfies (20.2), these definitions are no longer valid, or, more precisely, they are only definitions up to sign, since the electromagnetic field strength is two-valued.
    Even when the electric and magnetic charges of a single particle are only defined up to sign, the *relative* sign of the charges of two particles connected by a curve is well defined, provided that all along the curve the electromagnetic field strength is well defined, that is, that φ takes values in $\mathcal{R}$. Take a sphere Ω enclosing this curve, and consider the open set U bounded outside by Ω and inside by two smaller spheres Ω_1 and Ω_2 containing the particles (Figure 14). Since U is simply connected, we can take a continuous, single-valued branch for $F_{\mu\nu}$ on U, and compute the flux of the electric or magnetic field strength through Ω_1 and Ω_2. If the fluxes have the same sign, we say that the particles have charges of the same sign; otherwise, of opposite signs. The relative charge of a pair of particles may depend on the curve connecting them, but it does not change under a continuous deformation of the curve, so long as the curve does not leave the domain where $\varphi(x) \in \mathcal{R}$.
    Now consider a system consisting of a string and several particles, so that its state differs significantly from a vacuum state only in the neighborhood of

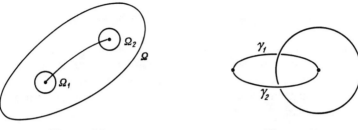

Figure 14   Figure 15

a closed curve and of several points. (We could also allow the curve to be disconnected: then we have several strings.) Where the state is close to a vacuum,
we think of it as being semiclassical, that is, we replace it with a classical field.
We assume that the electromagnetic field strength of this classical field changes
sign as we go through the string.

In this situation, it follows from the preceding discussion that the relative
sign of two charged particles does depend on the choice of the curve connecting
the particles—more specifically, if two curves γ_1 and γ_2 connecting the particles
form a loop that encircles the string, as in Figure 15, they define opposite
relative signs. This is because the field strength changes sign when one goes
around the loop.

Furthermore, if we have a time-dependent field, the relative sign of two fixed
charges, with respect to to a fixed curve γ, will nonetheless change if the curve
is crossed by the string as the system evolves. In any case, the relative sign
of the charges with respect to γ is undefined at the moment when the string
crosses γ.

It is reasonable to define the relative sign of two charges by using straight-
line segments whenever possible. Then, if we fix the sign of one particle, we
determine the sign of any other charge, unless the segment connecting the two
happens to be crossing a string at that moment. In this case, the relative signs
before and after the crossing differ.

All of this indicates that in these models charge is not conserved. However,
it is possible to define charge in such a way that conservation of charge is true in
a certain sense. To this end we span the string with a film, or, in other words, we
consider a surface whose boundary is the curve in whose neighborhood the field
differs considerably from a vacuum. If there are several strings, we assume that
these films do not intersect one another. We now consider the space obtained by
removing the films, as well as the neighborhoods where the electromagnetic field
strength $F_{\mu\nu}$ is undefined. The result is a simply connected set on which we can
select a continuous, single-valued branch for $F_{\mu\nu}$. We use this branch to define
the (electric or magnetic) charge of a particle; likewise we define the charge of a
string as $(4\pi)^{-1}$ times the flux of the field strength through the boundary of a
neighborhood of the spanning film. Maxwell's equations imply that the charge,
so defined, is conserved. The tensor $F_{\mu\nu}$, and therefore the charge of a particle,
changes sign when one goes through the film spanning a string. However, this

change in the particle's charge is accompanied by an appropriate change in the string's charge.

Obviously, conservation of charge in this sense is nonlocal: the charge of a particle depends on the choice of a film spanning the string, and its change when the film is traversed has no physical meaning. This resembles the situation in general relativity, where we can formulate a law of conservation of energy, but there is no covariant concept of energy density, so we cannot talk of energy localization.

The reasoning above applies equally well to electric and magnetic charges: both types are defined in the same way, in terms of flux.

Recall that, if $H = U(1)$, the magnetic and topological charges coincide. We now want to study the relationship between the the two when H is a one-dimensional disconnected group. Given a field inside a bounded region on whose boundary Γ the field is close to a vacuum, and assuming that Γ is a topological sphere, we define the topological charge of the field as the homotopy class of the obvious mapping $\Gamma \to \mathcal{R}$, that is, as an element of $\{S^2, \mathcal{R}\} = \{S^2, R\}$. If $H = U(1)$, the space R is simply connected and $\{S^2, R\} = \pi_2(R) = \mathbf{Z}$. But if H is a one-dimensional disconnected group, $R = G/H$ is not simply connected. (We assume that G itself is simply connected, so that $\pi_1(R) = \pi_0(H) = H/H_{\mathrm{con}}$.) This means that, in general, we cannot identify $\{S^2, R\}$ and $\pi_2(R) = \pi_1(H) = \mathbf{Z}$. We can obtain $\{S^2, R\}$ from $\pi_2(R)$ by identifying elements that can be mapped to one another under the action of $\pi_1(R)$ on $\pi_2(R)$ (T8.2). If $\alpha \in H$ conjugates each element of $U(1) \subset H$ to itself, that is, $\alpha u \alpha^{-1} = u$, the class of α in $\pi(R) = H/H_{\mathrm{con}}$ acts trivially on $\pi_2(R) = \mathbf{Z}$. If, on the contrary, α conjugates elements of $U(1) \subset H$ to their inverses, the class of α in $\pi(R) = H/H_{\mathrm{con}}$ maps each element of $\pi_2(R)$ to its opposite. Thus an element of $\{S^2, R\}$ is characterized by a non-negative integer; the topological charge is defined up to sign. This agrees with the fact that the magnetic charge, too, is only defined up to sign, because of the essential ambiguity in the magnetic field strength.

Until now we have assumed that the gauge group G is simply connected and that the unbroken symmetry group H is one-dimensional. It is not difficult to generalize our reasoning to the case where H is not one-dimensional, but the center of the Lie algebra $\mathcal{H}$ of H is. (As we saw in Chapter 14, the electromagnetic field strength is defined in terms of the generator of the center of $\mathcal{H}$.) This is the case for the standard models of grand unification, in which H is locally isomorphic to $SU(3) \times U(1)$. As we noted in Chapter 19, grand unification models may contain topologically nontrivial strings. However, they do not contain strings that change the magnetic charge of particles that go around them.

A discrete symmetry α giving the topological type of such a string would transform an electron into a positron. But in the Weinberg–Salam model, which is the low-energy limit of the usual grand unification models, there is no such symmetry.

We now study the case when the center $\mathcal{Z}$ of the Lie algebra of H has dimension $r > 1$. Using a formula like (15.1), we can assign an electromagnetic field strength tensor to each $h \in \mathcal{Z}$. We must use in the definition a function

$h(\varphi)$ such that $h(\varphi_0) = h$ and $h(T(g)\varphi) = \tau_g h(\varphi)$; as shown in Chapter 14, such a function exists (and is single-valued) if H is connected. However, in the situation that interests us here, where there exist topologically nontrivial strings, H is disconnected, so that $h(\varphi)$ and the electromagnetic field strength defined by means of it are generally multivalued. It is easy to see that the change in $h(\sigma\varphi(x))$ caused by going around a string is related to the string's topological class.

Suppose we go around the string following a closed curve along which φ takes values close to vacuum values, that is, $\varphi \in \mathcal{R}$. For simplicity, we assume that in fact these values lie in R; this does not change our results. The field φ on the curve x_t is of the form $T(g(t))\varphi_0$, where $0 \leq t \leq 1$ is the parameter along the curve, $g(t)$ is a continuous function of t with values in G and equal to the identity at 0, and φ_0 is the value of φ at the point x_0 (which is the same as x_1, since the curve is closed). By definition, the topological type of the string is determined by the element $\alpha = g^{-1}(1) \in H = H(\varphi_0)$. On the other hand, it is obvious that the continuous branch of $h(\varphi(x))$ on the curve is given by

$$h(t) = g(t)hg^{-1}(t), \qquad h(t) = h(\varphi(x_t)), \qquad h = h(\varphi_0).$$

(Strictly speaking, for groups other than matrix groups we must write $\tau_g h$ instead of ghg^{-1}.) When one goes around x_t, the value h of the function $h(\varphi(x))$ is replaced by $\alpha h\alpha^{-1}$. Thus, if $\alpha h\alpha^{-1} \neq h$, the field strength tensor changes when one goes around the string.

These results are of interest to theories that contain "mirror" particles. It is well known that theories that contain the standard set of elementary particles are not invariant under spatial reflection. But this invariance can be reinstated by introducing new particles obtained from the old by spatial reflection (mirror particles). Of course, the interaction of mirror particles with standard particles must be extremely weak, in order not to contradict experimental data.

It is easy to construct a grand unification theory that includes both ordinary and mirror particles. The connected part of the unbroken symmetry group H must be locally isomorphic to $SU(3) \times U(1) \times SU(3) \times U(1)$. This is because every ordinary particle has a mirror counterpart; therefore, in addition to the $SU(3)$ and $U(1)$ corresponding to gluons and photons, the unbroken symmetry group must include as factors an $SU(3)$ and a $U(1)$ for mirror gluons and photons. In addition, besides the connected part, H must contain a discrete symmetry α that interchanges ordinary and mirror particles. By the preceding discussion, this means that the theory contains topologically nontrivial strings corresponding to α.

In a theory with mirror particles, there are two essentially different electromagnetic field strength tensors, associated with ordinary photons and mirror photons, respectively—or, more precisely, with the generators h_1 and h_2 of the two copies of $U(1)$ that are direct factors in H. Upon going around a string with symmetry of type α, the ordinary and mirror field strengths are interchanged. In other words, both are branches of the same multivalued function. This follows from the equality $\alpha h_1\alpha^{-1} = h_2$.

We can say that in the presence of such strings the difference between ordinary and mirror particles is purely conventional. This is the same situation observed in theories with nonlocal conservation of electric charge. If there are no strings in a given subset of three-dimensional space (that is, if the part of that subset where $\varphi \in \mathcal{R}$ is simply connected), the subset has two single-valued electromagnetic field strength tensors, and one can distinguish between ordinary and mirror particles, at least when these particles participate in an electromagnetic interaction. But if an ordinary particle leaves this subset, goes around a string, and comes back, it becomes a mirror particle. We call such strings "Alice strings," since they behave like Alice's looking glass. If Alice strings exist, they can be found through astronomical observations. (The term "Alice string" was introduced in [57]. It prompted the creation of the term "Cheshire charge" for nonlocalized charge.)

It is natural to think that if topologically nontrivial strings exist, they must be extremely long. They would necessarily have appeared in the early stages of the universe. Strings that are not very long would have disappeared, or collapsed, during the evolution of the universe.

A cosmological analysis shows that extremely massive objects would contain ordinary and mirror particles in approximately equal numbers, while smaller objects must contain mostly particles of a single type. If a string happens to intersect the line from the earth to a galaxy (or some other astronomical object) consisting mostly of ordinary matter, the galaxy becomes invisible. Indeed, to establish whether a particle is made of ordinary or mirror particles, we must link the particle to a fixed standard object by means of a curve, and consider the branch of the field strength tensor that is continuous along this curve. Since light moves along geodesics, we can determine the type of the particles in a galaxy by connecting them with the earth with a geodesic. If a galaxy has different amounts of ordinary and mirror matter, its brightness will change as a string crosses in front of it. As the string moves, this phenomenon is observed for different objects along its visual path: the string generates a wave of brightness variation. (Unfortunately, for distant objects there exist less exotic explanations for the observed changes in brightness).

The motion of a string can have observable effects even if the objects behind it have about the same amount of ordinary and mirror matter. For example, the brightness of a quasar might change suddenly, since the brightness of the ordinary and mirror matter in a quasar vary independently. This might be the explanation for abrupt changes in brightness that have actually been observed; this phenomenon is difficult to explain otherwise.

We now discuss briefly the change in type of particle occurring when one goes around a string in an arbitrary model. If there are no strings, one can classify the existing particles in a standard way. In the presence of strings, the same classification is applicable only within a simply connected domain that does not contain strings. Suppose a particle leaves such a domain, circles around a string whose topological class is $\alpha \in H$, and returns to the domain. The group H acts in the obvious way on the set of stationary states of the system; one-particle states transform into one-particle states. The type of particle, however, may change under the action of discrete unbroken symmetries. It turns out that the change in type caused by going around a string is controlled by the action of the element $\alpha \in H$ giving the topological type of the string. To see this, consider a particle following a loop Γ that encircles a straight string but lies at a great distance from it. As a result of going once around Γ, the state vector of the particle is multiplied by $\mathrm{P}\exp\left(-\oint_\Gamma A_\mu\, dx^\mu\right)$. On the other hand, we showed in Chapter 19 that

$$\alpha = \left(\mathrm{P}\exp\left(-\oint_\Gamma A_\mu\, dx^\mu\right)\right)^{-1}$$

is the element of H that gives the topological type of the string, as we wished to show.

The assertions made earlier in this chapter about the change in electric charge and in mirror type as a particle goes around a string follows easily from the previous paragraph.

21. Nonlinear Fields

Classical field theory generally studies fields with values in $\mathbf{R}$ or in $\mathbf{R}^n$ (in the latter case we can think of a multicomponent field, or of n scalar-valued fields). We now turn to theories where the fields take values in a manifold with nontrivial topology, the so-called nonlinear fields.

As we know from Chapter 8, a local equilibrium state is described by a field with values in the degeneracy space R. The dynamics of such a space is fixed by the Lagrangian, whose form is determined to a great extent by the symmetry properties of the system. In quantum statistical physics, the Lagrangian can also be regarded as being quantized.

Quantum field theory can be thought of as quantum statistical physics at absolute zero, with ground states playing the role of equilibrium states. Thus, the degeneracy space R in quantum theory is the manifold of all ground states.

Consider, for example, the theory given by the Lagrangian

$$(21.1) \qquad \mathcal{L} = \tfrac{1}{2}\partial_\mu \varphi \, \partial^\mu \varphi - V(\varphi),$$

where $\varphi = (\varphi^1, \ldots, \varphi^n)$ is an n-component scalar field. The lowest energy is realized on constant fields that minimize $V(\varphi)$. In particular, if

$$V(\varphi) = \lambda((\varphi^1)^2 + \cdots + (\varphi^n)^2 - a^2)^2,$$

the minimum of $V(\varphi)$ is achieved at the set of points of $\mathbf{R}^n$ where $(\varphi^1)^2 + \cdots + (\varphi^n)^2 + a^2 = 0$, which is an $(n-1)$-dimensional sphere S^{n-1}.

The fields φ that minimize $V(\varphi)$ can be called classical ground states or classical vacuums. If there is only one classical vacuum, it corresponds to the quantum ground state in the semiclassical approximation, and its energy differs from the energy of the quantum ground state by corrections of the order of $\hbar$. If the classical vacuum is degenerate, the quantum corrections to the energies of different classical vacuums may be different. Then, in quantum theory, the degeneracy is lifted completely or partially. However, in the most interesting case, when the degeneracy is caused solely by symmetry, the quantum corrections to all classical vacuums are the same. Then in the semiclassical approximation every classical vacuum has a corresponding quantum ground state.

The analogs of states of local equilibrium are called *Goldstone fields*; they are fields that assume values in the manifold R of ground states. If the degeneracy of the ground states is entirely due to the action of the internal symmetry group G (that is, if G acts transitively on R), R can be identified with the quotient G/H, where H is the unbroken symmetry group. It is natural to assume that Goldstone fields vary slowly; indeed, since they energy differs little from that of the ground state, they cannot bear high gradients in either space or time, since such gradients would contribute greatly to the energy. For this reason, and using relativistic invariance, we can write the *Goldstone field Lagrangian* in the form

$$(21.2) \qquad \mathcal{L} = \tfrac{1}{2} g_{ij}(\varphi)\, \partial_\mu \varphi^i\, \partial^\mu \varphi^j,$$

where $(\varphi^1, \ldots, \varphi^n)$ are the local coordinates of the ground state manifold R and $g_{ij}(\varphi) = g_{ji}(\varphi)$ is a symmetric tensor field on R. We assume that the coordinates $(\varphi^1, \ldots, \varphi^n)$, and therefore the Lagrangian (21.2), are invariant under Lorentz transformations. More than that: (21.2) is the only expression that is Lorentz-invariant and quadratic in the derivatives $\partial_\mu \varphi^i$.

To justify the choice of the Goldstone field Lagrangian in the form (21.2), we note that one cannot construct a Lorentz-invariant expression involving only the first derivatives of the field and linear in these derivatives. On the other hand, an expression involving derivatives of order greater than one, or terms of order higher than two in the first derivatives, are small in comparison with (21.2)—except for linear terms in the second derivative, but these can be reduced to the form (21.2) by integration by parts. Of course, the Lagrangian (21.2) is approximate, and sometimes higher-order corrections are substantial.

If we use the semiclassical approximation to find the vacuum manifold, the same approximation can be used to find the Goldstone field Lagrangian. More precisely, we must consider the original Lagrangian, assuming that the field can only take values in R. For instance, for Lagrangian (21.1) with

$$V(\varphi) = ((\varphi^1)^2 + \cdots + (\varphi^n)^2 - a^2)^2,$$

the Goldstone field Lagrangian has the form

$$(21.3) \qquad \mathcal{L} = \tfrac{1}{2} \langle \partial_\mu \varphi, \partial^\mu \varphi \rangle,$$

where the values of φ are constrained by the equation $(\varphi^1)^2 + \cdots + (\varphi^n)^2 = a^2$, that is, they lie on a sphere of dimension $n - 1$. We introduce stereographic coordinates $(\pi^1, \ldots, \pi^{n-1})$ on the sphere, so that

$$\varphi^n = a\frac{\langle \pi, \pi \rangle - a^2}{a^2 + \langle \pi, \pi \rangle} \qquad \text{and} \qquad \varphi^i = \frac{2\pi^i a^2}{a^2 + \langle \pi, \pi \rangle} \quad \text{for } 1 \le i < n,$$

where $\langle \pi, \pi \rangle = (\pi^1)^2 + \cdots + (\pi^{n-1})^2$. Then (21.3) becomes

$$(21.4) \qquad \mathcal{L} = 2a^4\,\frac{\langle \partial_\mu \pi, \partial^\mu \pi \rangle}{(a^2 + \langle \pi, \pi \rangle)^2}.$$

Clearly (21.3) is invariant under the orthogonal group $SO(n)$, so that the same is true about (21.4).

The symmetric tensor field $g_{ij}(\varphi)$ gives a metric $ds^2 = g_{ij}(\varphi)\,d\varphi^i\,d\varphi^j$ on R. Since the energy corresponding to Lagrangian (21.1) must be positive, the tensor $g_{ij}(\varphi)$ is positive definite, so ds^2 is a Riemannian metric. Using this metric, we can write (21.2) as

$$(21.5) \qquad \mathcal{L} = \tfrac{1}{2}\langle \partial_\mu \varphi, \partial^\mu \varphi \rangle,$$

where the angle brackets stand for the scalar product corresponding to ds^2. If G acts transitively on R, that is, if $R = G/H$, the metric on R is invariant under G, so the tensor field $g_{ij}(\varphi)$ is completely determined by its value at a point φ_0 of R. Not only that, but the value at φ_0 is not arbitrary: it must be invariant under the stabilizer H at that point. In other words, g_{ij} must define an H-invariant scalar product on the tangent space $T_{\varphi_0}(R)$ to R at φ_0. If the representation of H in $T_{\varphi_0}(R)$ is irreducible, it is easy to show that there is a unique H-invariant scalar product (up to a factor). In this case the Lagrangian (21.2) is essentially unique, since a multiplicative factor does not affect the equations of motion.

If $G = \mathrm{SO}(n)$, $H = \mathrm{SO}(n-1)$ and $G/H = S^{n-1}$, the representation of H in the tangent space is the vector representation of $\mathrm{SO}(n-1)$, and so is irreducible. This means that for fields with values in S^{n-1}, there exists a unique $\mathrm{SO}(n)$-invariant Lagrangian of the form (21.2) (or (21.4) in stereographic coordinates), up to a factor.

The fields $\varphi^1, \dots, \varphi^n$ appearing in the Lagrangian (21.2) describe massless particles. To see this, we just have to expand the Lagrangian in the neighborhood of a coordinate-independent field $\varphi(x) = (\varphi_0^1, \dots, \varphi_0^n)$, which acts as a classical vacuum. This justifies the term "Goldstone field."

Recall that Goldstone's theorem states that if G is the continuous symmetry group and H is the group of unbroken symmetries, the theory contains n massless particles, where n is the number of violated symmetry generators, $n = \dim G - \dim H$. In the semiclassical approximation these particles correspond to the field $\varphi^1, \dots, \varphi^n$, which take values in $R = G/H$. Goldstone's theorem does not apply to gauge theories.

We now consider topological integrals of motion in theories with nonlinear fields. The energy functional corresponding to Lagrangian (21.2) has the form

$$(21.6) \qquad E = \frac{1}{2}\int \left(g^{ij}(\varphi)\pi_i\pi_j + g_{ij}(\varphi)\,\partial_a\varphi^i\,\partial_a\varphi^j \right) d^3x,$$

where $a = 1, 2, 3$, the $\pi_i(\mathbf{x}) = g_{ij}(\varphi)\dot\varphi^j$ are the generalized momenta, and $g^{ij}(\varphi)$ is the matrix inverse to $g_{ij}(\varphi)$. As usual when considering the topology of space fields with finite energy, we can assume that $\pi_i(\mathbf{x}) \equiv 0$; the energy functional then acquires the simple form

$$(21.7) \qquad E = \frac{1}{2}\int g_{ij}(\varphi)\,\partial_a\varphi^i\,\partial_a\varphi^j\,d^3x.$$

Fields with finite energy tend to the same limit as one goes off to infinity in different directions. Indeed, suppose that $\lim_{\lambda \to \infty}\varphi(\lambda\mathbf{n}) = \Phi(\mathbf{n})$, where $|\mathbf{n}| = 1$; then, passing to spherical coordinates in (21.7) we see that the energy in infinite, unless $\Phi(\mathbf{n})$ is a constant. Hence, in discussing fields of finite energy we assume that they tend to a finite limit as $\mathbf{x} \to \infty$:

$$(21.8) \qquad\qquad\qquad \lim_{x\to\infty} \varphi(\mathbf{x}) = \varphi_0.$$

The field $\varphi(\mathbf{x})$ can be regarded as a map from $\mathbf{R}^3$ into $R = G/H$. Condition (21.8) means that this map can be extended to a continuous map on the sphere S^3, taking the point at infinity to $\varphi_0 \in R$ (we recall that S^3 is obtained from $\mathbf{R}^3$ by adjoining a point at infinity).

We see that two Goldstone fields are separated by an infinitely high potential barrier if the corresponding maps $S^3 \to R$ are not homotopic to each other. In other words, the topological type of a Goldstone field is determined by the homotopy class of the map $S^3 \to R = G/H$. If R is simply connected, we can identify the set of such homotopy classes with $\pi_3(R) = \pi_3(G/H)$; see T14.2 for the calculation of this group. If $R = S^{n-1}$ (as is the case when $V(\varphi)$ is quadratic: see Lagrangian (21.4)), topologically nontrivial fields exist for $n = 3$ and $n = 4$, since $\pi_3(S^2) = \pi_3(S^3) = \mathbf{Z}$ (T10.2).

Our conclusions about the structure of the space of fields of finite energy remain valid when we add to the Lagrangian (21.2) terms containing higher derivatives. Furthermore, only in the presence of such terms can the minimum of the energy functional on the set of topologically nontrivial states be nonzero.

If the energy functional (21.7) has value E at $\varphi(\mathbf{x})$, it has value $\alpha^{-1}E$ at the field $\varphi(\alpha\mathbf{x})$. By increasing α and noticing that $\varphi(\alpha\mathbf{x})$ has same topological type as $\varphi(\mathbf{x})$, we conclude that there are fields of a given topological type with arbitrarily small energy.

The general assertions made above find their most important application in quantum chromodynamics. The basic constituents of quantum chromodynamics are fermion bispinor fields $\psi_i^a(x)$, called quark fields. Here $a = 1, 2, 3$ is the "color," and i is the "flavor." The common approach to constructing the Lagrangian is to begin with the free fermion Lagrangian, assuming that the quark mass is independent of the color. This Lagrangian is invariant under "rotations" in color space, that is, transformations of the form $\psi_i^a \to S_b^a \psi_i^b$, where $S_a^b \in \mathrm{SU}(3)$ is a unitary matrix of determinant one. Introducing, as usual, gauge fields with values in the Lie algebra of $\mathrm{SU}(3)$, we can construct a Lagrangian that is invariant under local gauge transformations with group $\mathrm{SU}(3)$. This Lagrangian, which contains quark fields and gauge fields (gluon fields), is known as the quantum chromodynamics Lagrangian (see Section 1.6). It can be used to describe strong interactions.

The masses of the two lightest quarks (up and down) are close to each other. If we consider them equal, the quantum chromodynamics Lagrangian is invariant under $\mathrm{SU}(2)$-rotations in flavor space, that is, transformations of the form $\psi_i^a \to V_i^j \psi_j^a$, with $i, j = 1, 2$ and $V_i^j \in \mathrm{SU}(2)$. If we assume that the two light quarks are massless, the symmetry group extends to $\mathrm{SU}(2) \times \mathrm{SU}(2)$, because we can rotate independently (in flavor space) the left-hand and the right-hand components of a quark field, in each case choosing a matrix from

SU(2). There is a spontaneous symmetry breaking that reduces $SU(2) \times SU(2)$ to SU(2), the group obtained by applying the same rotation to the left- and right-hand components of the field. The corresponding degeneracy space $R = SU(2) \times SU(2)/SU(2)$ can be identified with SU(2).

If G is any group, we can define an action of $G \times G$ on G by assigning to each element $(g_1, g_2) \in G \times G$ the map taking $g \in G$ into $g_1 g g_2^{-1}$. The pairs $(g, g) \in G \times G$, for $g \in G$, take the identity element to itself, so the quotient of $G \times G$ by the set of such diagonal pairs can be identified with G itself.

Topologically, SU(2) is equivalent to the sphere S^3, so that $\pi_3(SU(2)) = \pi_3(S^3) = \mathbf{Z}^2$. This means that the topological type of a Goldstone field is specified by an integer.

Under the usual identification of SU(2) with S^3 (T1.2), the action of $SU(2) \times SU(2)$ on SU(2) gives rise to the group of rotations of S^3. This implies that the Goldstone field Lagrangian in this situation is of the form (21.4), with $n = 4$. The corresponding Goldstone particles are identified with π *mesons*; if we ignore the mass of the two lightest quarks, the π mesons become massless.

If we ignore the mass of the three lightest quarks (up, down and strange), the symmetry group becomes $SU(3) \times SU(3)$. This breaks down spontaneously to the subgroup consisting of elements (g, g), for $g \in SU(3)$, which is obviously isomorphic to SU(3). The degeneracy space $R = SU(3) \times SU(3)/SU(3)$ is identified with SU(3). Since $\pi_3(SU(3)) = \mathbf{Z}$ (T10.2), the topological type of a Goldstone field is again specified by an integer.

To construct the Goldstone field Lagrangian, we take a metric on $R = SU(3)$ invariant under left and right translations. Such a metric is unique to within multiplication by a constant, and can be written as

$$(21.9) \qquad ds^2 = \operatorname{tr} dg\, dg^{-1} = -\operatorname{tr}(g^{-1}\, dg\, g^{-1}\, dg).$$

Indeed, the tangent space to SU(3) at the identity is identified with the Lie algebra $\mathfrak{su}(3)$. Thus, the desired metric is determined by the unique scalar product in $\mathfrak{su}(3)$ invariant under inner automorphisms $g \mapsto hgh^{-1}$. We obtain (21.9) if we recall that this scalar product is given, to within a factor, by $\langle a, b \rangle = \operatorname{tr} ab^\dagger = -\operatorname{tr} ab$. Using (21.9), we can write the Goldstone field Lagrangian (up to a constant factor) as

$$\mathcal{L} = \operatorname{tr} \partial_\mu g\, \partial^\mu g^{-1}.$$

We return to Goldstone fields in general. Note that the space of fields on which the action integral is defined may be disconnected. Of course, adding a constant to the integral changes nothing in the equation of motion. If the space of fields is disconnected, a different constant can be chosen for each component. This has important consequences. We think of a Goldstone field $\varphi(\mathbf{x}, t) = \varphi_t(\mathbf{x})$ as a path in the space $\mathcal{E}$ of fields $\varphi(\mathbf{x})$ on the spatial variable $\mathbf{x}$. We assume that $\varphi_t(\mathbf{x})$ satisfies (21.8), where φ_0 is fixed; then $\varphi_t(\mathbf{x})$, for t fixed, can be interpreted as a map $S^3 \to R$, and $\mathcal{E}$ as the space of spheroids in R (T8.1).

Given two Goldstone fields φ_1 and φ_2, consider the space $\mathcal{C}(\varphi_1, \varphi_2)$ of paths in $\mathcal{E}$ joining the two fields, that is, the space of of time-dependent Goldstone fields $\varphi_t(\mathbf{x})$ that equal $\varphi_1(\mathbf{x})$ at $t = t_1$ and $\varphi_2(\mathbf{x})$ at $t = t_2$. This space comes up naturally in the calculation of the evolution operator, for the amplitude of the transition from φ_1 to φ_2 in time $t_2 - t_1$ is represented as the functional integral of $\exp(i\hbar^{-1}S)$ over $\mathcal{C}(\varphi_1, \varphi_2)$ (Chapter 24). Allowing for the constant terms in the action integral S, we can represent this amplitude in the form

$$(21.10) \qquad \sum c_\alpha \int_{\mathcal{C}_\alpha(\varphi_1,\varphi_2)} \exp\left(\frac{i}{\hbar}S\right) \prod d\varphi(x,t),$$

where S is determined by the Lagrangian (21.2) (without the constant) and α indexes the components $\mathcal{C}_\alpha(\varphi_1, \varphi_2)$ of $\mathcal{C}(\varphi_1, \varphi_2)$. If $\varphi_1 = \varphi_2$, these components are in one-to-one correspondence with the elements of the fundamental group $\pi_1(\mathcal{E}, \varphi_1)$, and we can assume that α runs through this group. The group property of the evolution operator implies that $c_{\alpha\beta} = c_\alpha c_\beta$. If $\varphi_1 \neq \varphi_2$, we can reduce to the case $\varphi_1 = \varphi_2$, because $\mathcal{C}(\varphi_1, \varphi_2)$ is either empty or homotopically equivalent to $\mathcal{C}(\varphi_1, \varphi_1)$; to set up a homotopy equivalence, fix a path ν from φ_2 to φ_1, and associate to every path $\lambda \in \mathcal{C}(\varphi_1, \varphi_2)$ the path $\lambda\nu \in \mathcal{C}(\varphi_1, \varphi_1)$.

The group $\pi_1(\mathcal{E}, \varphi_1)$ is isomorphic to $\pi_4(R)$. We show this first if $\varphi_1(\mathbf{x}) \equiv \varphi_0$. In this case, a path $\varphi_t(\mathbf{x})$ can be seen as a map from the four-dimensional cube I^4 into E: for each t we consider the spheroid $\varphi_t(\mathbf{x})$ as a map from the cube I^3 into R taking the entire boundary of I^3 to φ_0. Thus we can identify four-dimensional spheroids in R with closed paths in $\mathcal{E}$ beginning and ending at φ_0. In the general case, when $\varphi_1(\mathbf{x})$ is not constant, we take into account that all connected components of $\mathcal{E}$ are homotopically equivalent, say by the map $\varphi \to \varphi + \rho$, where ρ is a fixed spheroid; this implies that the groups $\pi_1(\mathcal{E}, \varphi_1)$ are isomorphic, for different values of φ_1.

As already seen, if R is simply connected, we can assign to each field of finite energy an element of $\pi_3(R)$ giving the homotopy type of the field. If we include terms with higher derivatives in the Lagrangian, we can arrange it so that the energy on the set of topologically nontrivial fields is minimized for some field $\varphi(\mathbf{x})$—a topologically nontrivial soliton. In the semiclassical approximation a quantum particle corresponds to this soliton. To establish whether the particle is a boson or a fermion, we must consider the element $\alpha \in \pi_4(R)$ defined by the composition $\varphi(\lambda(\mathbf{x}))$ of the soliton φ (seen as a map $S^3 \to R$) with a homotopically nontrivial map $\lambda : S^4 \to S^3$. If the coefficient c_α in (21.10) corresponding to the element $\alpha \in \pi_4(R) = \pi_1(\mathcal{E})$ is 1, the particle is a boson; otherwise $c_\alpha = -1$ and the particle is a fermion. We will not prove this here, but we observe that for $R = S^3$ and $R = S^2$ we have $\pi_3(R) = \mathbf{Z}$ and $\pi_4(R) \doteq \mathbf{Z}_2$. In these cases, the composition of the map $S^3 \to R$ with a homotopically nontrivial map $\lambda : S^4 \to S^3$ determines a nonzero homomorphism from $\pi_3(R) \to \pi_4(R)$ (T7.4 and T10.2). This means that the particle associated with a soliton with an odd topological number is a fermion if $c_\alpha = -1$ for a nonzero $\alpha \in \pi_4(R)$.

22. Multivalued Action Integrals

In quantum mechanics and quantum field theory many quantities can be represented as functional integrals over possible histories: that is, along possible trajectories in quantum mechanics and over possible time dependencies of the fields in quantum field theory. The set of histories over which integration is performed and the form of the integrand depend on the quantity to be calculated: see Chapter 24. The important point to us is that the integrand includes the exponential of the classical action integral S multiplied by $i/\hbar$. This means that a functional integral may be meaningful even if S is a multivalued functional, so long as $\exp(iS/\hbar)$ is single-valued. In other words, several values of S correspond to the same integrand if they differ by an integral multiple of $2\pi\hbar = h$. This condition on multivalued functionals implies that the constants appearing in S may assume only a discrete set of values, that is, they are quantized. We give examples of this.

First we show that the quantization condition for the magnetic charge can be obtained from the requirement that the motion of a quantum particle in the field of this charge has meaning. Recall that the action integral S of a classical particle moving in an electromagnetic field according to the usual equations of motion

$$
(22.1) \qquad \frac{d^2 x^\mu}{ds^2} - e F^{\mu\nu} \frac{dx_\nu}{ds} = 0
$$

can be written in the form

$$
(22.2) \qquad S[x(\tau)] = -m \int ds - e \int A_\mu \, dx^\mu,
$$

where m is the mass and e the charge of the particle, ds is the space-time line element, A_μ the electromagnetic field potential, and the possible path $x^\mu(\tau)$ is determined how time x^0 and the space coordinates x^1, x^2, x^3 depend on the parameter τ, with $\tau_0 \leq \tau \leq \tau_1$. If $x(\tau)$ is a closed path in Minkowski space-time, that is, if $x^\mu(\tau_0) = x^\mu(\tau_1)$, we can use Stokes' theorem to rewrite (22.2) as

$$
(22.3) \qquad S[x(\tau)] = -m \int ds - \frac{e}{2} \int_\Gamma F_{\mu\nu} \, dx^\mu \wedge dx^\nu = -m \int ds - e \int_\Gamma F,
$$

where Γ is an arbitrary surface spanning the closed path $x(\tau)$ and $F = \frac{1}{2} F_{\mu\nu} \, dx^\mu \wedge dx^\nu$ is the two-form that corresponds to the electromagnetic field strength. Recall (T5.1) that F is closed, by Maxwell's equations.

If the electromagnetic field and the form F are defined on all of space-time, F is exact, that is, it can be written as $F = dA$, where $A = A_\mu\,dx^\mu$ is the electromagnetic field potential. But we are interested in the case when the electromagnetic field is not defined on all of space-time, only in some domain V. This occurs when a point magnetic charge is present, because the electromagnetic field is not defined at that point. We must exclude the trajectory of the charge in space-time from the domain of definition V of F.

If the domain of definition V of F is not all of space-time, F may not be exact. In this case we can talk about the field strength, but there is no globally defined field potential, and (22.2) is no longer valid. However, (22.3) is still meaningful, although it may be multivalued, depending of the choice of Γ. The difference between two values of S in this case is of the form $\int_\Gamma F$, where Γ is a closed surface—the closed surface obtained by combining two surfaces that span the loop $x(\tau)$, one of them with reverse orientation.

Thus, we see that the quantum mechanics of a particle moving in an electromagnetic field of strength F makes sense only if, for every closed surface Γ, the integral of F over Γ is a multiple of $2\pi\hbar/e$, for only then will the functional $\exp(iS/\hbar)$ be well defined. If all of Γ has the same time coordinate x^0, the integral $\int_\Gamma F$ reduces to the magnetic flux through Γ, for only the time components of $F_{\mu\nu}$ take part in the integral. By definition, the magnetic charge $\mathfrak{m}_\Gamma$ inside Γ equals this flux divided by 4π. We therefore get the following condition for magnetic charge quantization:

$$(22.4) \qquad\qquad \mathfrak{m}_\Gamma = \frac{\hbar}{2e}n,$$

where n is an integer (Chapter 14).

Condition (22.4) is not only necessary, but also sufficient for $\exp(iS/\hbar)$ to be single-valued if the domain of definition V is the complement of the world lines of punctual magnetic charges. For in this case any two-dimensional closed surface in V (every two-cycle) is homologous to a surface lying in a single time slice. The integral of F over two homologous surfaces is the same, so if (22.4) is satisfied the integral is a multiple of $2\pi\hbar$ for every closed surface.

Although the action integral (22.3) is defined only on closed paths, this restriction is not essential. For paths with different endpoints $x(\tau_0) = x_0$ and $x(\tau_1) = x_1$, we can still find a multivalued action integral that leads to the equations of motion (22.1). We must fix an arbitrary path $\tilde{x}(\tau)$ having the same endpoints, and use for Γ in formula (22.3) an oriented surface spanning the loop formed by x followed by $\tilde{x}$ in the reverse direction. Different choices of $\tilde{x}(\tau)$ lead to values of the integral that differ by a constant. Repeating the reasoning of the preceding paragraphs, we see that $\exp(iS/\hbar)$ is single-valued under the same condition (22.4) as in the case of closed loops.

We now discuss a general approach to the construction of multivalued action integrals for non-linear fields. We assume, for definiteness, that our fields $\varphi(\mathbf{x}, t)$ are defined in space-time, that is, that they depend on three space variables and on time. We let $\mathcal{M}$ denote the manifold where the fields take values. As noted

in Chapter 21, the simplest Lorentz-invariant action integral for such fields has
the form (21.5), and we can add to it more complicated terms containing higher
derivatives or higher powers of the first derivative. We assume that $\varphi(\mathbf{x}, t)$ tends
to a finite limit as $\mathbf{x} \to \infty$; as discussed in Chapter 21, this is necessary if the
field's energy is to be finite.

We let t vary from τ_0 to τ_1, and assume for now that $\varphi(\mathbf{x}, \tau_0) = \varphi(\mathbf{x}, \tau_1)$;
we will lift this restriction later. In this situation, the field $\varphi(\mathbf{x}, t)$ determines
a map $S^3 \times S^1 \to \mathcal{M}$. Indeed, for every value of t we have a map $\mathbf{R}^3 \to \mathcal{M}$
that can be extended to S^3 by assigning $\lim_{\mathbf{x} \to \infty} \varphi(\mathbf{x}, t)$ to the point at infinity.
The family of these maps $S^3 \to \mathcal{M}$ gives a map $S^3 \times [\tau_0, \tau_1] \to \mathcal{M}$, but we
can identify the endpoints of the interval $[\tau_0, \tau_1]$ because the field is the same
for these two values of t (in other words, we can assume that t takes values in
the circle). The map $S^3 \times S^1 \to \mathcal{M}$ thus obtained defines a closed, oriented
four-dimensional surface (cycle) in $\mathcal{M}$, which we denote by $Z(\varphi)$.

We now show how, given a closed 5-form ω on $\mathcal{M}$, we can construct a
multivalued functional S_ω associated with φ, and that S_ω can be included in
the expression for the action integral as a separate term. The definition of S_ω is

$$(22.5) \qquad S_\omega(\varphi) = \int_{\Gamma(\varphi)} \omega,$$

where $\Gamma(\varphi)$ is a five-dimensional oriented surface whose boundary is $Z(\varphi)$.

Such a surface only exists if $Z(\varphi)$ is null-homotopic. For this reason we must
either impose on $\mathcal{M}$ conditions that guarantee the existence of $\Gamma(\varphi)$, or restrict our
discussion to fields for which $Z(\varphi)$ is null-homotopic.

The value of S_ω may depend on the choice of $\Gamma(\varphi)$. If $\Gamma_1(\varphi)$ and $\Gamma_2(\varphi)$ are
both bounded by $Z(\varphi)$, we have

$$(22.6) \qquad \int_{\Gamma_1(\varphi)} \omega - \int_{\Gamma_2(\varphi)} \omega = \int_\Gamma \omega,$$

where Γ is a five-dimensional cycle (closed oriented surface) consisting of Γ_1,
together with Γ_2 with the opposite orientation. We say that ω is an *integral
form* if the integral of ω over any cycle is an integer (T5.2).

From (22.6) it follows that for an integral form ω any two values of S_ω differ
by an integer. When S_ω is included in an action integral S, we can construct
a quantum theory based on S if and only if ω is $2\pi\hbar$ times an integral form,
for then the multivaluedness of S_ω disappears when we take the exponential
$\exp(iS/\hbar)$.

The general construction above acquires physical meaning when we use a
non-linear field to describe quantum chromodynamics in the low-energy limit.
Recall that, in quantum chromodynamics, when we ignore the masses of the
three lightest quarks (up, down and strange), the SU(3) $\times$ SU(3) symmetry
breaks down to SU(3) symmetry. Thus the Goldstone fields, which play an
important role at low energies, take on values in $R =$ SU(3) $\times$ SU(3)/ SU(3),
which we identify with SU(3). Thus, a Goldstone field is a matrix field $g(x) =$
$(g_b^a(x))$ such that $g^{-1}(x) = g^\dagger(x)$ and $\det g(x) = 1$. The action integral for

Goldstone fields must be Lorentz-invariant and also invariant under the action of SU(3) × SU(3), that is, invariant under left and right translations in SU(3). The simplest functional with these properties is

$$(22.7) \qquad S = \text{const} \int \text{tr}(\partial_\mu g \, \partial^\mu g^{-1}) d^4 x$$

(Chapter 21). However, this functional has too many symmetries, including some that are broken in quantum chromodynamics, namely those of the form $g(x) \mapsto g^{-1}(x)$. To see why this symmetry is unwanted, write the Goldstone field as as

$$g(x) = \exp\left(\sum_{i=1}^{8} \pi_i(x)\lambda^i\right),$$

where $\lambda^1, \ldots, \lambda^8$ are the generators of SU(3) and the $\pi_i(x)$ are the fields of Goldstone mesons. Then $g(x) \mapsto g^{-1}(x)$ corresponds to the symmetry $\pi_i(x) \mapsto -\pi_i(x)$, which forbids transitions from a system with an even number of mesons into one with an odd number. However, the transition $K^+ K^- \rightarrow \pi^+ \pi^0 \pi^-$ is allowed in quantum chromodynamics. Thus, in the action integral for Goldstone fields there must be a term to break the symmetry $g(x) \mapsto g^{-1}(x)$. For this role we take (22.5), where ω is the 5-form on SU(3) given by

$$(22.8) \quad \omega = k\frac{-i}{240\pi^2} \text{tr}\left(g^{-1}\frac{\partial g}{\partial y^i}g^{-1}\frac{\partial g}{\partial y^j}g^{-1}\frac{\partial g}{\partial y^k}g^{-1}\frac{\partial g}{\partial y^l}g^{-1}\frac{\partial g}{\partial y^m}g^{-1}\right)$$
$$\times dy^i \wedge dy^j \wedge dy^k \wedge dy^l \wedge dy^m,$$

where k is any real number. This form is invariant under left and right translations in SU(3) (and in fact every 5-form in SU(3) with this property can be so written). This implies that ω is closed (T14.2), a fact that can also be verified directly. Furthermore, ω is integral if $k = 1$. Therefore we should choose $k = 2\pi n\hbar$, with n an integer, in order for $\exp(iS/\hbar)$ to be well defined, where S is the action integral.

Recall that, since $\pi_3(SU(3)) = \mathbf{Z}$, the topological type of a Goldstone field is given by an integer. In the semiclassical approximation, a quantum particle corresponds to a topologically nontrivial soliton $g(x)$. One can show that this particle is a fermion if the topological number of the soliton and the integer $n = k/(2\pi\hbar)$ are both odd. This makes it possible to identify solitons with baryons.

23. Functional Integrals

A *functional integral* is an integral over an infinite-dimensional space, usually a space of functions in one or several variables. One generally defines a functional integral as a limit of ordinary multiple integrals. For example, consider a functional $F(\varphi)$ on the space of functions $\varphi(t)$ of a single variable t, with $a \le t \le b$. We can restrict F to the space of continuous functions that are linear in each of the segments $[t_0, t_1], \ldots, [t_{N-1}, t_N]$, with $t_i = a + i(b-a)/N$. This space is finite-dimensional, since every such function is determined by its values at $a = t_0, t_1, \ldots, t_N = b$. If we call these values $\varphi_0, \ldots, \varphi_N$, the space is parametrized by $(\varphi_0, \ldots, \varphi_N)$, and we can consider the integral J_N of F with respect to $\varphi_1, \ldots, \varphi_N$. The integral of $F(\varphi)$ is naturally defined as the limit of the approximations J_N, as $N \to \infty$:

$$(23.1) \qquad J = \int F[\varphi] \prod_{a \le \tau \le b} d\varphi(\tau) = \lim_{N \to \infty} J_N.$$

However, in many important cases, the limit does not exist. Then it is expedient to isolate the divergent part of J_N and to interpret the limit of the remainder as the functional integral. For example, if there is a constant C for which $C^{-N} J_N$ has a limit, we define the functional integral as this limit. We merely have to redefine J_N, replacing $d\varphi_i$ by $C^{-1} d\varphi_i$.

This definition can be modified in several ways. Suppose we consider, for each piecewise linear function φ, not $F(\varphi)$ itself but an approximation to it. For example, if the functional is defined by an integral, we could approximate the integral by a Riemann sum using the same partition $t_0, \ldots, t_N$. It turns out that in many interesting cases the value of the functional integral depends on the choice of an approximation. Hence, strictly speaking, the expression $J = \int F(\varphi) \prod d\varphi(t)$ is only meaningful when we specify in what way we are passing to the limit.

Now suppose $\varphi(t)$ is defined on the whole t-axis, rather than just on an interval. Then, to construct a multiple-integral approximation to the functional integral, we must not only partition the axis into small segments, but also limit ourselves to a finite number of segments. Likewise, if $F[\varphi]$ is defined on functions in m variables, we construct the approximation by replacing $\mathbf{R}^m$ with a lattice, and determining J_N by integrating with respect to the values of φ at a finite number of lattice points (for example, those lying inside a cube).

In quantum field theory, the transition from a functional integral J to its finite-dimensional approximation J_N is closely related to the cutoff procedure.

Passing from functions on $\mathbf{R}^m$ to discrete functions on a lattice corresponds to the cutoff of high momenta (ultraviolet cutoff); ignoring all but a finite set of lattice points corresponds to a cutoff in the coordinates (infrared cutoff).

Let us consider Gaussian integrals, the simplest and most important examples of functional integrals. A Gaussian integral is an integral of the exponential of a quadratic form. In the finite-dimensional case we have

$$(23.2) \qquad \int \exp(-\tfrac{1}{2}\langle Ax, x\rangle)\, dx = (2\pi)^{N/2}(\det A)^{-1/2},$$

where the angle brackets denote some real-valued scalar product in an N-dimensional space, and A is a positive, self-adjoint linear operator; if (a_{ij}) is the matrix for A, we have $\langle Ax, x\rangle = a_{ij}x^i x^j$.

Now consider an infinite-dimensional Gaussian integral

$$(23.3) \qquad J = \int \exp(-\tfrac{1}{2}\langle Ax, x\rangle)\, dx,$$

where x takes values in an infinite-dimensional Hilbert space, and A operates on this space. We approximate (23.3) by the finite-dimensional integrals

$$(23.4) \qquad J_N = \int \exp(-\tfrac{1}{2}\langle A_N x, x\rangle)\, dx,$$

where x takes values in a finite-dimensional space, and A_N is an approximation for A on this space. For example, if A is a differential operator acting on functions in $\mathbf{R}^n$, we restrict to functions defined on a finite portion of a lattice, and choose for A_N the operator obtained by replacing derivatives with finite differences.

If $\det A_N$ tends to a finite limit as $N \to \infty$, it is natural to define the determinant of A as this limit: $\det A = \lim_{N\to\infty} \det A_N$. The J_N have no finite limit because of the factor $(2\pi)^{N/2}$; as explained above, we get rid of this divergent factor by replacing dx with $(2\pi)^{-1/2}dx$. We then define the Gaussian integral J as the limit of the redefined J_N, and this limit is $(\det A)^{-1/2}$.

It is reasonable to retain the formula

$$(23.5) \qquad J = \int \exp(-\tfrac{1}{2}\langle Ax, x\rangle)\, dx = (\det A)^{-1/2}$$

as the definition of an infinite-dimensional Gaussian integral even when the limit $\lim_{n\to\infty} \det A_N$ does not exist. In this case the (regularized) determinant of A can be defined in other ways. The simplest way, from the technical point of view, is by means of the ζ-function of A. We assume that the self-adjoint operator A is non-negative (that is, $\langle Ax, x\rangle \geq 0$) and that its spectrum is discrete. The ζ-function of A is given by

$$(23.6) \qquad \zeta(s) = \sum \lambda_k^{-s},$$

where the λ_k are the non-zero eigenvalues of A. This formula makes sense only for values of s at which the series converges; for other values $\zeta(s)$ is defined by analytic continuation. If $\zeta(s)$ can be analytically continued to the point $s = 0$, we define $\det A$ by the equation

$$(23.7) \qquad \ln \det A = -\frac{d\zeta(s)}{ds}\bigg|_{s=0},$$

which is motivated by the fact that

$$\ln \lambda_i = -\frac{d\lambda_i^{-s}}{ds}\bigg|_{s=0}.$$

If A is an elliptic operator of order r on an m-dimensional compact manifold (Chapter 26), the series converges for $\operatorname{Re} s > m/r$. The resulting function $\zeta(s)$ can be analytically continued into a meromorphic function of s, having no singularity at $s = 0$. Thus (23.7) can be used in this case to define $\det A$. For more details, see Chapter 27.

As an example, consider the functional integral

$$(23.8) \qquad J = \int \exp\left(-\frac{1}{2}\int ((\nabla\varphi)^2 + m^2\varphi^2)\,dx\right)\prod d\varphi(x)$$

$$= \int \exp\left(-\frac{1}{2}\int \varphi(-\Delta + m^2)\varphi\,dx\right)\prod d\varphi(x),$$

where $\varphi(x)$ is a field in $\mathbf{R}^4$, assumed to decay to zero at infinity, ∇ is the four-dimensional gradient, and Δ the four-dimensional Laplacian. By (23.5), this integral equals $(\det(-\Delta + m^2))^{-1/2}$. In order to compute the ζ-function of $A = -\Delta + m^2$, we use the identity

$$(23.9) \qquad \cdot \quad \zeta_A(s) = \sum \lambda_k^{-s} = \sum \frac{1}{\Gamma(s)}\int_0^\infty t^{s-1}\exp(-\lambda_k t)\,dt$$

$$= \frac{1}{\Gamma(s)}\int_0^\infty t^{s-1}\operatorname{tr}\exp(-At)\,dt.$$

The matrix entries $K(x, x_0, t) = \langle x|\exp(-At)|x_0\rangle$ of the operator $\exp(-At)$ satisfy the equation

$$(23.10) \qquad \frac{\partial}{\partial t}K(x, x_0, t) = -A_x K(x, x_0, t),$$

with initial condition $K(x, x_0, t = 0) = \delta(x - x_0)$. For $A = -\Delta + m^2$ this equation can easily be solved, to yield

$$(23.11) \qquad K(x, x_0, t) = \frac{1}{16\pi^2 t^2}\exp\left(-m^2 t - \frac{(x - x_0)^2}{4t}\right),$$

where we have used the fact that $\varphi(x)$ decays to zero at infinity. In particular,

$$(23.12) \qquad K(x, x, t) = \frac{1}{16\pi^2 t^2}\exp(-m^2 t).$$

In order to find the trace of $\exp(-At)$, we must integrate $K(x, x, t)$ with respect to dx; but this integral clearly diverges. To get a finite answer, we must introduce a coordinate cutoff, that is, assume that $\varphi(x)$ is defined only on a finite box of volume V. Then

$$(23.13) \qquad \operatorname{tr} \exp(-At) = V \frac{1}{16\pi^2 t^2} \exp(-m^2 t).$$

This allows us to compute the ζ-function, using (23.9):

$$(23.14) \qquad \zeta(s) = V \frac{1}{16\pi^2} (m^2)^{2-s} \frac{\Gamma(s-2)}{\Gamma(s)}.$$

Equation (23.9) can only be used when $\operatorname{Re} s > 2$, since only then does the integral in t converge. However, (23.14) allows us to analytically continue $\zeta(s)$ to the whole s-plane, except for the points $s = 1$ and $s = 2$, where ζ has simple poles. From (23.4) we see that

$$(23.15) \qquad \det(-\Delta + m^2) = -\zeta'(0) = V \frac{1}{32\pi^2} m^4 (-\tfrac{3}{2} + \ln m^2).$$

It is often interesting to know not only the regularized determinant of an operator, but also the asymptotic behavior of the approximating discrete operators. The determinant of $-\Delta + m^2$ becomes finite when we introduce ultraviolet cutoff (that is, when we pass to a lattice) and infrared cutoff (when we restrict to domain to a finite volume V). It is important to study the asymptotic behavior of the finite determinants when the cutoffs are lifted. This problem is certainly solvable, but it is technically much simpler to proceed instead by introducing what is called *cutoff in proper time*. The results do not change significantly. The determinant of a positive operator A cut off in proper time is the number $\det_\varepsilon A$ defined by

$$(23.16) \qquad \ln \det_\varepsilon A = \sum_i - \int_\varepsilon^\infty \frac{1}{t} \exp(-\lambda_i t)\, dt = \int_\varepsilon^\infty \frac{1}{t} \operatorname{tr} \exp(-At)\, dt.$$

For $A = -\Delta + m^2$, it follows from (23.13) that

$$(23.17) \qquad \ln \det_\varepsilon A \approx \frac{Vm^4}{32\pi^2} \left(-\frac{1}{m^4 \varepsilon^2} + \frac{2}{m^2 \varepsilon} + \ln \varepsilon + \ln m^2 - \tfrac{3}{2} - \Gamma'(1) \right),$$

and the error tends to zero as $\varepsilon \to 0$. We denote by $\ln \det' A$ the number obtained from (23.17) by discarding the part that diverges as $\varepsilon \to 0$. Just as we did in connection with $\det A = \exp(-\zeta'(0))$, we can consider $\det' A$ as the regularized determinant of A. Equation (23.17) implies that, for $A = \Delta + m^2$,

$$(23.18) \qquad \ln \det' A = \ln \det A - \frac{Vm^4}{32\pi^2} \Gamma'(1).$$

This is a particular case of a general relation that we will prove in Chapter 28. Substituting $x + a$ for x, we can reduce the functional integral

(23.19) $\int \exp(-\frac{1}{2}\langle Ax, x\rangle + \langle b, x\rangle)\, dx = (\det A)^{-1/2} \exp(-\frac{1}{2}\langle A^{-1}b, b\rangle)$

to (23.5). A reasoning similar to the one we used to study (23.5) gives

(23.20) $\int \exp(\frac{1}{2}i\langle Ax, x\rangle)\, dx = (\det A)^{-1/2}$

and

(23.21) $\int \exp(\frac{1}{2}i\langle Ax, x\rangle + i\langle b, x\rangle)\, dx = (\det A)^{-1/2} \exp(-\frac{1}{2}i\langle A^{-1}b, b\rangle),$

where A is a self-adjoint operator on a Hilbert space.

Gaussian integrals arise in quantum field theory when one studies Lagrangians that depend quadratically on the field. But they are also important in the study of more complicated Lagrangians, as they occur in the approximate calculation of functional integrals by the stationary-phase method or the Laplace method. We recall that, if $f(x)$ is a real-valued function of n real variables, the main contribution to the multiple integral $\int \exp(i\alpha^{-1} f(x))\, d^n x$ as $\alpha \to 0$ comes from neighborhoods of the critical points of $f(x)$. We will write the answer, in the case where there is a unique critical point x_0, by taking the Taylor series of $f(x)$ at x_0, keeping only terms of order at most two, and evaluating the resulting Gaussian integral. The asymptotic behavior of the solution is
(23.22)
$$\int \exp(i\alpha^{-1} f(x))\, dx \approx \exp(i\alpha^{-1} f(x_0))(-2\pi i\alpha)^{n/2} \left(\det \frac{\partial^2 f(x)}{\partial x^i\, \partial x^j} \right)^{-1/2} \Bigg|_{x=x_0}.$$

If there is more than one critical point, the asymptotic behavior is given by a sum of expressions like (23.22), one for each point.

Similarly, we can find the asymptotic behavior of the integral

$$\int \exp(-\alpha^{-1} f(x))\, dx,$$

where $f(x)$ is a real-valued function with an absolute maximum at x_0. The dominant asymptotic term in this case is

(23.23) $\exp(-\alpha^{-1} f(x_0))(2\pi a)^{n/2} \left(\det \frac{\partial^2 f(x)}{\partial x^i\, \partial x^j} \right)^{-1/2} \Bigg|_{x=x_0}.$

There are analogs of (23.22) and (23.23) for functional integrals, although the proof of these formulas is much more delicate. We can easily derive higher-order terms in (23.22) and (23.23): we just have to expand the exponential in a power series and include terms of order three or higher. For functional integrals, the approximations thus obtained correspond to results from perturbation theory.

The definition of a functional integral given above is not always adequate. Sometimes one must use a procedure known in quantum field theory as *renormalization*. Suppose the integrand in a functional integral depends on a parameter λ (possibly a vector $(\lambda^1, \ldots, \lambda^N)$). Then the approximating multiple integral J_N will also depend on λ. It may happen that, even if $J_N(\lambda)$ does not have a finite limit as $N \to \infty$ for λ fixed, one can choose a sequence λ_N of values of λ such that $J_N(\lambda_N)$ does converge to a nonzero finite limit as $N \to \infty$. This limit is then defined as the value of the functional integral. Of course, there can be many sequences λ_N with this property, leading to different limits $\lim_{N \to \infty} J_N(\lambda_N)$. In this case the functional integral is not well defined, but it can be seen as depending of a parameter μ (possibly a vector), which describes the passage to the limit. In quantum field theory the parameters $\lambda^1, \ldots, \lambda^N$ that appear in the integrand are called *bare charges* or *bare masses*; the parameters that describe the passage to the limit are called *physical charges* or *physical masses*.

24. Applications of Functional Integrals to Quantum Theory

We will now obtain expressions for several important physical quantities in terms of functional integrals. These expressions are often very useful in quantum mechanics and quantum field theory.

Consider the operator

(24.1)
$$\hat{A} = \sum_n a_n(q)\left(\frac{1}{i}\frac{\partial}{\partial q}\right)^n = \sum_n a_n(\hat{q})\hat{p}^n.$$

The function

(24.2)
$$A(p,q) = \sum_n a_n(q)p^n$$

is called the *symbol* (or *qp-symbol*) of $\hat{A}$. One can say that $\hat{A}$ is obtained from $A(p,q)$ by replacing the coordinate q by the coordinate operator $\hat{q}$, and the moment p by the moment operator $\hat{p} = i^{-1}(\partial/\partial q)$, with the condition that $\hat{q}$ is written to the left of $\hat{p}$. The transition from $A(p,q)$ to $\hat{A}$ can be seen as the quantization of the classical physical quantity $A(p,q)$. One can easily verify that the function $\tilde{f}(q) = \hat{A}f(q)$ can be written as

(24.3)
$$\tilde{f}(q) = \frac{1}{2\pi}\int A(p,q)f(q_1)e^{ip(q-q_1)}\,dp\,dq_1.$$

This formula allows us to obtain $\hat{A}$ from $A(p,q)$ even when $A(p,q)$ is not a polynomial in p. It also gives an expression for the matrix entries $\langle q_2 \mid \hat{A} \mid q_1 \rangle$ in terms of the symbol $A(p,q)$:

(24.4)
$$\langle q_2 \mid \hat{A} \mid q_1 \rangle = \frac{1}{2\pi}\int A(p,q_2)e^{-ip(q_1-q_2)}\,dp.$$

Equation (24.3) also implies that if $\hat{A}$, $\hat{B}$ and $\hat{C}$ are operators with $\hat{C} = \hat{A}\hat{B}$, their symbols are related by

(24.5)
$$C(p,q) = \frac{1}{2\pi}\int A(p_1,q)B(p,q_1)e^{-i(p_1-p)(q_1-q)}\,dp_1\,dq_1.$$

We use this formula to compute quantities of interest in quantum physics in terms of functional integrals. Specifically, we consider an operator $\hat{H}$ with symbol $H(p, q)$, and compute the matrix entries $\langle y \mid e^{-it\hat{H}} \mid x \rangle$ of the operator $e^{-it\hat{H}}$ in terms of a functional integral.

In the applications, $H(p, q)$ is the Hamiltonian of a classical mechanical system, $\hat{H}$ is the Hamiltonian operator of the corresponding quantum system, and $e^{-it\hat{H}}$ is the evolution operator for the system.

We show that

$$(24.6) \qquad \langle y \mid e^{-it\hat{H}} \mid x \rangle = \int \exp(iS) \prod dp(\tau) \, dq(\tau),$$

where

$$(24.7) \qquad S[p(\tau), q(\tau)] = \int_0^t \left(p(\tau) \frac{dq(\tau)}{d\tau} - H(p(\tau), q(\tau)) \right) d\tau.$$

The functional S can be interpreted as the classical action integral along the phase trajectory $(p(t), q(t))$. On the right-hand side of (24.6) we have a functional integral along the trajectory $(p(t), q(t))$, for $0 \leq \tau \leq t$, with boundary conditions $q(0) = x$ and $q(t) = y$. The meaning of this functional integral will be clarified by the proof.

To derive (24.6), we use the relation

$$(24.8) \qquad \exp(-it\hat{H}) = (\exp(-(it/N)\hat{H}))^N,$$

and notice that for N large we have $\exp(-(it/N)\hat{H}) \approx 1 - (it/N)\hat{H}$. Thus the symbol of the operator $\exp(-(it/N)\hat{H})$ is approximately equal to

$$\exp(-(it/N)H(p, q)),$$

and the error is of order N^{-2} as $N \to \infty$. Using (24.5) and (24.8), we get an approximate expression for the symbol of $\exp(-it\hat{H})$, which we call $G(p, q, t)$:

$$(24.9) \quad G(p, q, t)$$
$$\approx \frac{1}{(2\pi)^{(N-1)}} \int \exp\left(i \sum_1^N p_\alpha(q_\alpha - q_{\alpha-1}) - i\frac{t}{N} \sum_1^N H(p_\alpha, q_{\alpha-1}) \right) \prod_1^{N-1} dp_\alpha \, dq_\alpha,$$

where we have st $q_0 = q_N = q$ and $p_N = p$. The error in this estimate approaches zero as $N \to \infty$.

Combining (24.9) and (24.4) we get an approximation for $\langle y \mid e^{-it\hat{H}} \mid x \rangle$, which closely resembles (24.9), except that it contains an extra integral with respect to to $p_N = p$, and an extra factor $(2\pi)^{-1}$ coming from the extra integral; also, the boundary conditions are $q_0 = x$ and $q_N = y$. Thus,

$$(24.10) \quad \langle y \mid e^{-it\hat{H}} \mid x \rangle$$
$$\approx (2\pi)^{-N} \int \exp\left(i \left(\sum_i^N p_\alpha(q_\alpha - q_{\alpha-1}) - \frac{t}{N} \sum_1^N H(p_\alpha, q_{\alpha-1}) \right) \right) \prod_1^N dp_\alpha \prod_1^{N-1} dq_\alpha,$$

where $q_0 = x$ and $q_N = y$. The expression in the inner big parentheses is a Riemann sum for the integral in (24.7), where the trajectory $q(\tau)$ goes from x to y, and no boundary conditions are imposed on p. This leads us to set the limit of the right-hand side as $N \to \infty$ the functional integral of e^{iS} over the space of paths $(p(\tau), q(\tau))$ with boundary conditions $q(0) = x$ and $q(t) = y$. By convention, the factor $(2\pi)^{-N}$ is absorbed into the terms dp_α and dq_α.

Using a similar reasoning, we obtain

$$(24.11) \quad \langle y \mid e^{-tH} \mid x \rangle$$
$$= \exp\left(i \int_0^t p(\tau) \frac{dq(\tau)}{d\tau} d\tau - \int_0^t H(p(\tau), q(\tau)) d\tau \right) \prod dp(\tau) dq(\tau),$$

where, as before, the integral is over all paths $(p(\tau), q(\tau))$ with boundary conditions $q(0) = x$ and $q(t) = y$. This formula can also be obtained directly by substituting $i\hat{H}$ for $\hat{H}$ in (24.6), something we can do because we have not assumed that $\hat{H}$ is self-adjoint.

All the relations proved above hold practically unchanged for the case of an arbitrary (finite) number of degrees of freedom. Of course, one should then think of the variables p and q as vectors

$$p_\alpha = (p_{\alpha 1}, \ldots, p_{\alpha n}), \qquad q_\alpha = (q_\alpha^1, \ldots, q_\alpha^n),$$

where n is the number of degrees of freedom, and one must also replace $(2\pi)^{-1}$ by $(2\pi)^{-n}$ everywhere. In addition, formulas (24.6) and (24.11) have counterparts even in the case of an infinite number of degrees of freedom.

We define the symbol of the differential operator

$$(24.12) \qquad \hat{A} = \sum_k \sum_{i_1,\ldots,i_k} a^{i_1\ldots i_k}(q) i^{-k} \frac{\partial^k}{\partial q^{i_1} \ldots \partial q^{i_k}},$$

acting on the space of functions in the n variables $q^1, \ldots, q^n$, as the following function of the $2n$ variables $q^1, \ldots, q^n, p_1, \ldots, p_n$:

$$(24.13) \qquad A(p, q) = \sum_k \sum_{i_1,\ldots,i_k} a^{i_1\ldots i_k}(q) p_{i_1} \ldots p_{i_k}.$$

As before, $\hat{A}$ can be obtained from $A(p, q)$ by replacing the variables $q^1, \ldots, q^n$ by the position operators $\hat{q}^1, \ldots, \hat{q}^n$ (operators of multiplication by $q^1, \ldots, q^n$), and the variables $p_1, \ldots, p_n$ by the momentum operators $\hat{p}_1 = i^{-1}(\partial/\partial q^i), \ldots, \hat{p}_n$, making sure we write the $\hat{p}_i$ to the right of the $\hat{q}_i$.

Another concept, the *principal symbol* of $\hat{A}$, is often useful. The principal symbol of a differential operator of order r is defined as the sum of terms of order r in the p_i, in the expression (24.13).

We examine the important particular case where the Hamiltonian $H(p, q)$ is quadratic in the momenta. Then the integrals with respect to the p-variables in (24.9) and (24.10) are Gaussian integrals. Evaluating them by means of the standard formulas given in Chapter 23, and taking the limit as $N \to \infty$, we obtain expressions for $\langle y| \exp(-it\hat{H})|x\rangle$ and $\langle y| \exp(-t\hat{H})|x\rangle$ in terms of functional integrals along trajectories in q-space. For example, if

$$H = \frac{1}{2} \sum a^{ij}(q)p_i p_j + U(q),$$

where the $a^{ij}(q)$ form a positive definite matrix, the approximate expression (24.10) for $\langle y| \exp(-it\hat{H})|x\rangle$ is
(24.14)
$$\int \exp\left(i \sum_\alpha \frac{N}{2t} a_{ij}(q_{a-1})(q_a - q_{a-i})^i (q_a - q_{a-1})^j - \frac{t}{N} U(q_{a-1}) \right) \prod_\alpha \frac{dq_\alpha}{(\det a^{ij}(q_{a-1}))^{1/2}},$$

where $(a_{ij}(q))$ is the matrix inverse to $(a^{ij}(q))$.

This allows us to write $\langle y| \exp(-it\hat{H})|x\rangle$ as a functional integral

$$(24.15) \qquad \langle y| \exp(-it\hat{H})|x\rangle = \int \exp(iS[q]) \prod dq(\tau),$$

where

$$(24.16) \qquad S[q] = \int L(q(\tau), \dot{q}(\tau)) \, d\tau = \int (\tfrac{1}{2} a_{ij}(q)\dot{q}^i \dot{q}^j - U(q)) \, d\tau$$

is the classical action along the trajectory $q(\tau)$ in configuration space, and the integral is taken over trajectories satisfying the boundary conditions $q(0) = x$ and $q(t) = y$. The factor $(\det a(q))^{1/2}$ in (24.14) is accounted for by the introduction, in configuration space, of the volume element $dq = (\det a(q))^{1/2} \, dq^1 \ldots dq^n$, corresponding to the metric $ds^2 = a_{ij}(q) \, dq^i \, dq^j$.

A similar reasoning gives

$$(24.17) \qquad \langle y| \exp(-t\hat{H})|x\rangle = \int \exp(-S_{\text{eucl}}[q]) \prod dq(\tau),$$

where

$$(24.18) \qquad S_{\text{eucl}}[q] = \int (\tfrac{1}{2} a_{ij}(q)\dot{q}^i \dot{q}^j + U(q)) \, d\tau$$

is called the *Euclidean action*. (Apart from a sign, the Euclidean action can be obtained from the action (24.16) by replacing time t by imaginary time it.)

We remark that the functional integral (24.17) is much easier to deal with than (24.15), and therefore increasingly more popular. This is because the integrand in (24.15) has absolute value 1, so convergence depends on the fact that the integrand oscillates. The passage from (24.15) to (24.17) is known as *Euclidean rotation* or *Wick rotation*, and is related to analytic continuation in the time domain.

Usually the operator $\exp(-t\hat{H})$ can be defined for complex values of t in the half-plane $\operatorname{Re} t \geq 0$, and it varies continuously with t in this half-plane, and analytically in the open half-plane $\operatorname{Re} t > 0$. The matrix entries $\langle y| \exp(-t\hat{H})|x\rangle$ have the same analyticity properties.

Physical information can be extracted equally well from (24.15) and (24.17), although the operator $\exp(-it\hat{H})$ has a more immediate physical interpretation. For example, we have the relation

$$\langle y|\exp(-t\hat{H})|x\rangle = \sum_n \langle y|\exp(-t\hat{H})|n\rangle\langle n|x\rangle = \sum_n \exp(-\varepsilon_n t)\langle y|n\rangle\langle n|x\rangle,$$

where the $|n\rangle$ form a basis of eigenvectors for $\hat{H}$ and the ε_n are the corresponding eigenvalues. From this it follows that, if $\langle y|0\rangle \neq 0$ and $\langle 0|x\rangle \neq 0$, the asymptotic behavior of $\langle y|\exp(-t\hat{H})|x\rangle$ as $t \to \infty$ is defined by the energy ε_0 of the ground state $|0\rangle$ of the Hamiltonian $\hat{H}$:

$$\varepsilon_0 = -\lim_{t\to\infty}\frac{1}{t}\ln\langle y|\exp(-t\hat{H})|x\rangle.$$

Further, the partition function $Z_{(\beta)}$ of a quantum system with Hamiltonian $\hat{H}$ at a temperature $T = 1/\beta$ is given by

$$(24.19) \qquad Z_{(\beta)} = \operatorname{tr}\exp(-\beta\hat{H}) = \int \langle x|\exp(-\beta\hat{H})|x\rangle \, dx.$$

This becomes

$$(24.20) \qquad Z_{(\beta)} = \int \exp(-S_{\mathrm{eucl}}[q(\tau)]) \prod dq(\tau),$$

where $0 \leq \tau \leq \beta$ and the integral is taken over all possible closed trajectories, $q(0) = q(\beta)$. (One can say that the trajectory $q(\tau)$ is required to satisfy periodic boundary conditions, or that it is defined on the circle.)

The generating functional of the correlation function at temperature $T = 1/\beta$ can be written as

$$(24.21) \qquad G(J) = \frac{Z_{(\beta)}(J)}{Z_{(\beta)}},$$

where $Z_{(\beta)}(J)$ is the partition function in the presence of a source:

$$(24.22) \qquad Z_{(\beta)}(J) = \int \exp(-(S_{\mathrm{eucl}}[q] + Jq)) \prod dq(\tau).$$

(Here $Jq = \int J_i(\tau)q^i(\tau)\,d\tau$, and the function $J(\tau)$ is known as the *source*.)

Equation (24.21) can be considered the definition of the correlation function. To see that this definition coincides with the standard one, notice that

$$(24.23) \quad G(J) = \sum \frac{(-1)^n}{n!}\int_0^\beta d\tau_1 \cdots \int_0^\beta d\tau_n G_n^{i_1\ldots i_n}(\tau_1,\ldots,\tau_n\mid T)J_{i_1}(\tau_1)\ldots J_{i_n}(\tau_n),$$

where

$$(24.24) \quad G_n^{i_1\ldots i_n}(\tau_1,\ldots,\tau_n\mid T) = Z_{(\beta)}^{-1}\int q^{i_1}(\tau_1)\ldots q_{i_n}(\tau_n)\exp(-S_{\mathrm{eucl}}[q]) \prod dq(\tau),$$

and the integral is taken over closed trajectories.

We will momentarily show that, for $\tau_1 < \cdots < \tau_n$, we have

$$(24.25) \quad G_n^{i_1 \cdots i_n}(\tau_1, \ldots, \tau_n \mid T)$$
$$= Z_{(\beta)}^{-1} \operatorname{tr}(\exp(-\hat{H}(\beta - \tau_n))\hat{q}^{i_n} \exp(-\hat{H}(\tau_n - \tau_{n-1}))\hat{q}^{i_{n-1}} \ldots \hat{q}^{i_1} \exp(-\hat{H}\tau_1))$$
$$= Z_{(\beta)}^{-1} \operatorname{tr}(\exp(-\hat{H}\beta)\hat{q}^{i_n}(\tau_n) \ldots \hat{q}^{i_1}(\tau_1)) = \langle \hat{q}^{i_n}(\tau_n) \ldots \hat{q}^{i_1}(\tau_1) \rangle_T,$$

where $\hat{q}^i(\tau) = \exp(\hat{H}\tau)\hat{q}^i \exp(-\hat{H}\tau)$, and the angle brackets denote averaging with respect to the equilibrium state at temperature $T = \beta^{-1}$. The right-hand side of the last equality in (24.25) is the usual definition for the correlation function.

To prove (24.25), we impose additional conditions $q^{i_k}(\tau_k) = s_k$, for s_k fixed. Using (24.17), we integrate under these conditions, and then integrate over all possible values of the s_k.

From (24.24) it is clear that $G_n^{i_1 \cdots i_n}(\tau_1, \ldots, \tau_n \mid T)$ does not change if we permute the τ_k, so long as we apply the same permutation to the i_k. Thus, we can extend (24.25) to define $G_n^{i_1 \cdots i_n}(\tau_1, \ldots, \tau_n \mid T)$ for arbitrary $\tau_1, \ldots, \tau_n$: it is enough to apply a permutation that puts the τ_k in increasing order. Making T tend to zero, we obtain functions $G_n^{i_1 \cdots i_n}(\tau_1, \ldots, \tau_n)$ called *Euclidean Green's functions*, or *Schwinger functions*. They are expressed in terms of the products $\hat{q}^{i_n}(\tau_n) \ldots \hat{q}^{i_1}(\tau_1)$ averaged with respect to the ground state; the times $\tau_n, \ldots, \tau_1$ must be in decreasing order. By analytic continuation of the Euclidean Green's functions we obtain the ordinary Green's functions, which are the average with respect to the ground state of the products of Heisenberg operators, $\exp(i\hat{H}t)\hat{q}^k \exp(-i\hat{H}t)$, arranged in order of decreasing time.

Equations (24.15), (24.17) and (24.20) also hold in the case of infinitely many degrees of freedom; this can be checked by taking the limit of the finite case as the number of degrees of freedom tends to infinity. Note, however, that the divergences that arise in the infinite case usually lead to a situation where the functional integrals cannot be interpreted literally; a meaningful definition for these integrals must include a renormalization procedure.

Consider, for instance, the quantum partition function of the scalar field with action functional

$$(24.26) \qquad S = \int (\tfrac{1}{2}g^{\mu\nu}\, \partial_\mu\varphi\, \partial_\nu\varphi - U(\varphi))d^4x,$$
$$U(\varphi) = \tfrac{1}{2}m^2\varphi^2 + \tfrac{1}{4}\lambda\varphi^4,$$

assuming that we have introduced cutoff in the spatial variables $\mathbf{x} = (x^1, x^2, x^3)$. Here $g^{\mu\nu}$ is the metric tensor in Minkowski space-time, $\mathbf{x}$ lies in a three-dimensional cube W, and the field $\varphi(x) = \varphi(t, \mathbf{x})$ satisfies periodic boundary conditions in the three spatial variables. The partition function Z_V can be written in terms of a functional integral:

$$(24.27) \qquad Z_V = \int \exp(-S_{\text{eucl}}[\varphi]) \prod d\varphi(x),$$

where the integral is over the four-dimensional box V with coordinates $0 \leq t \leq \beta = 1/T$ and $\mathbf{x} \in W$, and the field φ is defined on V and satisfies periodic boundary conditions in all four coordinates. (Equivalently, we can say that φ is defined on the four-dimensional torus $T^4 = S^1 \times S^1 \times S^1 \times S^1$ obtained by identifying opposite walls of V.) The Euclidean action $S_{\text{eucl}}[\varphi]$ is obtained form (24.26) by replacing $g^{\mu\nu}$ by $\delta^{\mu\nu}$ and changing the sign of $U(\varphi)$.

The energy density of the ground state of a quantum system obtained by quantizing the classical theory with action integral (24.26) is

$$(24.28) \qquad \bar{\varepsilon}_0 = \lim_{V \to \infty} -\frac{\ln Z_V}{V}.$$

This is because, as $T \to 0$, the main contribution to the partition function $Z_V = \text{tr}\exp(-\beta H) = \sum_i \exp(-\beta\varepsilon_i)$ comes form the term $\exp(-\beta\varepsilon_0)$, which corresponds to the ground state energy ε_0 in the volume W.

If only the spatial dimensions of the box V tend to infinity, rather than all four dimensions, equation (24.28) gives the value of the free-energy density at temperature $T = \beta^{-1}$. The generating functional of the correlation functions in W at $T = \beta^{-1}$ is

$$(24.29) \qquad G_V(j) = \frac{Z_V(J)}{Z_V},$$

where $Z_V(J)$ is the partition function in the presence of a source and V has the same meaning as in (24.27). The functional integral for $Z_V(J)$ differs form the one for Z_V only by the addition of the term $\int_V J(x)\varphi(x)\,dx$ to the Euclidean action. As the dimensions of V tend to infinity, the functional $G_V(J)$ becomes the generating functional for the Euclidean Green's functions:

$$(24.30) \quad G(J) = \lim_{V\to\infty} G_V(J) = \sum_n \frac{(-1)^n}{n} \int G_n(x_1,\ldots,x_n)J(x_1)\ldots J(x_n)\,d^n x.$$

From (24.25) it follows that the Euclidean Green's functions have the form

$$G_n(x_1,\ldots,x_n) = \langle \hat\varphi(x_1)\ldots\hat\varphi(x_n)\rangle_0$$

for $x_i = (\tau_i, \mathbf{x}_i)$ with $\tau_1 > \cdots > \tau_n$, where the angle brackets denote averaging with respect to the ground state, and the symbol $\hat\varphi(x) = \hat\varphi(\tau,\mathbf{x})$ denotes a field operator that depends on the spatial coordinates $\mathbf{x}$ and imaginary time τ. By analytic continuation with respect to $\tau_1,\ldots,\tau_n$ one can obtain from the Euclidean Green's functions the ordinary Green's functions (averages, with respect to the ground state, of chronological products of Heisenberg operators).

As noted before, if there are infinitely many degrees of freedom one runs into ultraviolet divergences, so one cannot directly determine the functional integrals discussed above simply by taking the limit of the discrete case. One must first renormalize the bare mass m and the bare charge λ in (24.26), and also divide $J(x)$ by a number z depending on the scale of the lattice, or, alternatively, renormalize φ by multiplying it by z. Perturbation theory shows that the dependence of m, λ and z on the scale of the lattice can be chosen in such a way that, by isolating in a certain way the divergent numerical factors not depending on J, we obtain a finite (non-zero) limit for $Z_V(J)$. When we compute $G_V(J)$, the divergent factors cancel out. Thus the divergences in the

Green's functions can be eliminated by renormalizing the mass m, the charge λ and the field $\varphi(x)$.

It is important to emphasize that the considerations in this chapter are not mathematically rigorous. Along the lines of the reasoning above, one can give a justification of the use of functional integrals in quantum mechanics (i.e., a proof of (24.6) and (24.11)). However, the situation in quantum field theory is much more subtle. In particular, it appears that the results motivated by perturbation theory are not necessarily confirmed by rigorous consideration. This is true, in particular, of the theory with action functional (24.26) (the so-called φ_4^4 theory).

The theory of functional integrals in quantum field theory is far from being complete. For an exposition of many important results already obtained, see [82].

25. Quantization of Gauge Theories

The action integral in gauge theories is invariant under local gauge transformations. Such a large group of symmetries makes the Lagrangian degenerate, that is, the generalized velocities cannot be expressed uniquely in terms of the generalized momenta. This makes difficult the use of the Hamiltonian formalism in quantization. The difficulties are surmountable, but it is simpler to choose an alternative approach, based on functional integrals. Note that in the Hamiltonian formalism we can express physical quantities in terms of functional integrals whose integrand contains $\exp(iS)$ or $\exp(-S_{\mathrm{eucl}})$, where S is the action integral and S_{eucl} is the Euclidean action integral. Our strategy will consist in postulating, without resorting to the Hamiltonian formalism, that similar equalities hold in gauge theories.

For the sake of simplicity we consider the theory of a gauge field that does not interact with other fields; the effect of such interactions can be easily taken into account later. Recall that in a gauge theory the action integral has the form

$$(25.1) \qquad S = -\frac{1}{4g^2} \int g^{\mu\alpha} g^{\nu\beta} \langle \mathcal{F}_{\alpha\beta}, \mathcal{F}_{\mu\nu} \rangle \, d^4x,$$

where $g^{\mu\nu}$ is the metric tensor in Minkowski space. As usual, the strength $\mathcal{F}_{\alpha\beta}$ of the gauge field A_μ, which takes values in the Lie algebra $\mathcal{G}$ of the gauge group G, is defined by

$$\mathcal{F}_{\alpha\beta} = \partial_\alpha A_\beta - \partial_\beta A_\alpha + [A_\alpha, A_\beta],$$

and the angle brackets denote an invariant scalar product on $\mathcal{G}$ (if G is a matrix group, we can assume that $\langle a, b \rangle = -2\operatorname{tr}(ab)$).

By definition, the Euclidean action integral for Yang–Mills fields is obtained from the action integral (25.1) by replacing $g^{\mu\nu}$ with $\delta^{\mu\nu}$ (that is, by replacing the Minkowski metric with the Euclidean metric) and switching signs. Less formally, the Euclidean action is obtained form the pseudo-Euclidean action by replacing time with imaginary time.

It will also be convenient to consider the Euclidean action integral for gauge fields on arbitrary Riemannian manifolds. This is given by

$$(25.2) \qquad S_{\mathrm{eucl}} = \frac{1}{4g^2} \int g^{\mu\alpha} g^{\nu\beta} \langle \mathcal{F}_{\alpha\beta}, \mathcal{F}_{\mu\nu} \rangle \, dV = \frac{1}{4g^2} \int \langle \mathcal{F}_{\alpha\beta}, \mathcal{F}^{\alpha\beta} \rangle \, dV,$$

where $\mathcal{F}^{\alpha\beta} = g^{\alpha\mu} g^{\beta\nu} \mathcal{F}_{\mu\nu}$, and $dV = \sqrt{\det g_{ab}} \, d^4x$ is the volume element for the manifold. Notice that (25.2), like the action functionals discussed above, is invariant under the local gauge transformations

$$A_\mu(x) \mapsto A'_\mu(x) = g(x)A_\mu(x)g^{-1}(x) - \partial_\mu g(x)g^{-1}(x),$$

where $g(x)$ is a G-valued function.

Consider the functional integrals

(25.3) $$Z_V = \int \exp(-S_{\text{eucl}}[A]) \prod dA(x),$$

(25.4) $$Z_V(\Phi) = \int \Phi(A) \exp(-S_{\text{eucl}}[A]) \prod dA(x),$$

where V is a four-dimensional box over which the integral is taken, the gauge field $A_\mu(x)$ is defined in this box and satisfies periodic boundary conditions, and $\Phi(A)$ is an arbitrary gauge-invariant functional. For example, we can take $\Phi(A) = \sigma(x_1)\ldots\sigma(x_n)$, where

$$\sigma(x) = \frac{1}{4g^2} g^{\mu\alpha} g^{\nu\beta} \langle \mathcal{F}_{\alpha\beta}(x), \mathcal{F}_{\mu\nu}(x) \rangle$$

is the Euclidean action density at a point $x \in \mathbf{R}^4$. Another possible choice is $\Phi(A) = \operatorname{tr} b_\Gamma$, where b_Γ is the element of the gauge group that represents parallel transport around a loop Γ (T15.1):

$$b_\Gamma = \operatorname{P} \exp\left(-\int_\Gamma A_\mu \, dx^\mu\right).$$

We will call (25.3) the partition function in the four-dimensional box V, and the ratio of (25.4) to (25.3) the Euclidean Green's functional in V. Equation (24.28) implies that the limit of $V^{-1}(-\ln Z_V)$ as $V \to \infty$ should be interpreted physically as the energy density in the ground state (in fact, under the approach we are using this limit is the definition of that density). Similarly, in view of (24.29) and (24.30), it is natural to interpret the limit

(25.5) $$\lim_{V \to \infty} \frac{Z_V(\Phi)}{Z_V}$$

as a Euclidean Green's functional or Euclidean Green's function (dependent on the choice of $\Phi(A)$). For example, for $\Phi(A) = \sigma(x_1)\ldots\sigma(x_n)$, this limit can be interpreted physically as the average of the T-ordered product $\hat\sigma(x_1)\ldots\hat\sigma(x_n)$ with respect to the ground state. (Here the temporal coordinate of a point $x \in \mathbf{R}^4$ corresponds to imaginary time, T-ordered means ordered with respect to imaginary time, and $\hat\sigma(x)$ stands for the operator corresponding to $\sigma(x)$ in the operator formalism that we chose not to adopt.)

It is easy to construct analogs for Z_V and $Z_V(\Phi)$ for V an arbitrary complex Riemannian manifold: we use the same formulas (25.3) and (25.4), where $S_{\text{eucl}}[A]$ now denotes the Euclidean action integral on the manifold, given by (25.2). The case of a box with periodic boundary conditions corresponds to the manifold being a four-dimensional torus T^4, since we can identify opposite walls of the box.

One should bear in mind, however, that for a gauge field the term "periodic boundary conditions" can have two different meanings: we can require that the restrictions of the field to opposite sides of the box should coincide, or merely that they should be gauge-equivalent. In the first case the gauge field has the geometric meaning of a connection on the trivial principal bundle over the torus T^4, and in the second, that of a connection on an arbitrary principal bundle over T^4. Similarly, a gauge field on a manifold can be seen as a connection on an arbitrary principal bundle over the manifold (T15.3).

Naturally, the functional integrals (25.3) and (25.4), like any other functional integral, require a precise definition in terms of limits of finite-dimensional approximations. To provide such a definition, we consider a cubic lattice in $\mathbf{R}^4$. If a gauge field $A_\mu(x)$ is defined in $\mathbf{R}^4$, we can associate, to every oriented edge γ of the lattice, the element b_γ of the gauge group G that represents parallel transport along γ:

$$(25.6) \qquad b_\gamma = \mathrm{P}\exp\left(-\int_\gamma A_\mu dx^\mu\right).$$

A change in the orientation of γ replaces b_γ by its inverse. Next, under a gauge transformation with gauge function $g(x)$, the element b_γ transforms according to the simple law

$$(25.7) \qquad b'_\gamma = g(x_1)b_\gamma g^{-1}(x_0),$$

where x_0 and x_1 are the beginning and end points of γ (T15.2). This suggests that we take as the discrete analog of a gauge field a correspondence $\gamma \mapsto b_\gamma$ assigning to each oriented edge of the lattice an element of the gauge group G, in such a way that b_γ is replaced by its inverse when the orientation of γ is switched. As the discrete analogue of a gauge transformation we take (25.7), where $g(x)$ is a function defined on lattice vertices. To construct a discrete analog for the action integral that is invariant under transformations (25.7), we assign to each oriented two-face σ of the lattice the element $c_\sigma \in G$ representing parallel transport around the boundary of σ. For a cubic lattice, the faces are squares, and if the boundary of an oriented two-face σ consists of the oriented edges γ_1, γ_2, γ_3 and γ_4 (Figure 16), we have

$$(25.8) \qquad c_\sigma = b_{\gamma_4}b_{\gamma_3}b_{\gamma_2}b_{\gamma_1}.$$

The value of c_σ depends on the choice of a starting point for the boundary; for example, if we had started at the vertex between γ_3 and γ_4, we would have obtained

$$\tilde{c}_\sigma = b_{\gamma_3}b_{\gamma_2}b_{\gamma_1}b_{\gamma_4} = b_{\gamma_4}^{-1}c_\sigma b_{\gamma_4}.$$

In the case that interests us (see the next paragraph), the replacement of c_σ by a conjugate has no effect in subsequent computations, and therefore this ambiguity is not important.

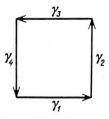

Figure 16

Under a gauge transformation, c_σ is replaced by a conjugate element in G:

$$c'_\sigma = g(x_0)c_\sigma g^{-1}(x_0),$$

where x_0 is the starting point of edge γ_1. If f is a function on G that is constant on each conjugate class, $f(kgk^{-1}) = f(g)$, the value of $f(c_\sigma)$ does not change under a gauge transformation. We can, therefore, construct a gauge-invariant functional by, say, attaching to each discrete gauge field the sum of the $f(c_\sigma)$ over all faces σ of the lattice.

It is easy to choose $f(c_\sigma)$ so that the discrete gauge theory thus obtained approximates the continuous gauge theory of interest. For instance, if G is a matrix group, we can make $f(g)$ proportional to $\mathrm{tr}(g-1)^2$, where 1 represents the identity. For a small square face σ with adjacent edges x^μ and x^ν, we have (see T15.2)

$$c_\sigma = 1 - a^2 \mathcal{F}_{\mu\nu},$$

where a is the length of the edges, so that

(25.9) $$f(c_\sigma) = \mathrm{const}\, a^4\, \mathrm{tr}\, \mathcal{F}_{\mu\nu}.$$

This means that, for an appropriate choice of the constant, the action $\sum_\sigma f(c_\sigma)$ approximates the action of a continuous gauge theory. Another possible choice for $f(g)$ is a constant times $\mathrm{tr}\, g$.

Once we have constructed an approximating discrete action, we can define a discrete analog for the functional integrals (25.3) and (25.4) simply by replacing $S_{\mathrm{eucl}}[A]$ with

(25.10) $$S^a(b) = \sum_\sigma f(c_\sigma).$$

In addition, integration over gauge fields is to be replaced by its discrete counterpart. We restrict ourselves to a finite lattice, so that there are only finitely many vertices and edges. Recall that a "discrete gauge field" is a map $\gamma \mapsto b_\gamma$, where the b_γ can be chosen independently of one another, under the sole condition that if γ and $\bar\gamma$ represent the same edge with opposite orientations, we have $b_{\bar\gamma} = b_\gamma^{-1}$. Thus the space of gauge fields is simply the direct product of finitely many copies of g, one per (non-oriented) edge. The invariant scalar product on the Lie algebra $\mathcal{G}$ of G gives an invariant metric and an invariant volume element on G, and therefore on the product $\prod_\gamma G$. Integration over gauge fields is with respect to this volume element.

The large group of symmetries that gauge fields possess complicates not only work in the Hamiltonian formalism, but also the evaluation of the functional integrals (25.3) and (25.4). Indeed, the perturbation theory approach to the calculation of a functional integral is based on discarding all but the quadratic part of the exponent, and evaluating the resulting Gaussian integral. However, since in gauge theories the quadratic part of the Lagrangian is degenerate because of the existence of an infinite-dimensional symmetry group, we cannot apply this procedure directly. The difficulty can be overcome if we note that the integral of a quantity that enjoys certain symmetries can be replaced by the integral of a related quantity over a subspace of the domain—for example, the integral of an even function over $\mathbf{R}$ is twice the integral over the positive axis only. This reduction yields integrals that can be dealt with as in the non-degenerate case.

To make this precise, we start with a general discussion of the integration of invariant functions of finitely many variables.

Let M be a Riemannian manifold and $f(x)$ a function on M, which is invariant under the action of a compact group G of isometries of the manifold: $f(gx) = f(x)$ for all $g \in G$ and all $x \in M$. We show that

$$(15.11) \qquad \int_M f(x)\,d\mu,$$

where $d\mu$ is the Riemannian volume element, can be reduced to an integral over a space of dimension lower than M if the orbits of G have positive dimension. For example, if $f(x)$ is a spherically symmetric function on $\mathbf{R}^3$, the integral of f can be reduced to a one-dimensional integral:

$$(15.12) \qquad \int f(|x|)\,d^3x = 4\pi \int_0^\infty f(r)r^2\,dr.$$

For now we assume that G acts freely (that is, all stabilizers H_x, for $x \in M$, are trivial), and that there is a submanifold N of M that intersects each orbit of G exactly once. Then (25.11) can be transformed into an integral over N. The simplest way to do this is to use the so-called *Faddeev–Popov trick*. Assume that N is locally defined by equations

$$(25.13) \qquad F^1(x) = \cdots = F^k(x) = 0,$$

which are assumed independent (that is, the differentials of $F^1, \ldots, F^k$ are linearly independent at each point $x \in M$). We can also write (25.13) as a single vector equation $F(x) = 0$, where $F = (F^1, \ldots, F^k)$ is a map from M into $\mathbf{R}^k$. Saying that the equations are independent means that the differential F_x of F at each point of $x \in M$ has rank k.

Define the function $W_F(x)$ by the equation

$$(25.14) \qquad W_F(x) \int \delta(F(gx))\,dg = 1,$$

where dg is the invariant element of volume on the group G, normalized so that the total volume of G is 1. Then

(25.15)
$$\int_M f(x)\,d\mu = \int_M W_F(x)f(x)\delta(F(x))\,d\mu,$$

where the presence of $\delta(F(x))$ on the right-hand side transforms the integral over M into an integral over N. To verify this equality, write (25.11) in the form

$$\int_M f(x)\,d\mu = \int_M d\mu \int_G dg\, f(x)W_F(x)\delta(F(gx)).$$

Replacing gx by $\tilde{x}$ and noting that both $f(x)$ and $W_F(x)$ are G-invariant, we get

$$\int_M f(x)\,d\mu = \int_M d\mu \int_G dg\, f(\tilde{x})W_F(\tilde{x})\delta(F(\tilde{x})).$$

Evaluating the integral over G, we see that

$$\int_M f(x)\,d\mu = V(G)\int_M W_F(x)f(x)\delta(F(x))\,d\mu,$$

which implies (25.15) because the volume $V(G)$ of G was assumed to be 1.

Although the argument above is for a finite-dimensional manifold and a compact group G of isometries, it can be extended to the infinite-dimensional case, with G not compact. It is often convenient to consider (25.15) as the defining equation for the functional integral when the integrand has an infinite-dimensional symmetry group. We apply this now to the integrals (25.3) and (25.4).

The integrands in these integrals are invariant under the group G^∞ of all local gauge transformations. However, this group does not act freely on the space of gauge fields $A_\mu(x)$, which here plays the role of the manifold M (T15.9). We therefore replace G^∞ by G_0^∞, the group of all local gauge transformations arising from functions $g(x)$ whose value at a fixed point x_0 is 1. When the fields are defined on Euclidean space-time, which we will assume for simplicity to the case, it is convenient to take the point at infinity for x_0, so that G_0^∞ consists of transformations arising from functions $g(x)$ such that $\lim_{x\to\infty} g(x) = 1$. Now G_0^∞ does act freely on the space of gauge fields (T15.9). We must therefore select a set N of gauge fields that intersects each orbit of G_0^∞ exactly once.

We do this by imposing on the fields the gauge condition $\partial_\mu A^\mu(x) = 0$. Strictly speaking, this only works in the abelian case; if G is non-abelian there may be no way to select, in a continuous way, a single representative for each orbit of G_0^∞ (T15.9). But even then we can use the gauge condition $\partial_\mu A^\mu(x) = 0$ to construct a perturbation theory, for the following reason: in such a theory, only gauge fields that differ little from a vacuum field (by which we mean a field that is gauge-equivalent to a zero field) play a significant role. However, two fields that are sufficiently close to being vacuum fields and that satisfy $\partial_\mu A^\mu(x) = 0$ cannot be linked by a gauge transformation arising from a function $g(x)$ with values near 1. This is because the nonlinear equation for $g(x)$ in this case differs by only a small term from a linear equation having no nontrivial solutions, and hence itself has no nontrivial solutions. (In the abelian case all equations are linear, and one need not assume that the fields are near a vacuum.)

When (25.15) is applied to (25.3) and (25.4), the role of $W_F(x)$ is played by the functional $W[A]$ defined by the equation

(25.16)
$$W[A]\int \delta(\partial^\mu A_\mu^g)\prod dg = 1,$$

where $A_\mu^g = gA_\mu g^{-1} - \partial_\mu g g^{-1}$, and the integral is over G_0^∞. As observed above, if a field A_μ satisfies $\partial_\mu A^\mu = 0$ and is sufficiently weak, we have $A_\mu^g = 0$ if and only if $g(x) \equiv 1$. Thus, in calculating the integral we can assume that $g(x)$ differs from 1 by an infinitesimal amount. Noting that

$$A_\mu^{1+\varepsilon}(x) = A_\mu(x) - \nabla_\mu \varepsilon(x),$$

where ε is an element of the Lie algebra of G_0^∞, we see that the computation of $W[A]$ reduces to that of the integral

(26.17)
$$\int \delta(\partial_\mu \nabla^\mu \varepsilon(x)) \prod d\varepsilon = \frac{1}{\det(\partial_\mu \nabla^\mu)}$$

over the Lie algebra of G_0^∞. Applying the Faddeev–Popov trick to (25.4), for example, we obtain

(26.18)
$$\int \Phi(A) \det(\partial^\mu \nabla_\mu) \exp(-S_{\mathrm{eucl}}) \delta(\partial^\mu A_\mu) \prod dA.$$

A perturbation theory applicable to this functional integral can be constructed in the standard way. To do this, it is convenient to replace A_μ with $\tilde{A}_\mu = g^{-1}A_\mu$, where g is the coupling constant. Then the factor g^{-2} vanishes from the action integral and the field strength $\mathcal{F}_{\mu\nu}$ is replaced with $\tilde{\mathcal{F}}_{\mu\nu} = \partial_\mu \tilde{A}_\nu - \partial_\nu \tilde{A}_\mu + g[\tilde{A}_\mu, \tilde{A}_\nu]$. Notice that the perturbation series expansion is in powers of g^2, rather than in powers of g, since it is g^2 that appears in the action integral. Hence, instead of g one often calls $g^2/(4\pi)$ the coupling constant, and denotes it by α. If necessary to distinguish it from the effective coupling constant, we will call α the bare coupling constant and denote it by α_{bare}.

The terms of the resulting perturbation series in powers of α_{bare} are generally subject to ultraviolet divergences. To eliminate them, we must introduce cutoff in momenta by discretizing, then lift the cutoff by passing again to the continuous limit, while making the coupling constant α_{bare} depend on the cutoff parameter. This dependence is usually fixed by the condition that the effective coupling constant at a certain point (the normalization point) must assume a given value α_0. The perturbation series can then be transformed into a series in powers of α_0. The terms of any series arising by this process and describing a physically meaningful quantity remain finite when the cutoff is lifted (assuming that the gauge field theory is renormalizable).

We now discuss in greater detail the passage from discrete to continuous gauge theories. For definiteness, we assume that the action theory of the discrete theory is given by (25.10), with $f(c) = \beta \operatorname{tr}(c^2 - 1)$. (This assumes the gauge group G is a matrix group; if not, we set $f(c) = \beta \operatorname{tr} \hat{c}$, where c is the operator corresponding to an element $c \in G$ in the adjoint representation.) If we choose the dependence of β on the scale a of the lattice in an appropriate way, this discrete action integral tends, as $a \to 0$, to the action integral of a continuous gauge theory. However, for such a choice of β as a function of a, the physical quantities computed from the discrete action integral have no meaningful limit.

Therefore, we take another approach, one motivated by perturbation theory. The dependence of β on a should be such that at least one physical quantity remains constant as a varies; it can then be expected that all other physically meaningful quantities will have a finite limit as $a \to 0$. The correlation functions of a discrete gauge model fall off exponentially with increasing distance, a fact that has been rigorously proved for β large. The rate of this decrease is characterized by the radius of correlation $r(\beta)$. In a continuous theory, the radius of correlation is the inverse of the mass of the lightest particle. (Recall that in our system of units, $\hbar = c = 1$, so that mass is expressed in units of inverse length.) Hence, in passing to a continuous limit, it is convenient to assume that $r(\beta)$ is independent of a; to see that one can force β to depend on a so that this condition is satisfied, notice that as $\beta \to 0$ the correlation radius gets very large in relation to a. If β is so chosen, all other physically meaningful quantities have a finite limit.

In this way we obtain all the physical quantities associated with a continuous gauge theory. Actually, although the initial Lagrangian of the continuous gauge theory depended on a dimensionless coupling constant, as a result of renormalization we get a whole family of theories, depending on a parameter that has the dimension of length, the fixed radius of correlation. It will be convenient to use the inverse of the correlation length, which is a mass m in our system of units, as the parameter.

Essentially all these theories are equivalent, that is, they differ only in the choice of scale. If, for example, we take a dimensionless physical quantity that depends on the vector of momenta $k = (k_1, \ldots, k_n)$, in our theories it will depend only on the ratios $k_1/m, \ldots, k_n/m$. In particular, the effective coupling constant α, considered as a function of k, actually depends only on k/m. (The effective coupling constant can be defined in different ways, but the assertion above is true in all cases since it depends only on dimension considerations.)

It is common in perturbation theory to express the effective coupling constant in terms of its value at a fixed point v, the normalization point. Thus, the coupling constant is a function of the momentum k, the normalization point v, and the value α_0 of the coupling constant at that point:

$$a(k/m) = \alpha(k, v, \alpha_0), \quad \text{with } \alpha_0 = \alpha(v/m).$$

We see that instead of one, we now have two parameters, v and α_0. In other words, a change in the normalization point v can be canceled by a corresponding change in α_0. This property is known as *renorm-invariance*.

We now show how to compute the effective coupling constant when it is small, by using perturbation theory techniques and renorm-invariance. Since in this continuous limit one should expect invariance under rotations, we assume that the effective coupling constant α is independent of the direction of the vector k, and therefore is a function of k^2/m^2. Conversely, k^2/m^2 can be expressed as a function of α. One easily verifies that the derivative of $\ln \alpha$ with respect to $\ln k^2$ depends only on k^2/m^2, and so can be seen as a function of α, known as the *Gell-Mann–Low function*:

(25.19)
$$\frac{\partial \ln \alpha}{\partial \ln k^2} = \beta(\alpha).$$

Obviously, $\beta(0) = 0$, since in the absence of interactions the effective coupling constant is zero for all values of k. This means that the expansion of $\beta(\alpha)$ in powers of α begins with linear term:

$$(25.20) \qquad \beta(\alpha) = b\alpha + \cdots .$$

Substituting (25.20) into (25.19) and ignoring terms of order higher than 1, we get

$$(25.21) \qquad \frac{d\ln\alpha}{\alpha} = \frac{d\alpha}{\alpha^2} = b\,d\ln k^2.$$

This in turn implies

$$(25.22) \qquad \alpha^{-1}(k^2, v^2, \alpha_0) = \alpha_0^{-1} - b\ln\frac{k^2}{v^2},$$

where v is the normalization point and $\alpha_0 = \alpha(v^2, v^2, \alpha_0)$ is the value of the effective coupling constant at that point. This equation is valid only when $\alpha(k)$ is small, since otherwise we cannot ignore the non-linear terms in (25.20).

The factor b in (25.20) can be calculated using perturbation techniques, or using Seeley coefficients (Chapter 32). For instance, for a gauge theory with $G = \mathrm{SU}(N)$ we get

$$(25.23) \qquad b = -\frac{11N}{12\pi}.$$

It turns out that b is negative whenever the gauge group G is non-abelian. This means that in such theories (25.22) holds when k^2 is large, so that α tends to 0 as k increases. This is known as asymptotic freedom. If b were positive, (25.22) could be used only for small values of k^2, and we would be able to derive no information on the behavior of the effective coupling constant for large values of k^2.

The arguments above can be applied, with minor changes, to any renormalizable theory where the action integral is scale-invariant, that is, does not change under the transformation $x \mapsto \lambda x$. The latter requirement implies, among other things, that the Lagrangian of the theory contain no mass terms. If the theory is asymptotically free, it remains so even if mass terms are added to the Lagrangian, since at high momenta the contribution of mass terms is negligible. Of course, in the process of adding mass terms the renormalizability must not be violated.

In the sequel we will need a group-invariant version of the Faddeev–Popov trick introduced above to reduce the integral of a function having certain symmetries to an integral of lower multiplicity. Assume, as before, that M is a Riemannian manifold acted on by a compact group G of isometries. Instead of taking a set N that intersects each orbit exactly once, we will work with the space of orbits itself, which we denote $W = M/G$. By removing from M, if necessary, a set of volume zero, we can assume that the stabilizers of all points in M are conjugate, and in particular that all orbits have the same dimension m.

We define the distance between two orbits as the minimum of the distance between two points, one on each orbit; this minimum is achieved and is non-zero if the orbits are distinct because G is compact. Further, in seeking a pair of closest points one of the points can be fixed arbitrarily, because, since G acts by isometries, all points in an orbit are equivalent. With this metric, W becomes a Riemannian manifold: if $x \in M$ is a point with orbit Gx, and dx is a vector orthogonal to Gx at x, the distance between Gx and $G(x + dx)$ is the length of the vector dx, because the geodesic in M connecting a pair of closest points on the two orbits is perpendicular to both orbits.

Using the Riemannian metric one can define on W a volume element $d\nu$. Denoting by $\lambda(x)$ the m-dimensional volume of the orbit Gx with respect to the Riemannian metric of M, we get a G-invariant weight function on M (since $\lambda(gx) = \lambda(x)$ for $g \in G$), and therefore a function on W, which we still denote $\lambda(x)$. If $f(x)$ is a G-invariant function on W, we easily see that

$$(25.24) \qquad \int_M f(x)\, d\mu = \int_W f(x)\lambda(x)\, d\nu,$$

where $d\mu$ is the element of volume on M. For example, for a spherically symmetric function on $\mathbf{R}^3$, formula (25.24) becomes (25.12), since $\lambda(r) = 4\pi r^2$ is the area of a sphere of radius r.

To use (25.24), we must be able to compute the volume $\lambda(x)$ of the orbits. Fix an invariant scalar product on the Lie algebra $\mathcal{G}$ of G, and normalize it so that the total volume of G with the induced metric is 1. For each point $x \in M$, define a linear operator $T_x : \mathcal{G} \to T_x M$, where $T_x M$ is the tangent space to M at x, describing the action of the Lie algebra at x—that is, T_x maps an infinitesimal transformation to the vector that it defines at x. The kernel of T_x is the Lie subalgebra $\mathcal{H}_x$ corresponding to the stabilizer H_x of x; passing to the quotient, we get a linear operator $\tilde{T}_x : \mathcal{G}/\mathcal{H}_x \to T_x M$. The operator $T_x^\dagger T_x$ is non-degenerate if and only if G acts with discrete stabilizers, while $\tilde{T}_x^\dagger \tilde{T}_x$ is always non-degenerate. We let $D(x)$ be the product of the non-zero eigenvalues of $T_x^\dagger T_x$, or, equivalently, the determinant of $\tilde{T}_x^\dagger \tilde{T}_x$. We claim that

$$(25.25) \qquad \lambda(x) = \frac{\sqrt{D(x)}}{V(H_x)},$$

where the volume $V(H_x)$ is with respect to the Riemannian metric induced from G. To verify (25.25), consider the quotient G/H_x, with the Riemannian metric induced from the one on G (the distance between two orbits being, as for M/G, the minimum of the distance between two points, one on each orbit). The quotient G/H_x is homeomorphic to the orbit Gx, under the map $g \mapsto gx$, for $g \in G$; this map is well defined because two elements g and gh, for $h \in H_x$, map x to the same point of the orbit. The differential of this map at the identity coincides with the operator $\tilde{T}_x$. The volume elements on Gx and G/H_x differ by a factor that depends only on the orbit, but not on x, and this factor is

$$K(x) = \det \tilde{T}_x = (\det \tilde{T}_x^\dagger \tilde{T}_x)^{1/2} = \sqrt{D(x)},$$

so that

$$(25.26) \qquad \lambda(x) = V(L(x)) = (D(x))^{-1/2} V(G/H_x).$$

To compute the volume of G/H_x, we must use (25.24) with $f(x) \equiv 1$ and $M = G$ (the space G/H_x can be thought of as the space of orbits of H_x acting on G from the right, all orbits having the same volume). We see that

$$V(G/H_x) = \frac{1}{V(H_x)}.$$

Using the assumption that all stabilizers are conjugate, and consequently have the same volume, we get

$$\int_M f(x)\, d\mu = \frac{1}{V(H)} \int_W f(x)(D(x))^{-1/2}\, d\nu.$$

To conclude this chapter, we apply this invariant version of the Faddeev–Popov trick to the calculation of the partition function of an electromagnetic field on a compact Riemannian manifold M. The Euclidean action integral is

(25.27) $\qquad S_{\text{eucl}} = \frac{1}{4g^2}(\mathcal{F}, \mathcal{F}) = \frac{1}{4g^2}(dA, dA) = \frac{1}{4g^2}(d^\dagger dA, A) = \frac{1}{2}(SA, A),$

where $S = (2g^2)^{-1}d^\dagger d$ and the field strength $\mathcal{F}$ is seen as a two-form, the exterior derivative of the one-form A, the potential. (For the scalar product of forms, see T6.9.) For notational simplicity, we set $2g^2 = 1$. The action (25.27) is invariant under gauge transformations $A \mapsto A + dg(x)$, so the quadratic form in it has an infinite order of degeneracy. Accordingly, the operator $S = d^\dagger d$ has infinitely many zero modes, each corresponding to a closed one-form, $dA = 0$. We impose on A the gauge condition $d^\dagger A = 0$, which singles out one representative in each class of gauge fields; in coordinates, the condition is $\nabla^\mu A_\mu = 0$, that is, the covariant divergence of the vector field A^μ is zero.

Thus we have a submanifold N of the space Γ_1 of all one-forms, and N can be identified with the space of orbits of Γ_1 under the action of the gauge group. Moreover, N is orthogonal to the orbits, since $(dg, A) = (g, d^\dagger A) = 0$ if $A \in N$. This means that the metric on the space of orbits coincides with the induced metric on N, since, as discussed above, the distance between neighboring orbits is measured along a normal. Therefore we can apply the invariant form of the Faddeev–Popov trick, with the role of the operator T in the paragraph preceding (25.25) being played by d. We get

(25.28) $\qquad Z = \int_{\Gamma_1} \exp(-S_{\text{eucl}})(A) \prod dA = \int_N \exp(-S_{\text{eucl}})(A)(\det d^\dagger d)^{1/2} \prod dA.$

Actually, we have not justified the use of (25.24) in the infinite-dimensional case. We should therefore consider the reasoning above as being of heuristic value only, and (25.24) as the defining equation for the integral on the left-hand side, which otherwise has no precise meaning. Also, we should include a factor of $V(H)^{-1}$ on the right-hand side of (25.28), where H is the stabilizer of the group of gauge transformations. In our case the stabilizer consists of functions such that $dg = 0$, that is, constant functions, so $H = \mathbf{R}$. Its volume is infinite, and for this reason we will ignore the factor $V(H)^{-1}$, as well as other divergent factors.

The operator S on N now has only a finite number of zero modes, namely, the number of linearly independent one-forms that satisfy simultaneously the conditions $dA = 0$ and $d^\dagger A = 0$. Such forms are called *harmonic*, and they form a space whose dimension is the first Betti number b^1 of M (T6.9). If $b^1 = 0$, that is, if M is acyclic in dimension one, S has no zero modes on N, and the integral of $\exp(-S_{\text{eucl}})$ over N equals $\sqrt{\det S}$; in this case N coincides with the space spanned by the eigenvectors of S with non-zero eigenvalues. We see, therefore, that

(25.29) $Z = (\det S)^{-1/2}(\det d^\dagger d)^{1/2} = \det(-\Delta_1)^{-1/2}\det(-\Delta_0),$

where $\Delta_1 = -(d^\dagger d + dd^\dagger))$ is the Laplace operator on one-forms and $\Delta_0 = -d^\dagger d$ is the scalar Laplacian (see the following chapter).

This expression for the partition function of an electromagnetic field can also be used when b^1 is non-zero, although the presence of harmonic one-forms, the zero modes of the operator $d^\dagger d + dd^\dagger$, leads to certain anomalies (Chapter 29).

26. Elliptic Operators

Recall that a second-order differential operator

$$(26.1) \qquad A = \sum a^{ij}(x)\partial_i\partial_j + \sum b^i(x)\partial_i + c(x)$$

is called *elliptic* if the quadratic form $\sum a^{ij}(x)p_ip_j$ is positive or negative definite for every x. More generally, recall that for a k-th order differential operator

$$(26.2) \qquad A = \sum_{i_1,\ldots,i_k} a^{i_1\ldots i_k}(x)\partial_{i_1} \ldots \partial_{i_k} + \text{lower order terms},$$

the *principal symbol* $\sigma(x,p)$ is the polynomial comprising the terms of highest degree, with the ∂_l replaced by ip_l (Chapter 24):

$$(26.3) \qquad \sigma(x,p) = i^k \sum_{i_1,\ldots,i_k} a^{i_1\ldots i_k}(x)p_{i_1} \ldots p_{i_k}.$$

We say that A is *elliptic* if $\sigma(x,p)$ vanishes only for $p = 0$.

We can also consider an operator of the form (26.2), with the $a^{i_1\ldots i_k}(x)$ being $r \times r$ matrices. In this case A acts on $\mathbf{R}^r$-valued functions (that is, r-component column vectors of functions), and the principal symbol is a matrix of polynomials in p. The operator is called elliptic if its principal symbol $\sigma(x,p)$ is non-degenerate except for $p = 0$.

As examples of elliptic operators, we consider the Laplacian

$$\Delta = (\partial_1)^2 + \cdots + (\partial_n)^2$$

and the Euclidean Dirac operator

$$\partial\!\!\!/ = i\gamma^j\partial_j,$$

where the γ^j are Hermitian matrices satisfying $\gamma^i\gamma^j + \gamma^j\gamma^i = 2\delta^{ij}$. In dimension $d = 2n$ or $d = 2n+1$ the matrices γ^i have order 2^n, and the Dirac operator can be seen as acting on spinor functions, or 2^n-component column vectors of functions. It is easy to check that the Dirac operator is invariant under orthogonal transformations: for every $g \in \mathrm{SO}(d)$, we have an associated transformation of spinor functions $\psi(x) \mapsto \psi'(x) = S_g\psi(g^{-1}x)$, where S_g is the image of g under the spinor representation of $\mathrm{SO}(d)$, which is 2^n-dimensional. Furthermore, we have

$$(\partial\!\!\!/\ \psi)' = \partial\!\!\!/\ \psi'.$$

The Laplace and Dirac operators can also be defined on Riemannian manifolds. If $g_{ij}(x)$ is the metric tensor and $g^{ij}(x)$ its inverse, and if ∇_i denotes the covariant derivative in the i-th coordinate direction, the Laplacian is

$$(26.4) \qquad\qquad \Delta = g^{ij}\nabla_i\nabla_j = \nabla^i\nabla_i = \nabla_i\nabla^i.$$

Thus $\Delta\varphi$ is the covariant divergence of the vector field with components $\nabla^i\varphi = \partial^i\varphi$. The Laplacian can also be written in the form $\Delta = -\delta d$, where d is the exterior derivative (which associates to a scalar function φ the one-form $d\varphi$), and $\delta = d^\dagger$ is its adjoint operator with respect to the scalar product in the space of forms. (For more on d, δ and the scalar product of forms, see T6.9.) The operator δ associates with a one-form $A_i dx^i$ the scalar function $\nabla^i A_i$, the covariant divergence of the vector field with components A_i.

We can generalize the Laplacian so it acts on k-forms, rather than functions. The definition is

$$(26.5) \qquad\qquad \Delta_k = -(\delta d + d\delta),$$

where the exterior derivative d takes r-forms to $(r+1)$-forms, and the adjoint operator $\delta = d^\dagger$ takes $(r+1)$-forms to r-forms (T6.9).

On an open set $U \subset \mathbf{R}^n$, a k-form $\omega = \omega_{i_1\ldots i_k}(x)dx^{i_1} \wedge \cdots \wedge dx^{i_k}$ is specified by its coefficient functions $\omega = \omega_{i_1\ldots i_k}(x)$, for all k-tuples $i_1 < \cdots < i_k$. Thus we can see Δ_k as acting on $d(k)$-tuples of real-valued functions, where $d(k) = C_n^k$ is the dimension of the space of k-forms. We can also see a k-form as a section of the trivial $\mathbf{R}^{d(k)}$-bundle over U. For a general n-dimensional manifold M, it is not possible to identify k-forms with $d(k)$-tuples of functions, since the functions depend on the local coordinate system. But we can still see k-forms as sections of a vector bundle with fiber $\mathbf{R}^{d(k)}$ and base M, and the Laplacian acts on the space of such sections.

This situation generalizes to elliptic operators acting on the space of sections $\Gamma(\xi)$ of a vector bundle $\xi = (E, M, \mathbf{R}^d)$ over an n-dimensional manifold M. The total space E is formed by gluing together the total spaces of a number of trivial bundles $\xi_i = (U_i \times \mathbf{R}^d, U_i, \mathbf{R}^d)$, where the U_i are open sets in $\mathbf{R}^n$, and the gluing maps take fibers to fibers, acting linearly on each fiber. By restriction, a section φ of ξ gives rise to sections φ_i of the local trivializations $U_i \times \mathbf{R}^d$, each of which can be seen as an $\mathbf{R}^d$-valued function on U_i. We can talk about differentiable operators on each space $\Gamma(\xi_i)$ of sections of $U_i \times \mathbf{R}^d$. To define a differentiable operator on $\Gamma(\xi)$, we must specify a differentiable operator on each $\Gamma(\xi_i)$, subject to the obvious compatibility conditions. A differentiable operator on $\Gamma(\xi)$ is *elliptic* if its restriction to each of the $\Gamma(\xi_i)$ is elliptic—in other words, if it is elliptic in any coordinate system.

In order to define the Dirac operator on an n-dimensional Riemannian manifold M, we work as follows. At each point of M we fix an (ordered) orthonormal basis $(e_1, \ldots, e_n)$. We denote the coordinates of these vectors by e_a^m; the superscript m is known as the world index and the subscript a as the internal index.

The orthonormality condition implies that $g_{mn}(x)e_a^m e_b^n = \delta_{ab}$. Any vector A^m can be expressed in this basis as $A^m = A^a e_a^m$. For a covector (a vector with subscripts instead of superscripts) we set $A_m = e_m^a A_a$, where the e_m^a are the entries of the matrix inverse to e_a^m. The transition from world indices to internal indices is carried out in a similar way for any other tensor. In particular, the covariant derivative satisfies

$$(29.6) \qquad \nabla_a A^b = e_a^m e_n^b \nabla_m A^n = \partial_a A^b + \omega_a{}^b{}_c A^c,$$

where the $\omega_a{}^b{}_c$ are called the *Ricci rotation coefficients*, and satisfy $\omega_a{}^b{}_c = -\omega_a{}^c{}_b$. We see that, for a fixed, we can regard $\omega_a{}^b{}_c$ as an element ω_a of the Lie algebra $\mathfrak{so}(n)$ of skew-symmetric matrices. Given an s-dimensional representation T of $SO(n)$, the element $\omega_a \in \mathfrak{so}(n)$ maps to an $s \times s$ matrix $\hat{\omega}_a$ by the adjoint representation.

We say that an s-component quantity Φ defined on a Riemannian manifold M *transforms according to* the representation T of $SO(n)$ if, when we change from the coordinate system $e_1, \ldots, e_n$ to the coordinate system $\tilde{e}_a = g_a^b e_b$ (where the g_a^b form an orthogonal matrix), Φ transforms according to the law

$$\tilde{\Phi} = T(g)\Phi.$$

If Φ is a field that transforms according to T, we can define the covariant derivative of Φ as

$$\nabla_a \Phi = \partial_a \varphi + \hat{\omega}_a \Phi.$$

A quantity that transforms according to the spinor representation of $SO(n)$ is called a *spinor*. By the preceding equation, we can write the Dirac operator on spinors in the standard form

$$\slashed{\nabla} = i\gamma^a \nabla_a.$$

So far we have assumed that an orthonormal basis can be consistently defined at all points of M. This is not always possible, but we can always cover M with domains U_α inside each of which there is such a field of orthonormal bases. In the intersection $U_\alpha \cap U_\beta$, we have two coordinate systems, related by $\tilde{e}_a^{(\alpha)}(x) = g(x, \alpha, \beta)_a^b e_b^{(\beta)}(x)$, where the $g(x, \alpha, \beta)_a^b$ form an orthogonal matrix (dependent on x). A field that transforms according to the representation T is defined by its expression in coordinates when restricted to each U_α, and these restrictions must satisfy the compatibility conditions

$$\Phi^{(\alpha)}(x) = T(g(x, \alpha, \beta))\Phi^{(\beta)}(x).$$

For a spinor field this definition needs refining, because the spinor representation is two-valued. For the spinor field to be well defined we need to select in each overlap $U_\alpha \cap U_\beta$ a branch of the two-valued function $T(g(x, \alpha, \beta))$, in such a way that on triple overlaps $U_\alpha \cap U_\beta \cap U_\gamma$ the choice is consistent, that is,

$$T(g(x, \alpha, \gamma)) = T(g(x, \alpha, \beta))\, T(g(x, \beta, \gamma)).$$

Once such a choice is made, we say that we gave given the manifold a *spinor structure*.

Not every manifold can be given a spinor structure. A necessary and sufficient condition for this to be possible is that the so-called two-dimensional *Stiefel class* w_2 vanishes. If the manifold is not simply connected, it may have inequivalent spinor structures—the set of such structures is in one-to-one correspondence with the set of homomorphisms $\pi_1(M) \to \mathbf{Z}_2$, or, which is the same, with the first homology group $H^1(M, \mathbf{Z}_2)$. When talking about the Dirac operator, we will always assume that the manifold has been given a spinor structure.

The spinor representation of SO(n) is reducible if n is even. Indeed, in this case the matrix $\gamma^{n+1} = -i^{n/2}\gamma^1 \ldots \gamma^n$ is Hermitian and anticommutes with all the γ^i; its eigenvalues are ± 1 because its square is the identity. The space of spinors then splits into a direct sum of subspaces consisting, respectively, of *right spinors* (those satisfying $\gamma^{n+1}\psi = \psi$) and *left spinors* (those satisfying $\gamma^{n+1}\psi = -\psi$). These subspaces are invariant under any operator of the spinor representation. Since γ^{n+1} and $\partial\!\!\!/$ anticommute, $\partial\!\!\!/$ transforms left spinors into right spinors and vice versa.

Using the Euclidean Dirac operator, we define the Euclidean action functional for Dirac's equation as

$$(26.7) \qquad S_{\text{eucl}}[\psi] = (\psi, \partial\!\!\!/\, \psi) - im(\psi, \psi),$$

where the scalar product of two spinor fields, $(\psi_1, \psi_2) = \int \langle \psi_1(x), \psi_2(x) \rangle$, is based on the scalar product $\langle\,,\,\rangle$ invariant with respect to the spinor representation of SO(n). (Since the γ^i are Hermitian matrices, we can take as the invariant scalar product the sum of products of components of one spinor with complex conjugates of components of the other.)

Equation (26.7) still makes sense if ψ represents a k-component spinor field (that is, a field with values in $\mathbf{C}^k$), with the Euclidean action being a sum of terms corresponding each to one component. The action functional is invariant under unitary transformations in *isotopic space*, that is, under transformations $\psi^a \mapsto u^a_b \psi^b$, where u^a_b is a unitary matrix and $1 \leq a, b \leq k$, the *isotopic indices*, refer to the component of the spinor field.

This allows us to introduce, in a standard way, the Euclidean action of fermions interacting with a gauge field. Namely, if we localize with respect to the subgroup $T(G) \subset U(k)$ of the internal symmetry group, where G is the gauge group (a compact Lie group), the Euclidean action of a fermion field in an external gauge field $A_\mu(x)$ is

$$(26.8) \quad S_{\text{eucl}}[\psi] = (\psi, i\gamma^\mu(\partial_\mu + t(A_\mu))\psi) - im(\psi, \psi) = (\psi, \nabla\!\!\!\!/\, \psi) - im(\psi, \psi).$$

Here, as usual, A_μ takes values in the Lie algebra $\mathcal{G}$ of G, and t is the representation of $\mathcal{G}$ corresponding to the representation T of G. In order to obtain the total Euclidean action functional incorporating the interaction of the fermion and gauge fields, we must add to (26.8) the Euclidean action functional of the gauge field.

The operator $\slashed{\nabla} = i\gamma^\mu \nabla_\mu = i\gamma^\mu(\partial_\mu + t(A_\mu))$ is called the *Dirac operator in an external gauge field*. Like, the usual Dirac operator, it is elliptic, because the introduction of the gauge field does not alter the principal symbol. In (26.8), for simplicity of notation, we considered the field on a subset of Euclidean space. The passage to the case of a field defined on a Riemannian manifold presents no difficulties. In particular, one can define the Dirac operator in the presence of a gauge field on a Riemannian manifold by means of the formula

$$i\gamma^a(\partial_a + \hat{\omega}_a + t(A_a)),$$

where $A_a = e_a^\mu A_\mu$.

In Equation (26.8), if n is even and $m = 0$, the symmetry group contains also the chiral transformations $\exp(\alpha\gamma^{n+1})$. Then the homomorphism T of the gauge field G can act on the group generated by unitary transformations of isotopic space together with chiral transformations. In other words, the generators of the representation t of the Lie algebra $\mathcal{G}$ may include the matrix γ^{n+1}.

27. The Index and Other Properties of Elliptic Operators

We now enumerate, mostly without proof, the most important properties of elliptic operators. Let A be an elliptic operator on a compact manifold M, that is, an elliptic operator on the space of sections $\Gamma = \Gamma(\xi)$ of a vector bundle $\xi = (E, M, \mathbf{R}^n, p)$ over M. The *kernel* of A, denoted Ker A, is the set of solutions of the equation $Af = 0$; the dimension $l(a)$ of Ker A is also called the number of *zero modes* of A. It can be proved that all eigenvalues of A have finite multiplicity, and in particular that $l(A)$ is finite. It can also be shown that the *image* of A, that is, the set of $g \in \Gamma$ such that $g = Af$ for some $f \in G$, is a subspace of finite codimension $r(A)$ in Γ; in other words, that set is the space of solutions of a finite set of linear equations.

Suppose that M has a Riemannian metric, and the fibers of ξ have a scalar product that varies differentiably over the base (in other words, ξ is a $U(n)$-bundle: see T9.4). Then the space Γ of sections has a natural scalar product,

$$(27.1) \qquad (f_1, f_2) = \int \langle f_1(x), f_2(x) \rangle \, dV,$$

where the angle brackets denote the scalar product on the fibers, and dV is the element of volume given by the Riemannian metric on M. The scalar product on Γ allows us to define the operator $A^\dagger$ adjoint to A. If $g = Af$ is in the image of A and h is in the kernel of $A^\dagger$, we have $(g, h) = (f, A^\dagger h) = 0$. This means that any $g \in \mathrm{Im}\, A$ satisfies the equations

$$(27.2) \qquad (g, h_1) = \cdots = (g, h_r) = 0,$$

where $h_1, \ldots, h_r$, for $r = l(A^\dagger)$, are the zero modes of $A^\dagger$. With a bit more trouble one can show that (27.2) is sufficient to characterize the vectors in the image of A, so that

$$(27.3) \qquad r(A) = l(A^\dagger).$$

The *index* of an elliptic operator is the difference

$$(27.4) \qquad \mathrm{index}\, A = l(A) - r(A).$$

Because of (27.3), the index can be defined as the difference between the number of zero modes of A and $A^\dagger$:

(27.5) $\text{index } A = l(A) - l(A^\dagger).$

It turns out that the index does not change under continuous changes in the operator (so long as the operator remains elliptic all the time). This fact, which will follow from the properties of elliptic operators stated below, suggests that the index of an elliptic operator is computable by topological methods. This is, indeed, the case: that so-called *Atiyah–Singer theory* allows one to reduce the calculation of the index to a topological problem. We will not discuss this theory any further; instead we will show how one can compute the index by means of more elementary, although lengthier, calculations. We note in passing that the invariance of the index under continuous changes implies that the index is determined entirely by the principal symbol: two operators of order r that differ only by terms of order lower than r can be continuously deformed into one another without changing the terms of highest order, which are the ones that determine ellipticity.

To obtain a formula that enables us to calculate the index, we start by remarking that

(27.6) $\text{index } A = l(A^\dagger A) - l(AA^\dagger).$

Indeed, if $Af = 0$, clearly $A^\dagger A f = 0$. If, conversely, $A^\dagger A = 0$, we get $(Af, Af) = (f, A^\dagger A f) = 0$, so that $l(A^\dagger A) = l(A)$. Similarly, $l(AA^\dagger) = l(A^\dagger)$.

$A^\dagger A$ and $AA^\dagger$ are non-negative elliptic operators, and to understand them better we turn our attention to this class of operators. Let B be a non-negative elliptic operator. We can then consider the equation

(27.7) $\dfrac{\partial g}{\partial t} = -Bg,$

called the *heat equation* by analogy with the usual case, where B is a constant multiple of the Laplace operator. The methods used in the study of the ordinary heat equation suggest that (27.7) has, for $t \geq 0$ and any smooth initial condition, a unique solution, which depends smoothly on the parameter t and on $x \in M$. We denote by $\exp(-Bt)$ the operator that assigns to each initial condition $g \in \Gamma$ the element $g(t) \in \Gamma$ obtained at time t by solving (27.7) with initial condition g. (If we assume that the exponential of an operator obeys the usual rules of differentiation, we can easily verify that $g(t) = \exp(-Bt)g$ satisfies (27.7) with initial condition $g(0) = g$.) It can be shown that the matrix entries $\langle y| \exp(-Bt)|x \rangle$ are smooth functions for $t \geq 0$, which implies that $\exp(-Bt)$ has finite trace. Moreover, the matrix entries of the operator $R_1 \exp(-Bt)R_2$, where R_1 and R_2 are arbitrary differential operators, are also smooth functions, and therefore $R_1 \exp(-Bt)R_2$, too, has finite trace.

In studying non-negative elliptic operators B, it is very useful to consider the asymptotic behavior of $\text{tr}\exp(-Bt)$ as $t \to +0$ or $t \to +\infty$. The latter is manifested when $\text{tr}\exp(-Bt)$ is represented as a sum over the eigenvalues:

(27.8) $\text{tr}\exp(-Bt) = \sum_i \exp(-t\lambda_i) = l(B) + \displaystyle\int_{\lambda_i > 0} \exp(-t\lambda_i),$

and it follows immediately that $\operatorname{tr}\exp(-Bt)$ approaches $l(B)$ with an error that tends to zero exponentially as $t \to \infty$.

The asymptotic behavior as $t \to +0$ is given by

$$(27.9) \qquad \sum_k \Phi_k(B)t^k,$$

where k takes on the values $-n/r$, $-(n-2)/r$, $-(n-4)/r$, and so on; here n is the dimension of M and r is the order of B. The $\Phi_k(B)$ are called *Seeley coefficients*: we will consider them in more detail below in the case where B has order two.

A generalization of (27.9) can also be obtained for $\operatorname{tr}R\exp(-tB)$, where R is a differential operator of order s. In this case the asymptotic behavior as $t \to +0$ is given by

$$(27.10) \qquad \sum -k\Psi_k(R|B)t^{-k},$$

where k takes on the values $(n-2l)/r$, $(n-2l-2)/r$, and so on, l being the integer part of $s/2$. The numbers $\Psi_k(R|B)$ are also called Seeley coefficients.

We now show that the index of an operator can be expressed in terms of Seeley coefficients. First, note that $A^\dagger A$ and $AA^\dagger$ have the same positive eigenvalues, with the same multiplicities, because $A^\dagger Af = \lambda f$ implies $(AA^\dagger)(Af) = \lambda Af$. Using this on (27.6), we get

$$(27.11) \qquad \operatorname{index} A = \operatorname{tr}\exp(-tA^\dagger A) - \operatorname{tr}\exp(-tAA^\dagger)$$

for any $t > 0$. Taking the limit as $t \to +0$, we see that

$$(27.12) \qquad \operatorname{index} A = \Phi_0(A^\dagger A) - \Phi_0(AA^\dagger).$$

Using (27.11) or (27.12), we prove easily the index invariance property mentioned above. For the right-hand side of (27.11) changes continuously with A, while the left-hand side is an integer, which therefore has to remain constant so long as A varies continuously.

We turn briefly to the calculation of the Seeley coefficients and of the index. We start with the case where B is a scalar elliptic operator of order two on an n-dimensional manifold, and we write B in the form (26.1), in some local coordinate system. Then the asymptotic behavior of $\langle x|\exp(-tB)|y\rangle$ as $t \to 0$ is

$$(27.13) \qquad \langle x|\exp(-tB)|y\rangle \approx \exp\left(\frac{-S(x,y)}{t}\right)\sum_k A_k(x,y)t^k,$$

where k takes on the values $-n/2$, $-n/2+1$, $-n/2+2$, and so on. One easily ascertains that $S(x,y)$, for y fixed, satisfies the equation

$$(27.14) \qquad a^{ik}(x)\frac{\partial S}{\partial x^i}\frac{\partial S}{\partial x^k} + S = 0.$$

A solution of this equation is given by

$$(27.15) \qquad S(x, y) = \tfrac{1}{4}\rho(x, y)^2,$$

where $\rho(x, y)$ is the distance from x to y in the Riemannian metric $ds^2 = -a_{ik}\, dx^i\, dx^k$, the a_{ik} forming a matrix inverse to the matrix of the a^{ik}. Choosing $S(x, y)$ to be of this form and setting

$$(27.16) \qquad A_{-n/2}(x, y) = (4\pi)^{-n/2}\sqrt{\det(-a_{ik})},$$

we satisfy the initial condition

$$(27.17) \qquad \lim_{t \to +0} \langle x| \exp(-tB)|y\rangle = \delta(x - y).$$

For the remaining functions $A_k(x, y)$ in (29.13), we obtain a system of equations that is to be solved by expansion in powers of $x - y$ (we are still fixing y). Once we know the right-hand side of (29.13), we can easily find the asymptotic behavior of $\operatorname{tr}(\hat{R}\exp(-tB))$ as $t \to +0$, where $\hat{R}$ is a differential operator. In particular, we consider the case where R has order zero, that is, is an operator of multiplication by a function $R(x)$. Then

$$(27.18) \qquad \operatorname{tr}(\hat{R}\exp(-Bt)) = \int_M R(x)\langle x| \exp(-tB)|x\rangle\, dV.$$

Thus, computing the Seeley coefficients boils down to finding the asymptotics of the expression $\langle x| \exp(-tB)|x\rangle$, which, by (20.13), has the form

$$(27.19) \qquad \langle x| \exp(-tB)|x\rangle = \sum \Psi_k(x)t^k,$$

where $\Psi_k(x) = A_k(x, x)$. Setting $\Psi_k(\hat{R}|B) = \int R(x)\Psi_k(x)\, dV$, we can write

$$\operatorname{tr}(\hat{R}\exp(-Bt)) = \sum \Psi_k(\hat{R}|B)t^k.$$

In this way, we can obtain the value of $A_k(x, x)$ with a finite number of arithmetic operations; notice that in the end we didn't have to evaluate any integrals. The same method can be used for matrix (rather than scalar) operators of second order whose principal symbol is a multiple of the identity matrix—for example, the square of the Dirac operator. The only difference is that for matrix operators the integrand in (29.18) should be replaced by the trace $\operatorname{tr}(R(x)\langle x| \exp(-tB)|x\rangle)$. Although conceptually there is no difficulty, in practice the calculations are quite complicated.

These calculations have been carried out in great generality for the first few Seeley coefficients. Here we give the expressions for $\Psi_{-2}(x)$, $\Psi_{-1}(x)$ and $\Psi_0(x)$, in the case of a four-dimensional manifold. Recall that our elliptic operator has the form (26.1), where $b^i(x)$ and $c(x)$ are matrix functions and the $a^{ij}(x)$ are scalar functions. (Of course, the a^{ij} form a matrix with respect to the spatial indices i, j, but in isotopic space each a^{ij} is a scalar, or, if you prefer, a multiple of the identity matrix.) By the ellipticity assumption, the quadratic form $-a^{ij}(x)p_i p_j$

is positive or negative definite; we assume without loss of generality that it is positive definite. We define a Riemannian metric on M by the formula $ds^2 = -a_{ij}(x)\, dx^i\, dx^j$, where the matrix of the a_{ij} is inverse to the matrix of the a^{ij}. It is easy to see that one can find a gauge field $A_i(x)$ on M such that the elliptic operator under consideration takes the form

$$(27.20) \qquad \hat{B} = a^{ij}\nabla_i\nabla_j - E(x),$$

where the ∇_i represent the covariant derivative with respect to the Riemannian metric $-a_{ij}(x)$ and the gauge field $A_i(x)$. When ∇_j acts on a field Φ^a having only an isotopic index a, the result is

$$\psi_j^a = \partial_j\Phi^a + (A_j)_b^a\Phi^b.$$

When ∇_j acts on ψ_i^a, a field with both a spatial index and an isotopic index, the result is

$$\partial_j\psi_i^a - \Gamma_{ji}^k\psi_k^a + (A_j)_b^a\psi_i^b,$$

where the Γ_{ji}^k are the Christoffel symbols for the metric $-a_{ij}$. A calculation shows that, for an operator of the form (27.20), the Seeley coefficients are given by

$$(27.21) \quad \Psi_{-2} = (4\pi)^{-2},$$

$$(27.22) \quad \dot{\Psi}_{-1} = -(4\pi)^{-2}(E + \tfrac{1}{6}R),$$

$$(27.23) \qquad \Psi_0 = (4\pi)^{-2}(-\tfrac{1}{30}\nabla^2 R + \tfrac{1}{72}R^2 - \tfrac{1}{180}R_{ij}R^{ij}$$
$$+ \tfrac{1}{180}R_{ijkl}R^{ijkl} + \tfrac{1}{6}RE + \tfrac{1}{2}E^2 - \tfrac{1}{6}\nabla^2 E + \tfrac{1}{12}\mathcal{F}_{ij}\mathcal{F}^{ij}),$$

where R_{ijkl} is the Riemann tensor, R_{ij} the Ricci tensor, R the scalar curvature, and $\mathcal{F}_{ij}$ the strength of the gauge field A_i.

The same relations apply to an operator on an n-dimensional manifold if we replace $(4\pi)^{-2}$ with $(4\pi)^{-n/2}$ and the left-hand sides Ψ_{-2}, Ψ_{-1} and Ψ_0 with $\Psi_{-n/2}$, $\Psi_{-n/2+1}$ and $\Psi_{-n/2+2}$. One can also obtain the coefficients Φ_k from these formulas, by taking the trace and integrating over x.

We now show how equations (27.12) and (27.23) can be used in the computation of the index of the Euclidean Dirac operator in a gauge field. We consider that we have a spinor field that transforms according to the representation T of the gauge group G. The definition of the Dirac operator in a gauge field differs from the ordinary definition only in that the usual derivative is replaced by the covariant derivative $\nabla_\mu = \partial_\mu + t(A_\mu)$, where t is the representation of the Lie algebra $\mathcal{G}$ of G that corresponds to T: see Chapter 26.

We assume the dimension to be even. Then the Dirac operator $\slashed{\nabla}$ splits into two operators L and $L^\dagger$, the first of which maps left spinors to right spinors, and the second the other way around. Since the Dirac operator is self-adjoint and elliptic, L and $L^\dagger$ are adjoint to one another and elliptic. The zero modes of L are called the *left zero modes* of $\slashed{\nabla}$, and the zero modes of $L^\dagger$ are the *right zero modes* of $\slashed{\nabla}$. Thus, the index of L is the difference between the number of left and right zero modes of $\slashed{\nabla}$.

We compute the index of L using (27.12). For simplicity, we assume we are in a compact four-dimensional manifold with a flat metric—the four-torus, for example. Then the square of the Dirac operator is given by

$$
\begin{aligned}
\nabla^2 &= (i\gamma^\mu(\partial_\mu + t(A_\mu)))^2 = -\gamma^\mu\gamma^\nu\nabla_\mu\nabla_\nu \\
&= -\tfrac{1}{2}\gamma^\mu\gamma^\nu(\nabla_\mu\nabla_\nu + \nabla_\nu\nabla_\mu) - \tfrac{1}{2}\gamma^\mu\gamma^\nu(\nabla_\mu\nabla_\nu - \nabla_\nu\nabla_\mu) \\
&= -\nabla_\mu\nabla^\mu - \tfrac{1}{2}\gamma^\mu\gamma^\nu t(\mathcal{F}_{\mu\nu}).
\end{aligned}
$$

The action of ∇^2 on left and right spinors coincides with the action of $L^\dagger L$ and $LL^\dagger$, respectively. Computing the Seeley coefficients by means of (27.23), we get

$$
\text{index } L = \Phi_0(L^\dagger L) - \Phi_0(LL^\dagger) = -\frac{1}{32\pi^2}\int \text{tr}(t(\mathcal{F}_{\mu\nu})t(\tilde{\mathcal{F}}^{\mu\nu}))\, d^4x,
$$

where $\tilde{\mathcal{F}}^{\mu\nu} = \tfrac{1}{2}\varepsilon^{\mu\nu\rho\sigma}\mathcal{F}_{\rho\sigma}$.

Now we make the assumption that G is a simple group. Then

$$
2\,\text{tr}(t(a)t(b)) = -\alpha_T\langle a, b\rangle,
$$

where the number α_T is the so-called *Dynkin index* of the representation T (T14.2), and is an integer if the invariant scalar product $\langle\,,\,\rangle$ on the Lie algebra $\mathcal{G}$ is normalized in an appropriate way (T14.2). If we set

(27.24)
$$
q(A) = \frac{1}{32\pi^2}\int \langle\mathcal{F}_{\mu\nu}, \tilde{\mathcal{F}}^{\mu\nu}\rangle\, d^4x,
$$

we obtain

(27.25)
$$
\text{index } L = \alpha_T\, q(A).
$$

We call $q(A)$ the *topological number* of the gauge field A (T15.4). This number does not change if A varies continuously; this is shown in T15.4, and also follows from (27.24), since the index of an operator satisfies the same property.

The same arguments can be applied to the Dirac operator on an arbitrary compact four-dimensional Riemannian manifold, and they lead to the formula

$$
\begin{aligned}
\text{index } L &= -\frac{1}{32\pi^2}\int dV\, (\text{tr}(t(\mathcal{F}_{\mu\nu})t(\tilde{\mathcal{F}}^{\mu\nu})) - \tfrac{1}{24}\varepsilon^{\mu\nu\alpha\beta}R_{\mu\nu\lambda\rho}R^{\lambda\rho}_{\alpha\beta}) \\
&= \alpha_T q(A) + \tfrac{1}{8}\tau(M),
\end{aligned}
$$

where $\tau(M) = (96\pi^2)^{-1}\int dV\, \varepsilon^{\mu\nu\alpha\beta}R_{\mu\nu\lambda\rho}R^{\lambda\rho}_{\alpha\beta}$, called the *signature of M*, is an integer that does not depend on the metric on M (this again follows from the fact that the index does not change as the operator changes continuously).

28. Determinants of Elliptic Operators

As noted in Chapter 23, the (regularized) determinant of a non-negative operator B can be defined in terms of the ζ-function of B by the formula $\ln \det B = -\zeta'(0)$. This definition can be used whenever the ζ-function $\zeta(s) = \sum \lambda_i^{-s}$ can be analytically continued to the point $s = 0$. We now show that for a non-negative elliptic operator, the poles of the analytic continuation of $\zeta(s)$ coincide with the values that k takes in the asymptotic expression (27.9) for $\operatorname{tr} \exp(-tB)$, namely, $s = -n/r$, $-(n-2)/r$, and so on, and that the residue of $\zeta(s)$ at $s = k$ is $\Gamma^{-1}(k)\Phi_{-k}(B)$, where Γ is the usual Γ-function. Furthermore, $\zeta(s)$ is analytic at $s = 0$, and $\zeta(0) = \Phi_0(B) - l(B)$. Because of this we can define $\det B$ using the ζ-function.

To study the behavior of the ζ-function we use the formula

$$(28.1) \qquad \lambda_i^{-s} = \frac{1}{\Gamma(s)} \int_0^\infty t^{s-1} \exp(-t\lambda_i)\, dt.$$

Summing over all non-zero eigenvalues of B, we get

$$(28.2) \qquad \zeta(s) = \frac{1}{\Gamma(s)} \int_0^\infty t^{s-1} (\operatorname{tr} \exp(-tB) - l(B))\, dt.$$

(We used this formula in Chapter 27 for B positive.) For (28.2) to be valid it is necessary that the integral converge. It always does converge as $t \to +\infty$, by the remark following (27.8); it also converges as $t \to +0$ if $\operatorname{Re} s > n/r$. We now write $\operatorname{tr} \exp(-tB)$ in the form

$$(28.3) \qquad \operatorname{tr} \exp(-tB) = \sum_{k \geq 0} \Phi_{-k}(B) t^{-k} + \rho(t),$$

where $|\rho(t)|$ is bounded by a constant times t as $t \to +0$, by (27.9). Splitting (28.2) into an integral over $[0, 1]$ and one over $[1, \infty)$, we find that

$$(28.4) \quad \zeta(s) = \frac{1}{\Gamma(s)} \left(\sum_{k>0} \frac{\Phi_{-k}(B)}{s - k} + \frac{\Phi_0(B) - l(B)}{s} \right.$$

$$\left. + \int_0^\infty (\operatorname{tr} \exp(-tB) - l(B)) t^{s-1}\, dt + \int_0^1 \rho(t) t^{s-1}\, dt \right).$$

Clearly, the right-hand side defines an analytic continuation for $\zeta(s)$ on the half-plane $\operatorname{Re} s \geq 0$. From (28.4) we can easily read out the position of the poles and the residues. The singularity at $s = 0$ turns out to be removable, since $\lim_{s \to 0} s\Gamma(s) = 1$.

To analytically continue $\zeta(s)$ to the left half-plane, one needs to use more terms in the asymptotic expansion (27.9); here, however, we have no need for this continuation.

From (28.4) follows the following expression for the determinant of a non-negative elliptic operator:

$$(28.5) \quad \ln \det B = \sum_{k>0} \frac{\Phi_{-k}(B)}{k} + \Gamma'(1)(\Phi_0(B) - l(B))$$

$$- \int_1^\infty (\operatorname{tr} \exp(-tB) - l(B)) \frac{dt}{t} - \int_1^\infty \left(\operatorname{tr} \exp(-tB) - \sum_{k>0} \Phi_{-k}(B) t^{-k} \right) \frac{dt}{t}.$$

Another definition for the determinant of a non-negative elliptic operator is suggested by the formula

$$(28.6) \qquad \ln \det A - \ln \det B = - \int_0^\infty (\operatorname{tr} \exp(-tA) - \operatorname{tr} \exp(-tB)) \frac{dt}{t},$$

valid for finite-dimensional positive operators. To prove this formula, express both sides in terms of the eigenvalues of A and B and use the analog of (28.6) for numbers.

For the operators we are interested in, the integral $\int_0^\infty t^{-1} \operatorname{tr} \exp(-Bt) \, dt$ diverges near zero. To obtain a finite result, we replace the lower limit by some positive number $\varepsilon > 0$. We say that

$$(28.7) \qquad \det_\varepsilon B = \exp\left(-\sum_i \int_\varepsilon^\infty \exp(-t\lambda_i) \frac{dt}{t} \right),$$

where the λ_i are the strictly positive eigenvalues of B, is obtained from the (infinite) determinant of B by *cutoff in proper time*. From (28.7) we have

$$(28.8) \qquad \ln \det_\varepsilon B = - \int_\varepsilon^\infty (\operatorname{tr} \exp(-tB) - l(B))) \frac{dt}{t}.$$

Since we know the asymptotic behavior of $\operatorname{tr} \exp(-tB)$ as $t \to 0$, we can also find the behavior of $\ln \det_\varepsilon B$ as $\varepsilon \to 0$. The divergent part of $\ln \det_\varepsilon B$ can be written as

$$(28.9) \qquad -\sum_{k>0} \frac{\Phi_{-k}(B)\varepsilon^{-k}}{k} + (\Phi_0(B) - l(B)) \ln \varepsilon.$$

We denote by $\ln \det' B$ the "finite part" of $\ln \det_\varepsilon B$, that is, the limit of the difference between $\ln \det_\varepsilon B$ and the expression in (28.9). We have

$$(28.10) \quad \ln \det' B = - \int_0^\infty \left(\operatorname{tr} \exp(-tB) - \sum_{k>0} \Phi_{-k}(B) t^{-k} \right.$$

$$\left. - \theta(1-t)\Phi_0(B) - \theta(t-1)l(B) \right) \frac{dt}{t}.$$

Comparing (28.5) and (28.10), we conclude that the regularized determinant $\det B$, defined by means of the ζ-function, does not differ significantly from $\det' B$, defined by cutoff in proper time:

$$(28.11) \qquad \ln \det' B = \ln \det B - \Gamma'(1)(\Phi_0(B) - l(B)).$$

So far we have considered only the determinants of non-negative elliptic operators. The determinant of an arbitrary elliptic operator A can be defined as

$$(28.12) \qquad \ln \det A = \tfrac{1}{2} \ln \det A^\dagger A.$$

To conclude this section, we derive a relation between the determinants of operators that differ only by a numerical factor. Notice that the ζ-functions of αB and B satisfy

$$(28.13) \qquad \zeta_{\alpha B}(s) = \alpha^{-s} \zeta_B(s).$$

For large s this follows form the obvious formula $\lambda'_k = \alpha \lambda_k$ relating the eigenvalues of αB with those of B; for other values of s it follows by analytic continuation. Differentiating (28.13) with respect to s and setting $s = 0$, we get

$$(28.14) \quad \ln \det \alpha B = \ln \det B + (\ln \alpha) \zeta_B(0) = \ln \det B + (\ln \alpha)(\Phi_0(B) - l(B))$$

because of (28.4). Combining this with (28.12), we get

$$(28.15) \qquad \ln \det \alpha A = \ln \det A + (\ln \alpha)(\Phi_0(A^\dagger A) - l(A)).$$

We now use these formulas to study how the determinant of the Dirac operator in a gauge field $A_\mu(x)$ is affected when we apply a scaling transformation to the gauge field, that is, a transformation $A_\mu(x) \mapsto A^{(\lambda)}_\mu(x) = \lambda A_\mu(\lambda x)$. For concreteness, assume the operator operates on fermion fields in four-dimensional Euclidean space. The transformation W_λ that takes the fermion field $\psi(x)$ into the field $\psi^{(\lambda)} = \lambda^2 \psi(\lambda x)$ preserves the scalar product, and conjugates the operators $\nabla\!\!\!\!/ = \partial\!\!\!/ + t(A_\mu)$ and $\lambda \nabla\!\!\!\!/^{(\lambda)} = \lambda(\partial\!\!\!/ + t(A^{(\lambda)}_\mu))$, so that they are unitarily equivalent. (Here, as usual, t is a representation of the Lie algebra of the gauge group into the space where the fermion fields take values.) A formal application of (28.14) gives

$$(28.15) \qquad \ln \det \nabla\!\!\!\!/^{(\lambda)} = \ln \det(\lambda^{-1} \nabla\!\!\!\!/) = \ln \det \nabla\!\!\!\!/ - (\ln \lambda) \zeta_{\nabla\!\!\!/}(0),$$

whence we get, by (28.15),

$$(28.17) \qquad \ln \det \nabla\!\!\!\!/^{(\lambda)} = \ln \det \nabla\!\!\!\!/ + \ln \lambda \left(l(\nabla\!\!\!\!/) - \frac{\alpha_T}{96\pi^2} \int d^4 x \langle \mathcal{F}_{\mu\nu}, \mathcal{F}^{\mu\nu} \rangle \right).$$

The application of (28.15) to operators defined in all of $\mathbf{R}^n$ is not quite justified because of our previous assumption that we were working in a compact manifold. To be rigorous, we must introduce spatial cutoff, say, by going over to fields that are periodic in each coordinate, with period R (that is, fields on the four-dimensional torus). The gauge field should be thought of as concentrated in a bounded region of space. In (28.17), then, the cutoff parameter for the left-hand side is λR, while for the right-hand side it is λ. This can also be applied to the case $A_\mu(x) = 0$, which allows one to determine how the determinant depends on the cutoff parameter. Consider the difference between the logarithm of the determinant of the Dirac operator in a field A_μ and the same quantity in the absence of the field, and let $\rho(A)$ be the limit of this quantity as $R \to \infty$, that is, as the spatial cutoff is lifted. We have

$$(28.18) \qquad \rho(A^{(\lambda)}) - \rho(A) = -\frac{\alpha_T}{96\pi^2} \ln \lambda \int \langle \mathcal{F}_{\mu\nu}, \mathcal{F}^{\mu\nu} \rangle \, d^4x.$$

This formula can be seen as a more accurate version of (28.17).

The arguments above can be applied to other operators as well—for example, to the Laplacian in a gauge field.

29. Quantum Anomalies

One talks about an anomaly in quantum field theory when a relation that holds on the classical level fails to do so after quantization.

For example, it can happen that a symmetry of the classical Lagrangian is not a symmetry of the corresponding quantum theory. In particular, the conformal invariance of the classical action does not always lead to the same property after quantization.

We consider the simplest example: a scalar field in general relativity. The Euclidean action functional can be written as

$$(29.1) \qquad \mathcal{S}_g[\varphi] = \frac{1}{2} \left(\int g^{\mu\nu} \, \partial_\mu \varphi \, d_\nu \varphi \, dV + \frac{1}{6} \int R\varphi^2 \, dV \right).$$

Here $g_{\mu\nu}$ is the Riemannian metric on the four-dimensional manifold M (when we pass to the Euclidean action, the pseudo-Riemannian metric of general relativity becomes Riemannian), R is the scalar curvature, and $dV = \sqrt{g} \, d^4x$ is the element of volume. We assume the manifold M is compact (this corresponds to the introduction of spatial cutoff). One can verify that the functional (29.1) is conformally invariant, that is, it does not change if one replaces the metric $g_{\mu\nu}(x)$ with the new metric $\rho(x)g_{\mu\nu}(x)$, which is conformally equivalent to the first. As explained in Chapter 24, the quantities that arise in quantum theory can be expressed by means of functional integrals whose integrands include $\exp(-\mathcal{S})$. We look at the partition function Z_g arising from (29.1), which is the functional integral

$$(29.2) \qquad Z_g = \int \exp(-\mathcal{S}_g[\varphi]) \prod d\varphi$$

over the space of fields $\varphi(x)$, equipped with the scalar product

$$(29.3) \qquad \langle \varphi, \varphi' \rangle = \int \varphi(x) \varphi'(x) \, dV.$$

The integral in (29.2) is a Gaussian integral:

$$(29.4) \qquad \mathcal{S}_g[\varphi] = \langle S_g \varphi, \varphi \rangle = \langle \varphi, S_g \varphi \rangle,$$

where

$$(29.5) \qquad S_g = -\Delta_0 + \tfrac{1}{6} \hat{R},$$

Δ_0 being the scalar Laplacian (with respect to the metric $g_{\mu\nu}$) and $\hat{R}$ the operator of multiplication by the function $R(x)$. Thus, Z_g is related to the determinant of the elliptic operator S_g:

$$(29.6) \qquad Z_g = (\det(-\Delta_0 + \tfrac{1}{6}\hat{R}))^{-1/2}.$$

To study whether the conformal invariance of the classical theory is preserved after quantization, we make the substitution $\varphi(x) \mapsto \rho(x)^{-1/2}\varphi(x)$ in the functional integral. We know that the integrand remains unchanged under this substitution if, at the same time, we replace the metric $g_{\mu\nu}(x)$ by $\rho(x)g_{\mu\nu}(x)$. We might expect, then, that the partition function Z_g does not change either under the (conformal) change in the metric. But this is not the case: although the integrand does not change, the scalar product (29.3) used in the definition of the functional integral does. (Roughly speaking, the volume element in the space of functions changes.) It is exactly this effect that leads to the conformal non-invariance of the quantum theory.

We show now how to compute the variation in the partition function Z_g under an infinitesimal conformal change in the metric. The calculation is based on general facts about the variation of a Gaussian functional integral under changes in the scalar product.

Consider the functional integral

$$(29.7) \qquad Z = \int \exp(-\mathcal{S}[f])\, df,$$

where $\mathcal{S}[f]$ is a non-negative quadratic functional; this integral can be seen as the partition function for the functional $\mathcal{S}$. Assume that the integral in (29.7) is with respect to a scalar product that depends on a parameter u. Then the value of the integral also depends on u; we denote this value by Z_u.

To compute the variation of Z_u with respect to u, we first write the quadratic functional $\mathcal{S}$ in the form

$$(29.8) \qquad \mathcal{S}[f] = \langle S_u f, f \rangle_u,$$

where S_u is an operator that is self-adjoint with respect to the scalar product $\langle\,,\,\rangle_u$. As explained in Chapter 23, the evaluation of (29.7) reduces to the calculation of the determinant of the operator S_u:

$$(29.9) \qquad Z_u = (\det S_u)^{-1/2}.$$

Differentiating (29.8) with respect to u, we get

$$(29.10) \qquad \frac{dS_u}{du} = -B_u S_u,$$

where B_u is the operator that describes the change in the scalar product with respect to u:

$$\frac{d}{du}\langle f, g \rangle_u = \langle B_u f, g \rangle_u = \langle f, B_u g \rangle_u.$$

To compute the variation Z_u, we compute first the variation $\operatorname{tr} \exp(-S_u t)$, using (29.10). Clearly,

$$(29.11) \quad \frac{d}{du} \operatorname{tr} \exp(-S_u t) = \operatorname{tr}\left(-\frac{dS_u}{du} t \exp(-S_u t)\right)$$

$$= \operatorname{tr}(t B_u S_u \exp(-S_u t)) - t \frac{d}{dt} \operatorname{tr}(B_u \exp(-S_u t)).$$

This equation allows us to find the variation of the determinant of S_u cutoff in proper time (see (28.8)). We get
(29.12)
$$\frac{d}{du} \ln \det_\varepsilon S_u = -\frac{d}{du} \int_\varepsilon^\infty \frac{dt}{t} \operatorname{tr} \exp(-S_u t) = -\int_\varepsilon^\infty \left(\frac{d}{du} \operatorname{tr} \exp(-S_u t)\right) \frac{dt}{t}$$

$$= \int_\varepsilon^\infty \frac{d}{dt} \operatorname{tr}(B_u \exp(-S_u t)) \, dt = \operatorname{tr}(B_u \exp(-S_u)t)\Big|_\varepsilon^\infty.$$

The asymptotic behavior of the right-hand side as $t \to \infty$ is determined by the zero modes of S_u. Indeed, by decomposing with respect to a basis of eigenfunctions ψ_n of S_u, we get

$$(29.13) \quad \operatorname{tr} B_u \exp(-S_u t) = \sum_n \langle \varphi_n | B_u | \varphi_n \rangle \exp(-\lambda_n t),$$

so that, if all the eigenvalues λ_n of S_u are positive, $\operatorname{tr} B_u \exp(-S_u t)$ tends to zero as $t \to \infty$. But if S_u has zero modes, we have

$$(29.14) \quad \lim_{t \to \infty} \operatorname{tr} B_u \exp(-S_u t) = \sum_{\lambda_n = 0} \langle \varphi_n | B_u | \varphi_n \rangle = \operatorname{tr} B_u \Pi(S_u),$$

where the operator $\Pi(S_u)$ projects a function onto the kernel of S_u, that is, $\Pi(\sum_n c_n \varphi_n) = \sum_{\lambda_n = 0} c_n \varphi_n$.

The asymptotic behavior of the right-hand side of (29.12) as $t \to +0$ is given by (27.10) for elliptic operators. Recall that the regularized determinant $\ln \det' S_u$ is defined by eliminating the divergent part of $\ln \det_\varepsilon S_u$ as $\varepsilon \to 0$. Thus $(d/du) \ln \det' S_u$ is also the non-divergent part of $(d/du) \ln \det_\varepsilon S_u$, and we get

$$(29.15) \quad \frac{d}{du} \ln \det' S_u = -\Psi_0(B_u|S_u) + \operatorname{tr} B_u \Pi(S_u).$$

(Here we have assumed that $\operatorname{tr}(B_u \exp(-S_u t))$ behaves like $\sum_k \Psi_{-k}(B_u|S_u)t^{-k}$ as $t \to +0$. This is the case, in particular, if S_u is an elliptic operator on a compact manifold and B_u is a differential operator.)

Although it does lead to correct results, the proof above cannot be considered rigorous, since the variation in the divergent part of $\ln \det_\varepsilon S_u$ could contribute to the finite part. But for elliptic operators on compact manifolds, it is fairly easy to give a rigorous proof, based on (27.10) and (28.10). One must also use the fact that (29.11) implies that

$$(29.16) \qquad \frac{d}{du}\Phi_k(S_u) = -k\Psi_k(B_u|S_u)$$

where the $\Phi_k(S_u)$ are the coefficients of the asymptotic behavior of $\mathrm{tr}\exp(-S_u t)$ as $t \to +0$.

We can also write (29.15) in the form

$$(29.17) \qquad \delta\ln\det' S = -\Psi_0(\delta B|s) + \mathrm{tr}(\delta B\Pi(S)),$$

where δB is the operator describing infinitesimal variations in the scalar product:

$$\delta\langle\varphi,\varphi'\rangle = \langle\delta B\varphi,\varphi'\rangle.$$

Equation (29.17) holds even if we replace $\det'$ with $\det$, that is, when the regularized determinant is defined by means of the ζ-function, rather than by cutoff in proper time.

From (29.17) we get a formula for the variation of the partition function (29.7) with respect to the scalar product:

$$(29.18) \qquad \delta\ln Z = \tfrac{1}{2}(\Psi_0(\delta B|S) - \mathrm{tr}(\delta B\Pi(S))).$$

We use this formula to calculate the variation in the partition function (29.2) caused by conformal changes in the metric. As noted before, the variation is due to the change in the scalar product (29.3). The change in the scalar product due to an infinitesimal conformal change in the metric,

$$(29.19) \qquad \delta g_{\mu\nu}(x) = \delta\rho(x)g_{\mu\nu}(x),$$

is given by

$$(29.20) \qquad \delta\langle\varphi,\varphi'\rangle = \langle\varphi,\delta\rho\varphi'\rangle,$$

taking into account that the field changes by $\delta\varphi(x) = -\tfrac{1}{2}\delta\rho(x)\varphi(x)$. Thus, the operator δB in (29.18) is the operator of multiplication by $\delta\rho(x)$. Using Equation (27.23), we obtain an expression for $\Psi_0(\delta B \mid -\Delta_0 + \tfrac{1}{6}\hat{R})$. Since the operator $-\Delta_0 + \tfrac{1}{6}\hat{R}$ has no zero modes, we get

$$(29.21) \qquad \delta\ln Z = \frac{1}{360(4\pi)^2}\int \delta\rho(x)(R^{\gamma\delta}_{\alpha\beta}R^{\alpha\beta}_{\gamma\delta} - R_{\alpha\beta}R^{\alpha\beta} - \Delta_0 R)\,dV.$$

Thus, the partition function (29.2) of the conformally invariant functional (29.1) changes under a conformal change of the metric. This fact is known as *conformal anomaly*.

The variational derivative of the action functional with respect to the metric tensor $g_{\mu\nu}$ equals, by definition, the energy-momentum tensor

$$(29.22) \qquad T^{\mu\nu}(x) = \frac{2}{\sqrt{g}}\frac{\delta S}{\delta g_{\mu\nu}(x)}.$$

Using this, we can obtain an expression for the energy-momentum tensor averaged over the ground state:

$$(29.23) \qquad \langle T^{\mu\nu}(x)\rangle = \frac{2}{\sqrt{g}}\frac{\delta \ln Z}{\delta g_{\mu\nu}(x)}.$$

Since the action functional (29.1) is invariant under conformal changes of the metric, the trace of the energy-momentum tensor in the classical theory is zero:

$$(29.24) \qquad T^{\mu}_{\mu}(x) = g_{\mu\nu}(x)T^{\mu\nu}(x) = \frac{2}{\sqrt{g}}\frac{\partial S}{\partial g_{\mu\nu}(x)}g_{\mu\nu}(x) = 0.$$

Here we have used the fact that the variation of S under a conformal change in the metric (29.19) is given by

$$(29.25) \qquad \delta S = \int \frac{\partial S}{\partial g_{\mu\nu}(x)}g_{\mu\nu}(x)\,\delta\rho(x)\,dV.$$

A similar formula holds for any functional. Using the counterpart of (29.25) for the functional $\ln Z$, together with (29.23) and (29.21), we get

$$(29.26) \qquad \langle T^{\mu}_{\mu}\rangle = \frac{1}{180(4\pi)^2}(R_{\alpha\beta\gamma\delta}R^{\alpha\beta\gamma\delta} - R_{\alpha\beta}R^{\alpha\beta} - \nabla^2_0 R).$$

Thus, the quantity $T^{\mu}_{\mu}(x)$, which is zero in the classical theory, becomes nonzero after quantization. Sometimes just this phenomenon is understood as a conformal anomaly.

We conclude by computing the conformal anomaly for the electromagnetic field. As we saw in Chapter 25, the partition function in this case is

$$(29.27) \qquad Z = (\det(-\nabla^2_1))^{-1/2}\det(-\nabla^2_0),$$

where $\nabla^2_1 = -(d^\dagger d + dd^\dagger)$ is the Laplace operator on one-forms and $\nabla^2_0 = -d^\dagger d$ is the scalar Laplacian. We want to find the variation in the partition function due to a conformal variation in the metric. The calculation is based on a general result, which we will use often in what follows.

Consider a quadratic functional $S[f]$ defined on a space Γ_1 and invariant under transformations of the form $f \mapsto f + Tg$, where $T : \Gamma_0 \to \Gamma_1$ is a linear operator. An example of such an operator is the action integral for the electromagnetic field, which is invariant under gauge transformations, with $T : g \mapsto dg$ mapping functions into one-forms (covector fields). After fixing scalar products in Γ_1 and Γ_0, we can write S in the form $S[f] = \langle Sf, f\rangle$, where S is a self-adjoint operator, and define the operator $T^\dagger$ adjoint to T. Arguing as we did for the electromagnetic field in Chapter 25, we find that it is reasonable to define the partition function for S by the formula

$$(29.28) \qquad Z = (\det S)^{-1/2}(\det T^\dagger T)^{1/2} = (\det(S + TT^\dagger))^{-1/2}\det T^\dagger T.$$

In any case, this definition is reasonable if $S + TT^\dagger$ and $T^\dagger T$ are elliptic operators, which we will assume from now on.

We study the variation of the partition function (29.28) arising from an infinitesimal change in the scalar products on Γ_0 and Γ_1, described by the operators δB_0 and δB_1 (that is,

$$(29.28') \qquad \delta\langle f, g\rangle_0 = \langle \delta B_0 f, g\rangle_0 = \langle f, \delta B_0 g\rangle_0$$

for the scalar product on Γ_0, and likewise for the scalar product on Γ_1). Then the variation in the partition function can be expressed as

$$(29.29) \qquad \delta \ln Z = \tfrac{1}{2}\beta(\delta B_1|\square_1) - \tfrac{1}{2}\beta(\delta B_0|\square_0),$$

where we have set

$$(29.30) \qquad \beta(B|A) = \Psi_0(B|A) - \mathrm{tr}(\Pi(B)A), \quad \square_1 = S + TT^\dagger, \quad \square_0 = T^\dagger T.$$

The proof of (29.29) is basically the same as that of (29.18). First we must study the variations in S and $T^\dagger$ with respect to the scalar products. It is easy to see that

$$(29.31) \qquad \delta S = -(\delta B_1)S,$$

and, by taking the variation of the equation $\langle T^\dagger f, g\rangle_0 = \langle f, Tg\rangle_1$, that

$$(29.32) \qquad \delta T^\dagger = T^\dagger \delta B_1 - (\delta B_0)T^\dagger.$$

From these two equations we get

$$(29.33) \qquad \delta\square_1 = -(\delta B_1)S + TT^\dagger \delta B_1 - T\delta B_0 T^\dagger,$$
$$(29.34) \qquad \delta\square_0 = T^\dagger \delta B_1 T - \delta B_0 T^\dagger T.$$

Using the relations

$$(29.35) \qquad \exp(-t\square_1)T = T\exp(-t\square_0),$$
$$(29.36) \qquad \exp(-t\square_0)T^\dagger = T^\dagger \exp(-t\square_0),$$
$$(29.37) \qquad \exp(-t\square_1)S = S\exp(-t\square_1),$$

which follow from the equations $\square_1 T = T\square_0$, $\square_0 T^\dagger = T^\dagger\square_1$ and $\square_1 S = S\square_1$, we obtain

$$
\begin{aligned}
(29.38) \quad &\delta(-\tfrac{1}{2}\mathrm{tr}(\exp(-t\square_1) + \mathrm{tr}\exp(-t\square_0))) \\
&= \tfrac{1}{2}t\,\mathrm{tr}((-\delta B_1 S + TT^\dagger \delta B_1 - T\delta B_0 T^\dagger)\exp(-t\square_1)) \\
&\quad - t\,\mathrm{tr}((T^\dagger \delta B_1 T - \delta B_0 T^\dagger T)\exp(-t\square_0)) \\
&= \tfrac{1}{2}t\,\mathrm{tr}((-\delta B_1 S - \delta B_1 TT^\dagger)\exp(-t\square_1)) + \tfrac{1}{2}t\,\mathrm{tr}(\delta B_0 T^\dagger T\exp(-t\square_0)) \\
&= \tfrac{1}{2}t\frac{d}{dt}(\mathrm{tr}(\delta B_1 \exp(-t\square_1)) - \mathrm{tr}(\delta B_0 \exp(-t\square_0))).
\end{aligned}
$$

Now we note that

$$(29.39) \quad \delta(-\tfrac{1}{2}\ln\det_\varepsilon \square_1 + \ln\det_\varepsilon \square_0) = -\int_\varepsilon^\infty \delta(-\tfrac{1}{2}\mathrm{tr}\exp(-t\square_1) + \mathrm{tr}\exp(-t\square_0))\frac{dt}{t}.$$

Substituting (29.38) into (29.39), we obtain the integral of a derivative, which we then evaluate, completing the reduction of the study of the variation to the study of the asymptotic behavior of the integrand.

We now apply (29.29) to the study of the variations of the partition function (29.27) under conformal changes in the metric, with ∇_1^2 and ∇_0^2 playing the role of $\square_1$ and $\square_0$. An infinitesimal conformal variation of the metric, $g_{\mu\nu} \mapsto (1 + \sigma(x))g_{\mu\nu}$, gives rise to variations in the scalar products on the spaces Γ_1 and Γ_0 of one-forms and functions, given respectively by $\delta B_1 = \hat{\sigma}$ and $\delta B_0 = 2\hat{\sigma}$, where $\hat{\sigma}$ is the operator of multiplication by the function $\sigma(x)$. We see that

$$\delta \ln Z = \tfrac{1}{2}\beta(\hat{\sigma}|\square_1) - \tfrac{1}{2}\beta(2\hat{\sigma}|\square_0)$$
$$= \tfrac{1}{2}(\Psi_0(\hat{\sigma}|\square_1) - \mathrm{tr}(\hat{\sigma}\Pi(\square_1))) - (\Psi_0(\hat{\sigma}|\square_0) - \mathrm{tr}(\hat{\sigma}\Pi(\square_0))).$$

The number of zero modes of Δ_i, as we know, is the i-th Betti number b^i of the manifold M. If M is acyclic in dimension 1, we have $b^1 = 0$ and $b^0 = 1$. The zero modes of ∇_0^2 are the constant functions. Clearly,

$$\mathrm{tr}\,\hat{\sigma}\Pi(\nabla_0^2) = \frac{1}{V}\int \sigma(x)\,dV = \tfrac{1}{2}\delta \ln V,$$

where V is the volume of M. Using the expression for the Seeley coefficients demonstrated in Chapter 27, we get

$$\delta \ln\left(\frac{Z}{\sqrt{V}}\right) = -\frac{1}{180(4\pi)^2}\int dV\,\sigma(\tfrac{13}{2}R_{\mu\nu\rho\sigma}R^{\mu\nu\rho\sigma} - 44R_{\mu\nu}R^{\mu\nu} + \tfrac{25}{2}R^2 - 9\nabla_0^2 R).$$

We have considered here only four-dimensional theories as examples. However, the general theorems proved above can be used to obtain important results in other dimensions as well. In particular, they are very useful in the explicit calculation of two-dimensional determinants arising in string theory and in two-dimensional nonlinear theories.

Consider, for example, the determinant of the scalar Laplacian Δ_0 on the two-dimensional sphere S^2 with an arbitrary Riemannian metric $g_{\mu\nu}$. The determinant of this operator is connected with the partition function corresponding to the Euclidean action functional

$$(29.40) \qquad S_g(\varphi) = \frac{1}{2}\int g^{\mu\nu}\partial_\mu\varphi\partial_\nu\varphi\,dV.$$

This action functional remains unchanged if we replace the metric $g_{\mu\nu}(x)$ by $\rho(x)g_{\mu\nu}(x)$, but the inner product $\langle\varphi,\varphi'\rangle$ changes. We can study the corresponding change in $Z = (\det\Delta_0)^{-1/2}$ by means of (29.18). The variation in Z under an infinitesimal variation $\delta g_{\mu\nu}(x) = \delta\rho(x) = g_{\mu\nu}(x)$ is given by

$$(29.41) \qquad \delta \ln Z = \frac{1}{48\pi}\int \delta \ln \rho(x)R(x)\,dV - \frac{1}{2V}\int \delta \ln \rho(x)\,dV.$$

(Here we used (27.29) and the fact that Δ_0 has one zero mode $\varphi = V^{-1/2}$, where V is the area of S^2 with respect to the metric $g_{\mu\nu}$.) Every two-dimensional Riemannian manifold homeomorphic to S^2 is conformally equivalent to S^2 with the standard metric $\bar{g}_{\mu\nu}$; in other words, it is isometric to S^2 with the metric $g_{\mu\nu}(x) = \rho(x)\bar{g}_{\mu\nu}(x)$. Integrating (29.41) and using the relations

$$(29.42) \qquad \bar{R} = \rho(R - \Delta \ln \rho)$$

and

$$(29.43) \qquad R = \rho^{-1}(\bar{R} + \bar{\Delta} \ln \rho),$$

where R and Δ denote the curvature and the Laplacian with respect to the metric $g_{\mu\nu}$ and $\bar{R}$, $\bar{\Delta}$ are the corresponding quantities for the metric $\bar{g}_{\mu\nu}$, we obtain an expression for Z.

An interesting three-dimensional example is given by the action

$$(29.44) \qquad S(A) = \frac{1}{2} \int A \wedge dA = \int \varepsilon^{\alpha\beta\gamma} A_{\alpha} \partial_{\beta} A_{\gamma} \, d^3 x,$$

where A denotes a one-form on the compact three-dimensional manifold M. This functional is invariant with respect to gauge transformations $A \mapsto A + d\lambda$, where λ is an arbitrary function (0-form). It is also independent of the choice of a Riemannian metric on M; therefore one can conjecture that the corresponding partition function Z also does not depend on the metric. This is not obvious (since the calculation of Z involves a metric-dependent gauge-condition), but it turns out to be true. The proof can be based on a slight modification of the results above.

Namely, note that we can apply in this case the general formula (29.28) for the calculation of the partition function, but it is not convenient to use (29.28′) because $S + TT^*$ is not elliptic. We will represent the partition function in the form

$$(29.45) \qquad Z = (\det(S^2 + TT^{\dagger}))^{-1/4}(\det(T^{\dagger}T))^{3/4},$$

which is equivalent to (29.28) and useful when $S^2 + TT^{\dagger}$ and $T^{\dagger}T$ are elliptic. For the functional (29.44), $T^{\dagger}T$ coincides with the scalar Laplacian Δ_0, and $S^2 + TT^{\dagger}$ is the Laplacian $\Delta_1 = -(dd^{\dagger} + d^{\dagger}d)$ on the space of one-forms; we obtain

$$(29.46) \qquad Z = (\det \Delta_1)^{-1/4}(\det \Delta_0)^{3/4}.$$

The arguments used in the proof of (28.29), applied to the partition function represented in the form (29.45), lead to the formula

$$(29.47) \qquad \delta \ln Z = \tfrac{1}{2}\beta(\delta B_1 | S^2 + TT^{\dagger}) - \tfrac{1}{2}\beta(\delta B_0 | T^{\dagger}T).$$

This formula allows us to calculate the variation of (29.46) under an infinitesimal change in the metric of M.

For the functional (29.44), the Seeley coefficients Ψ_0 vanish. (Ψ_0 vanishes for every elliptic differential operator on an odd-dimensional manifold.) Hence $\delta \ln Z$ can be expressed in terms of the zero modes of Δ_1 and Δ_0. It is well-known that the number of zero modes of δ_1 coincides with the i-th Betti number b_i of M. Suppose that $b_1 = 0$ and that M is connected, so that $b_0 = 1$ and Δ_0 has only one zero mode $\varphi_0 = V^{-1/2}$, where V is the volume of M. We get

$$\delta \ln Z = \tfrac{1}{2}\langle \delta B_0 \varphi_0, \varphi_0 \rangle = \tfrac{1}{2} \ln \delta V.$$

This means that the expression

$$(29.48) \qquad V^{-1/2}Z = V^{-1/2}(\det \Delta_1)^{-1/4}(\det \Delta_0)^{3/4}$$

does not depend on the Riemannian metric on M. We thus obtain a nontrivial invariant of the three-dimensional smooth manifold M. A somewhat more general invariant can be obtained if we regard (29.44) as a functional on one-forms taking values in the fibers of a locally flat vector bundle over M.

The invariants just described constitute what is known in mathematics as the Ray–Singer torsion, which is a smooth version of the Reidemeister torsion. It can be defined for any m-dimensional manifold M by

$$(29.49) \qquad \log \operatorname{Tor} M = \frac{1}{2} \sum_{0 \leq k \leq m} (-1)^k k \ln \det \Delta_k,$$

where $\Delta_k = -(dd^\dagger + d^\dagger d)$ denotes the Laplacian on the space of k-forms. This multidimensional torsion, too, can be obtained from considerations based on quantum field theory. One can consider, for example, the functional (29.44), where A denotes an n-form on a $(2n + 1)$-dimensional compact manifold. (In this case we cannot use the standard Faddeev–Popov trick to calculate the partition function, because the gauge group $A \mapsto A + d\lambda$, where λ is an $(n-1)$-form, does not act freely. However, the definition of the partition function can be modified to cover this case; see [55].)

The idea of obtaining invariants of different mathematical objects by considering physical quantities (the partition function or correlation functions) arising from appropriate action functionals is quite general. It is the central idea of topological quantum field theory. One of its most important generalizations is connected with a nonabelian generalization of (29.44).

The one-form $A = A_\alpha dx^\alpha$ in (29.44) can be considered as an electromagnetic field on the three-dimensional manifold M; the equations of motion corresponding to (29.44) have the form

$$(29.50) \qquad \varepsilon^{\alpha\beta\gamma} F_{\beta\gamma} = 0,$$

where $F_{\alpha\beta}$ denotes the electromagnetic field strength: $F_{\alpha\beta} = \partial_\alpha A_\beta - \partial_\beta A_\alpha$. We would like to replace the electromagnetic field by a gauge field A that takes values in the Lie algebra $\mathfrak{g}$ of a compact Lie group G, and construct an action functional leading to the equations of motion (29.50), where $F_{\alpha\beta} = \partial_\alpha A_\beta - \partial_\beta A_\alpha + [A_\alpha, A_\beta]$ is the strength of the gauge field. It is easy to check that such an action functional can be written as

$$(29.51) \qquad S(A) = \int \varepsilon^{\alpha\beta\gamma} A_\alpha^a F_{\beta\gamma}^a \, d^3x - \frac{1}{3} \int \varepsilon^{\alpha\beta\gamma} f_{abc} A_\alpha^a A_\beta^b A_\gamma^c \, d^3x,$$

where A_α^a and $F_{\beta\gamma}^a$ are components of A_α and $F_{\beta\gamma}$ with respect to an orthonormal basis in $\mathfrak{g}$, and f_{abc} denotes the structure constants of $\mathfrak{g}$ with respect to this basis. (We have fixed an invariant scalar product on $\mathfrak{g}$.)

This action functional, called the *Chern–Simons functional*, is invariant with respect to infinitesimal gauge transformations $A_\mu \mapsto A_\mu + \nabla_\mu \lambda$, where λ is a function taking values in $\mathfrak{g}$, and $\nabla_\mu \lambda = \partial_\mu \alpha + [A_\mu, \lambda]$. It does not depend on the metric on M; therefore its partition function should give an invariant of the smooth structure on M. The same is true for appropriately chosen correlation functions. The physical quantities associated with the action functional (29.51) were calculated by Witten [79], who proved that these quantities are closely related to the Jones polynomial of knots.

30. Instantons

We have seen that several quantities in quantum field theory can be expressed in terms of functional integrals whose integrands contain either $\exp(iS)$, where S is the action integral, or $\exp(-S_{\text{eucl}})$, where S_{eucl} is the Euclidean action integral. For the approximate evaluation of these integrals, one can use the stationary-phase method in the first case, and the Laplace method in the second. The application of these methods can be seen as an application of the semiclassical approximation. Indeed, if we momentarily abandon our convention that $\hbar = 1$, so the exponentials involve $i\hbar^{-1}S$ and $\hbar^{-1}S_{\text{eucl}}$ instead of iS or S_{eucl}, we can establish the asymptotic behavior as $\hbar \to 0$ by applying the stationary-phase method or the Laplace method.

We will discuss here the Laplace method, whose first step is to find a global minimum for S_{eucl}. The next step is to replace S_{eucl} near this minimum with its quadratic part, and then calculate the resulting Gaussian integral.

The same procedure can be applied to local minima of S_{eucl} as well. At first glance the contribution to the functional integral near a local minimum is insignificant, because it is dampened by the factor $\exp(-\hbar^{-1}\Delta S)$, where ΔS is the difference between the global and the local minimum. But this contribution is nonetheless interesting, because it leads to corrections that cannot be obtained by perturbation theory. In any case, for certain physical quantities the contribution of the local minima can be decisive.

The local minima of S_{eucl} are known as *instantons*. Wee are mainly interested in instantons in gauge theories, but we start our discussion with instantons in a simple quantum-mechanical model. Consider the Hamiltonian

$$(30.1) \qquad H = \tfrac{1}{2}p^2 + \lambda(q^2 - a^2)^2$$

and the corresponding Euclidean action

$$(30.2) \qquad S_{\text{eucl}} = \int \left(\frac{1}{2}\left(\frac{dq}{d\tau}\right)^2 + \lambda(q^2 - a^2)^2 \right) d\tau.$$

The potential energy $U(q) = \lambda(q^2 - a^2)^2$ has two minima, $q = \pm a$, and thus there are two classical ground states, or classical vacuums. The same minima (more precisely, the constant functions $q(\tau) \equiv a$ and $q(\tau) \equiv -a$) minimize the functional (30.2). As explained in Chapter 9, a classical vacuum can be used to find a quantum ground state by first replacing the potential energy with its quadratic part, and then allowing for deviations from the quadratic approximation by using perturbation techniques. Using this procedure, we can obtain from the two classical vacuums two ground states for the Hamiltonian (30.1), both having the same energy.

It is well known, however, that the energy levels of the one-dimensional Hamiltonian (30.1) are non-degenerate. This apparent contradiction is resolved if we notice that, near each classical vacuum, we have not a stationary state but only a quasistationary one, thanks to the possibility of tunneling from one well to the other. Formally, this translates into the fact that the perturbation series diverges, and the energy of the ground state is not analytic in $\hbar$. The possibility of tunneling leads to a situation in which, instead of two ground states that can be transformed into one another by replacing q with $-q$, we have a ground state with an even wave function and an excited state with an odd wave function. The energies of the two states differ by an exponentially small amount

$$(30.3) \qquad \Delta E = \hbar\omega\sqrt{\frac{2\omega^3}{\pi\lambda\hbar}}\,\exp\left(-\frac{\omega^3}{12\lambda\hbar}\right),$$

where $\omega = 2a\sqrt{2\lambda}$.

We now revert to our usual system of units, where $\hbar = 1$, and investigate the Hamiltonian (30.1) using the Euclidean action integral. The essential contribution to this functional integral come from the global minima of the Euclidean action, corresponding to the classical vacuums $q(\tau) \equiv \pm a$. But, in addition to these, the functional (30.2) also has local minima, with value $\frac{4}{3}\sqrt{2\lambda}\,a^3$ and achieved by the functions

$$(30.4) \qquad q(\tau) = \pm s(\tau - c),$$

where

$$(30.5) \qquad s(\tau) = a\tanh(a\sqrt{2\lambda}\,\tau).$$

We are considering that the functions $q(\tau)$ in (30.2) are defined on the whole real line. Strictly speaking, we should first assume that these functions are defined on a finite interval, and then make the length of the interval go to infinity.

Notice that the functional (30.2) has appeared before, in a different notation, as the energy functional (9.4) of the one-dimensional field theory with Lagrangian (9.3); in that context the functions (30.4) have the meaning of topologically nontrivial local minima of the energy functional. They are global minima in the class of topologically nontrivial functions (functions that have different limits at $+\infty$ and $-\infty$), and thus they are instantons. To distinguish between them, we call one—say, the one with the + sign in (30.4)—an instanton, and the other an *anti-instanton*. Using the functions (30.4), which are stationary points of the functional (30.2), we can construct almost-stationary points by matching instantons and anti-instantons, that is, by setting, say,

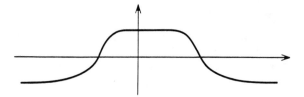

Figure 17

$$q(\tau) = \begin{cases} s(\tau - c_1) & \text{for } \tau \le b, \\ -s(\tau - c_2) & \text{for } \tau \ge b, \end{cases}$$

where $c_1 \ll b \ll c_2$ (Figure 17).

Similarly, one can construct almost-stationary points of the functional S_{eucl} consisting of several alternating instantons and anti-instantons. These are sometimes called *multi-instanton solutions*. When computing the functional integral, one must take into account all these stationary and almost-stationary points of S_{eucl}. It turns out that the answer can be reproduced by means of elementary quantum mechanical methods. We will not detail these calculations; we note only that the value of S_{eucl} on the functions (30.4) coincides with the exponent in (30.3).

The physical meaning of instantons is obvious from the discussion above. They are solutions of the classical equations of motion with an imaginary time variable and describe the tunneling between potential wells (here between $q = -a$ and $q = a$). In particular, if one wants to find the transition probability from one classical vacuum to the other (the matrix entry $\langle -a| \exp(-H\tau)|a\rangle$ as $\tau \to \infty$), one must consider instantons, because the topologically trivial solutions do not contribute to the corresponding functional integral. Classical vacuums only mix because of instantons.

Of course, the use of instantons in the example above is only of methodological interest, since the problem can be solved by essentially simpler methods. But instantons also arise in more complicated problems, and then it makes sense to use them. Such is the case, for example, with the two-dimensional Georgi–Glashow model (Chapter 11). The Euclidean action integral for this two-dimensional model coincides, to within notation, with the energy functional for the three-dimensional Georgi–Glashow model. This means that the local minima of the energy functional can be interpreted as instantons in the two-dimensional model. A topologically nontrivial minimum of the energy functional (magnetic monopoles) can be interpreted as a *topologically nontrivial instantons*, that is, an instanton that cannot be connected with a classical vacuum by means of a continuous family of fields having finite Euclidean action.

We now turn to instantons in gauge theories. We start by studying the space of gauge fields in $\mathbf{R}^4$ for which the Euclidean action integral (25.2) is finite. At infinity, such fields resemble purely gauge fields, that is, they have the asymptotics $g^{-1}(x)\partial_\mu g(x)$ as $x \to \infty$, where $g(x)$ is a function with values in the gauge group.

To prove this statement rigorously, we must insist that the fields behave well at infinity. Then the finiteness of the Euclidean action integral implies that $\mathcal{F}_{\mu\nu}(x)$ tends to zero fast enough, so that the field can be considered to be a gauge field to within a specified precision.

To each field A_μ that has the asymptotic behavior $g^{-1}(x)\partial_\mu g(x)$ as $x \to \infty$ we can assign a homotopy class of maps from a sphere of very large radius into the gauge group G, the class being realized by the function g.

We now assume that G is a *simple, non-abelian, compact Lie group*. Then the set of homotopy classes of maps $S^3 \to G$ is in one-to-one correspondence with the integers:

$$(30.6) \qquad \{S^3, G\} = \pi_3(G) = \mathbf{Z}$$

(T14.2). Thus, a field with a finite action integral has an associated integer, its topological number. (If $G = \mathrm{SU}(2)$, which is homeomorphic to S^3, this integer is the degree of the mapping $S^3 \to S^3$.) The topological number does not change under continuous variations of the field, so long as the Euclidean action integral remains finite. Thus, the space of gauge fields with finite Euclidean action integral splits into components, one for each topological number. It turns out that the topological number q of a field A_μ can be expressed analytically by

$$(30.7) \qquad q = \frac{1}{64\pi^2} \int d^4x \, \varepsilon^{\mu\nu\rho\sigma} \langle \mathcal{F}_{\mu\nu}, \mathcal{F}_{\rho\sigma} \rangle = \frac{1}{32\pi^2} \int d^4x \, \langle \mathcal{F}_{\mu\nu}, \tilde{\mathcal{F}}^{\mu\nu} \rangle,$$

where $\tilde{\mathcal{F}}^{\mu\nu} = \frac{1}{2}\varepsilon^{\mu\nu\rho\sigma}\mathcal{F}_{\rho\sigma}$ is the antisymmetric tensor dual to $\mathcal{F}_{\rho\sigma}$, and the angle brackets denote the invariant scalar product on the Lie algebra $\mathcal{G}$, normalized as explained in T14.2. (For $G = \mathrm{SU}(n)$ we have $\langle a, b \rangle = -\operatorname{tr} ab$.)

We show (30.7) for $G = \mathrm{SU}(2)$; for the general case, see T15.4. If $G = \mathrm{SU}(2)$, an element of the Lie algebra can be seen as a three-dimensional vector; in particular, $A_\mu = (A_\mu^1, A_\mu^2, A_\mu^3)$ and $\mathcal{F}_{\mu\nu} = (F_{\mu\nu}^1, F_{\mu\nu}^2, F_{\mu\nu}^3)$. It is easy to verify that

$$(30.8) \qquad \langle \mathcal{F}_{\mu\nu}, \tilde{\mathcal{F}}^{\mu\nu} \rangle = F_{\mu\nu}^a \tilde{F}_a^{\mu\nu} = \partial_\mu K^\mu,$$

where

$$(30.9) \quad K^\mu = 2\varepsilon^{\mu\nu\rho\sigma} A_\nu^a (\partial_\rho A_\sigma^a + \tfrac{1}{3}\varepsilon_{abc} A_\rho^b A_\sigma^c) = \varepsilon^{\mu\nu\rho\sigma}(A_\nu^a F_{\rho\sigma}^a - \tfrac{1}{3}\varepsilon_{abc} A_\nu^a A_\rho^b A_\sigma^c).$$

Using this, we can easily transform the four-dimensional integral (30.7) into an integral over a sphere at infinity. It is easy to check that for a field A_μ that behaves at infinity like $g^{-1}\partial_\mu g$, the integral over this sphere reduces to the standard expression for the degree of the map $g : S^3 \to \mathrm{SU}(2)$, where $\mathrm{SU}(2)$ is identified with S^3. The terms containing $F_{\rho\sigma}^a$ in the right-hand side of (30.9) can be ignored at infinity.

We now use (30.7) to find a lower bound for the action integral for fields with a fixed topological charge q. We have

$$(30.10) \qquad 0 \leq \int d^4x \, (\mathcal{F}_{\mu\nu} \pm \tilde{\mathcal{F}}_{\mu\nu})^2 = 2 \int d^4x \, ((\mathcal{F}_{\mu\nu})^2 \pm \langle \mathcal{F}_{\mu\nu}, \tilde{\mathcal{F}}^{\mu\nu} \rangle).$$

From this it follows that

$$(30.11) \qquad S \geq \frac{8\pi^2}{g^2} |q|,$$

and clearly equality holds if and only if

$$(30.12) \qquad \mathcal{F}_{\mu\nu} = \tilde{\mathcal{F}}_{\mu\nu} \quad \text{for } q \geq 0$$

or

$$(30.13). \qquad \mathcal{F}_{\mu\nu} = -\tilde{\mathcal{F}}_{\mu\nu} \quad \text{for } q \leq 0$$

Fields satisfying (30.12) are called *self-dual*, and those satisfying (30.13) are *anti-self-dual*. Obviously, self-dual and anti-self-dual fields are extremals of the Euclidean action integral. (We could also have used Bianchi's identity (T15.2) to prove directly that (30.12) and (30.13) imply Euler's equations for the functional (25.2), that is, the Euclidean equations of motion.)

We will see that there exist self-dual fields with any non-positive charge q. This implies that the smallest value of the Euclidean action integral on fields with topological charge q is exactly $(8\pi^2/g^2)|q|$.

To construct a self-dual field with $q = 1$ for $G = \mathrm{SU}(2)$, we recall that the map

$$(30.14) \qquad n \mapsto u(n) = n^0 + i \sum_{j=1}^{3} \sigma^j n^j,$$

where $n = (n^0, n^1, n^2, n^3) \in \mathbf{R}^4$ and $\sigma^1, \sigma^2, \sigma^3$ are the Pauli matrices, is a homeomorphism between S^3 and $\mathrm{SU}(2)$, and therefore has degree 1. Hence it is natural to look for a solution of the form

$$(30.15) \qquad A_\mu = \alpha(|x|) u^{-1}\left(\frac{x}{|x|}\right) \partial_\mu u\left(\frac{x}{|x|}\right).$$

For $\alpha(x) = \lambda^2(x^2 + \lambda^2)^{-1}$, where λ is an arbitrary real number, we indeed get a solution for (30.12). Translations preserve self-duality, so we obtain a five-parameter family of solutions to (30.12), with $q = 1$:

$$(30.16) \qquad A_\mu = \frac{\lambda^2}{(x-a)^2 + \lambda^2} u^{-1}\left(\frac{x-a}{|x-a|}\right) \partial_\mu u\left(\frac{x-a}{|x-a|}\right),$$

where u is defined by (30.14).

Under a spatial reflection a self-dual field is transformed into an anti-self-dual field, so (30.16) gives a five-parameter family of solutions to (30.13) with $q = -1$. It turns out that the fields just described exhaust (to within gauge equivalence) all solutions with $q = \pm 1$. Self-dual fields with $q = 1$ are often called *instantons*, and self-dual fields with $q > 1$ are called q-instanton solutions. Anti-self-dual fields with $q = -1$ are called anti-instantons. (However, according to the definition given at the beginning of this chapter, all these fields are instantons.)

The parameter λ gives the size of an instanton (or anti-instanton), and the vector a the position of its center.

To construct self-dual fields with $q = 1$ in the case $G = \mathrm{SU}(N)$, we need only note that there is a natural inclusion of $\mathrm{SU}(2)$ into $\mathrm{SU}(N)$—for an element of $\mathrm{SU}(2)$ given by a 2×2-matrix A, we simply form the block matrix with A and I_{n-2} as blocks along the diagonal, where I_{n-2} is the rank-$(n-2)$ identity matrix. Likewise, the Lie algebra of $\mathrm{SU}(2)$ is naturally identified with a Lie subalgebra of $\mathrm{SU}(N)$. We now interpret (30.16) as a field taking values in the Lie algebra of $\mathrm{SU}(N)$; this field is clearly self-dual, with $q = 1$. Every solution to the self-duality equation for $G = \mathrm{SU}(n)$ is gauge-equivalent to a field of the form (30.16).

We can construct in an analogous way instantons with $q = 1$ for G an arbitrary simple non-abelian compact Lie group. It is enough to use the existence of an inclusion of $\mathrm{SU}(2)$ into G that gives rise to an isomorphism between $\pi_3(\mathrm{SU}(2)) = \mathbf{Z}$ and $\pi_3(G)$ (see T14.2). Using such an inclusion, we can interpret (30.16) as a solution to the self-duality equation for G, with $q = 1$.

A broader class of self-dual fields can be obtained using the substitution

$$(30.17) \qquad A_a^0 = \partial_a \ln f \quad \text{and} \quad A_a^i = (\varepsilon_{aik}\partial_k - \delta_{ai}\partial_0) \ln f \text{ for } i, k = 1, 2, 3,$$

where f is an arbitrary function satisfying the four-dimensional Laplace equation $(\partial_0^2 + \nabla^2)f = 0$. In particular, if we set

$$(30.18) \qquad f(x) = \sum_{i=1}^{q+1} \frac{\lambda_i^2}{(x - x_i)^2},$$

we get a self-dual field having singularities at $x_1, \ldots, x_{q+1}$. It is easy to see that these singularities can be removed by a gauge transformation, so that the field has a finite strength and a finite Euclidean action integral, and has topological charge q.

Up to this point we have studied instantons in gauge theories in Euclidean space. It is also natural to consider the extremals of the Euclidean action integral on a Riemannian manifold. As noted earlier, in defining functional integrals for the Euclidean Green's functions we must introduce spatial cutoff. If we impose periodic boundary conditions, spatial cutoff becomes equivalent to considering gauge fields on a four-dimensional torus. Passing to fields on the four-sphere S^4, too, can be seen as a form of spatial cutoff. The reasoning in the preceding paragraphs can be applied to instantons on general Riemannian manifolds, with the difference that the tensor dual to $\mathcal{F}_{\alpha\beta}$ should be taken as

$$(30.19) \qquad \tilde{\mathcal{F}}^{\alpha\beta} = \tfrac{1}{2}g^{-1/2}\varepsilon^{\alpha\beta\gamma\delta}\mathcal{F}_{\gamma\delta},$$

where g, as usual, denotes the determinant of the metric tensor $g_{\alpha\beta}$. Note that (30.12) and (30.13) are invariant under a conformal transformation of the metric, $g_{\alpha\beta}(x) \mapsto g'_{\alpha\beta}(x) = \rho(x)g_{\alpha\beta}(x)$.

Using stereographic projection from the four-sphere S^4 to $\mathbf{R}^4$, we can associate to any gauge field in $\mathbf{R}^4$ a gauge field on S^4, defined in the complement of the north pole. Under the condition, imposed above, that our fields on $\mathbf{R}^4$ behave at infinity like pure gauge fields $g^{-1}(x)\partial_\mu g(x)$, the corresponding fields on S^4 minus the north pole can be continuously extended to the whole sphere (for more details, see T15.3). For this reason the study of gauge fields on $\mathbf{R}^4$ having finite Euclidean action integral can be conveniently reduced to the study of the corresponding fields on S^4. Moreover the metrics of S^4 and $\mathbf{R}^4$ are conformally equivalent under stereographic projection, so the question of self-duality (or anti-self-duality) is also equivalent in $\mathbf{R}^4$ and S^4.

To specify a gauge field on S^4, we must give two gauge fields $A_\mu^{(1)}$ and $A_\mu^{(2)}$ in two coordinate systems whose domains together cover the whole sphere, and the fields must be gauge-equivalent in the intersection of the domains of the coordinate systems. Geometrically, a gauge field on S^4 is a section of the fiber bundle obtained from two direct products by pasting them together using the function that establishes the gauge equivalence between $A_\mu^{(1)}$ and $A_\mu^{(2)}$. The topological type of this bundle uniquely defines the topological number of the global gauge field (T15.5).

In Chapter 31 we show that for $G = \mathrm{SU}(2)$ and $q \geq 1$ there exists a $(8q-3)$-dimensional family of gauge-inequivalent self-dual fields on S^4, and hence on $\mathbf{R}^4$, with topological number q. Similarly, for $q \leq -1$ the dimension of the space of gauge-inequivalent anti-self-dual fields is $8|q| - 3$.

To clarify the physical meaning of the number $8q - 3$, we consider a field that coincides with a purely gauge field far from the points $a_1, \ldots, a_q$, and near these points has the form

$$(30.20) \qquad A_\mu(x) = V_i \frac{\lambda_i^2}{(x - a_i)^2 + \lambda^2} u^{-1}\left(\frac{x - a_i}{|x - a_i|}\right) \partial_\mu u\left(\frac{x - a_i}{|x - a_i|}\right) V_i^{-1},$$

with $V_i \in \mathrm{SU}(2)$, and $|x - a_i| \ll |a_i - a_j|$ and $\lambda_i \ll |a_i - a_j|$. (Roughly speaking, such a field consists of q distant single-instanton solutions (30.16), each affected by a gauge rotation with a matrix $V_i \in \mathrm{SU}(2)$). This field is an approximate solution for the self-duality equation. It depends on $8q$ parameters ($4q$ coming from the a_i, plus q from the λ_i, plus $3q$ from the V_i), but three of them are inessential, since multiplication of all the V_i by the same matrix $V \in \mathrm{SU}(2)$ does not affect the field up to gauge equivalence. It can be shown that the exact solutions to the self-duality equations lie near these approximate solutions. This result is by no means trivial, and in fact one can easily construct approximate extremals for the Euclidean action by combining distant instanton and anti-instanton solutions, without there being any corresponding exact extremals.

If $G = \mathrm{SU}(N)$, for $N > 2$, we can use the inclusion $\mathrm{SU}(2) \subset \mathrm{SU}(N)$ as explained above and look at (30.20) as an approximate solution to the self-duality equation. In fact, we can consider each V_i in (30.20) as an element of $\mathrm{SU}(N)$, so the equation represents a field consisting of q distant single-instanton solutions, each affected by a gauge rotation with a matrix $V_i \in \mathrm{SU}(N)$. The resulting set of approximate solutions to the self-duality equation has $4Nq$ parameters. Indeed, each V_i contributes $N^2 - 1$ parameters, since $\dim \mathrm{SU}(N) = N^2 - 1$. But every matrix V taken from $\mathrm{SU}(2) \subset \mathrm{SU}(N)$ commutes with an $(N-2)^2$-dimensional family of matrices (block matrices with the rank-two identity matrix at the top left and an arbitrary $(N-2) \times (N-2)$ matrix at the bottom right). Thus, although at first glance the field (30.20) depends on $(N^2 - 1)q + 5q$ parameters, we must subtract $(N-2)^2 q$, leaving $4Nq$. It can be shown that for each such approximate solution there is an exact solution near it, so the space of solutions to the self-duality equation with topological number q has dimension $4Nq$. Furthermore, every solution to the self-duality equation with topological number q can be transformed into one of the fields obtained in this way by means of a gauge transformation in G_0^∞. (Recall that G_0^∞ is the group of gauge transformations determined by functions $g(x)$ that equal unity at a fixed point x_0. On $\mathbf{R}^4$ we generally take $x_0 = \infty$, so the condition is that $g(x) \to 1$ as $x \to \infty$).

Our restriction to gauge transformations in G_0^∞ means that we have excluded global gauge transformations. If we think of gauge-equivalent fields as being identical, the dimension of the space of solutions to the self-duality equation decreases. If the topological number q is at least $\frac{1}{2}N$, this dimension, which we denote by r_q, is $4Nq - (N^2 - 1)$. Indeed, let H_A be the group of gauge transformations that fix a generic instanton A. By identifying together instantons that are G^∞-equivalent rather than only those that are G_0^∞-equivalent, the number of parameters for A drops by $\dim G - \dim H_A$. One can verify that, for $q \geq \frac{1}{2}N$, only the trivial gauge transformation fixes A, so $\dim H_A = 0$ and the dimension of the space of solutions drops by $\dim G = \dim \mathrm{SU}(N) = N^2 - 1$. For $1 \leq q < \frac{1}{2}N$ one can show that A is gauge-equivalent to an instanton with values in the Lie algebra of $\mathrm{SU}(2q) \subset \mathrm{SU}(N)$, so that $\dim H_A = (N - 2q)^2$, the dimension of the subgroup of $\mathrm{SU}(N)$ consisting of elements that commute with $\mathrm{SU}(2q)$ (this subgroup is isomorphic to $U(N - 2q)$). We conclude that in this case $r_q = 4q^2 + 1$.

We now show that instantons in a gauge theory can also be associated with tunneling between classical vacuums. We use the gauge condition $A_0 = 0$. Then a field of lowest energy (a classical vacuum) is given by $A_i(\mathbf{x})$ such that $\mathcal{F}_{ij} = 0$ for $i, j = 1, 2, 3$. Such a field is purely a gauge field, that is, it can be written as $A_i(\mathbf{x}) = g^{-1}(\mathbf{x}) \partial_i g(\mathbf{x})$. We will assume that the function $g(\mathbf{x})$ has a limit as $\mathbf{x} \to \infty$:

$$\text{(30.21)} \qquad\qquad \lim_{\mathbf{x} \to \infty} g(\mathbf{x}) = g_0.$$

Then $g(\mathbf{x})$ can be considered as a map from the sphere $S^3 = \mathbf{R}^3 \cup \{\infty\}$ into G.

Condition (30.21) can be explained by passing to the limit as the size of the sphere goes to infinity. It corresponds to imposing zero boundary conditions on a field in a finite volume. Periodic boundary conditions would correspond to a function $g(\mathbf{x}) : T^3 \to G$ on the three-torus.

We can associate to every function $g(\mathbf{x})$ satisfying (30.21), and, consequently, to every classical vacuum, a topological number, giving the homotopy class of the corresponding map $S^3 \to G$. (Here we are using the assumption that

G is a simple, non-abelian, compact Lie group, which implies that $\{S^3, G\} = \mathbf{Z}$). Instantons correspond to tunneling between classical vacuums with different topological numbers. To see this, we impose the gauge condition $A_0 = 0$ and the condition that our fields have the asymptotic behavior $A_\mu \sim g^{-1}(x)\partial_\mu g(x)$ at infinity, where $g(x) = g(\mathbf{x}, t)$ tends to 1 for $\mathbf{x} \in \mathbf{R}^3$ and $t \to -\infty$, and also for $t \in \mathbf{R}$ and $x \to \infty$; and $g(x)$ tends to $g(\mathbf{x})$ as $t \to \infty$. (These conditions can be ensured by the application of an appropriate gauge transformation.) If the instanton $A_\mu(\mathbf{x}, t)$ satisfies these conditions, it describes a tunneling process between the classical vacuums $A_\mu = 0$ and $A_\mu = g^{-1}\partial_\mu g$. In particular, (30.16) is gauge-equivalent to a field that describes tunneling between a classical vacuum of topological number 0 to one of topological number 1.

As noted at the beginning of this chapter, instantons must be taken into account when we compute the functional integrals (25.3) and (25.4) by the Laplace method. It is important, however, that in calculating the instanton contribution we can substitute a Gaussian integral for the integral over directions orthogonal to the space (manifold) of instantons, so that the Laplace method reduces the overall integral to an integral over the instanton manifold. Taking into account gauge invariance and using the Faddeev–Popov trick, one can reduce the integral over the infinite-dimensional instanton manifold to an integral over the space of gauge-inequivalent instantons, which is finite-dimensional. For instance, for $G = \mathrm{SU}(2)$ the contribution to (25.4) of instantons with topological number $q = 1$ can be written as

$$(30.22) \qquad \int \Phi(\lambda, a)\rho(\lambda, a)\, d\lambda\, da,$$

where $\Phi(\lambda, a)$ is obtained by substituting the field (30.16), with parameters λ and a, into the functional $\Phi(A)$, and

$$(30.23) \qquad d\mu = \rho(\lambda, a)\, d\lambda\, da$$

is some measure on the instanton manifold. The measure $d\mu$ can be expressed in terms of infinite-dimensional determinants. The problem of computing a measure on the instanton manifold is discussed further in Chapter 32; here we note only that for $q = 1$ the measure $d\mu$ can be derived from the requirement that the answer be renorm-equivalent.

It turns out that for $G = \mathrm{SU}(2)$ and $q = 1$ the instanton contribution is given, to within a factor, by the expression

$$(30.24) \qquad \int g^{-8}\Phi(\lambda, a)\exp\left(-\frac{8\pi^2}{g^2(\lambda)}\right)\frac{d\lambda}{\lambda^5}\, d^4a,$$

where $g(\lambda)$ is the effective coupling constant corresponding to the scale λ (or to the momentum $p = 1/\lambda$), and is given by the formula

$$(30.25) \qquad \frac{8\pi^2}{g^2(\lambda)} = \frac{8\pi^2}{g^2} - \frac{22}{3}\ln\frac{\lambda}{\lambda_0},$$

where λ_0 is the normalization point. The measure $\lambda^{-5} d\lambda\, d^4 a$ on the instanton manifold is invariant under translations and expansions—in fact, under all conformal transformations—of $\mathbf{R}^4$.

The integral (25.4) diverges, but this divergence can be eliminated if we divide (25.4) by (25.3) and renormalize the quotient. In calculating the q-instanton contribution, we divide (25.4) by the contribution to (25.3) of topologically trivial fields; this contribution is computed by the Laplace method. The factor g^{-8} in (30.24) could be replaced by $g(\lambda)^{-8}$, but this is not necessary since we are limiting ourselves to the main approximation term in the Laplace method, and the difference between g^{-8} and $g(\lambda)^{-8}$ only plays a role in the next order of approximation.

We now discuss briefly the provenance of the term g^{-8}. In the formal calculation of the integral (25.4) by the Laplace method, it turns out that integration along each direction orthogonal to the instanton manifold yields a factor g^{-1}. The total number of these factors is

$$(30.26) \qquad \dim \mathcal{E}_q - (r_q + \dim G^\infty - h_q),$$

where $\dim \mathcal{E}_q$ and $\dim G^\infty$ are the (infinite) dimensions of, respectively, the space of all fields with topological number q and the group of all gauge transformations; r_q is the number of parameters on which depends the general solution of the self-duality equation with topological number q (with gauge-equivalent fields considered as the same); and h_q is the dimension of the subgroup of G^∞ consisting of elements that fix a given instanton with topological number q. Formula (30.26) holds for any G, not just for SU(2). If $G = $ SU(2), we have $r_q = 8q - 3$ and $h_q = 0$ for $g \geq 1$.

As discussed above, in calculating the instanton contribution to (25.4) it is necessary to divide by the contribution of instantons with topological number $q = 0$ (classical vacuums) to (25.3); this contribution contains the factor g^{-1} with multiplicity $\dim \mathcal{E}_0 - \dim G^\infty + 3$ (this is formula (30.26) with $r_0 = 0$ and $h_0 = 3$). Setting $\dim \mathcal{E}_q = \dim \mathcal{E}_0$, we see that in the the q-instanton contribution the factor q^{-1} appears with multiplicity $8q$. If $G = $ SU(N), as we saw above, we have $r_q = 4Nq - (N^2 - 1)$ and $h_q = 0$ for $q \geq \frac{1}{2}N$; while for $1 \leq q < \frac{1}{2}N$ we have $r_q = 4q^2 + 1$ and $h_q = (N - 2q)^2$, and for $q = 0$ we of course have $r_0 = 0$ and $h_0 = N^2 - 1$. We conclude that the q-instanton contribution to (25.4), after the normalization discussed above, has the factor q^{-1} always with multiplicity $4Nq$. For $q = 1$ we get (30.24), where g^{-8} should be replaced with g^{-4N} and the effective coupling constant $g^2(\lambda)$ is defined by

$$(30.27) \qquad \frac{8\pi^2}{g^2(\lambda)} = \frac{8\pi^2}{g^2} - \frac{11}{3} N \ln \frac{\lambda}{\lambda_0}.$$

The single-instanton contribution (30.24) clearly diverges for $\Phi(A) \equiv 1$. This means that (30.24) can reasonably be used only when the factor $\Phi(\lambda, a)$ suppresses the contribution from large-scale instantons. This could be expected because, as λ tends to infinity, the effective coupling constant ceases to be small. But, in spite of this divergence, much effort has been put into analyzing the contribution of instantons to the partition function (25.3).

In general, as we will show in Chapter 32, the q-instanton contribution for $q \geq 1$ can be written as

$$(30.28) \qquad I_q = g^{-\sigma} \exp\left(-\frac{8\pi^2}{g^2}q\right) \int_{R_q} \Phi(A) \frac{\lambda(A)}{\lambda(0)} \, d\mu_0,$$

where the integral is over the manifold of instantons with topological number q (gauge-equivalent fields being considered as the same), the measure $d\mu_0$ comes from the natural metric on the instanton manifold, σ is a constant depending on G and equal to $4N$ for $G = \mathrm{SU}(N)$, and $\lambda(A)$ is defined as

$$(\det \Delta_1^A)^{-1/2} \det \Delta_0^A,$$

where $\Delta_0^A = d_A^\dagger d_A$ and $\Delta_1^A = \frac{1}{2}d_A^\dagger(1 - *)d_A + d_A d_A^\dagger$ denote the Laplacian operators acting, respectively, on functions and on one-forms with values in the Lie algebra of the gauge group (here d_A is the covariant derivative and $*$ denotes taking the dual). The determinants of Δ_0^A and Δ_1^A are regularized via cutoff in proper time or with the help of the ζ-function (see Chapters 23 and 28); zero modes are disregarded in this calculation.

From (30.28), using a reasoning similar to the one at the end of Chapter 28 in order to find out the dependence of the determinants on the scale of the instanton (there is no dependence on the position, by translation invariance), we can obtain an explicit expression for the single-instanton contribution. We will not work out this computation here, because Chapter 32 contains a more general discussion, based on a similar idea.

In quantizing a gauge theory we begin with the Euclidean action integral (25.2). Without changing the classical equations of motion, one can add to this functional the following term:

$$(30.29) \qquad \frac{1}{2}i\theta(\mathcal{F}, *\mathcal{F}) = \frac{1}{4}i\theta \int d^4x \, \langle \mathcal{F}_{\mu\nu}, \mathcal{F}_{\alpha\beta} \rangle \varepsilon^{\mu\nu\alpha\beta}.$$

Indeed, as we know, this so-called θ-term differs from the topological charge only by a factor, and does not change under continuous variations of the gauge field, so that it does not contribute to the equations of motion. We can derive the same result by noting that the integrand in (30.29) is a total divergence. However, after quantization the θ-term has sizable effects. This is especially evident when we calculate the instanton contribution: the presence of the θ-term in the expression for the contribution from fields with topological number q (in particular, in the expression for the q-instanton contribution) leads to an additional factor $\exp(-16\pi^2 q \theta i)$. Thus, a new parameter θ appears in the theory after quantization.

31. The Number of Instanton Parameters

The main object of this chapter is to calculate the number of parameters on which the general q-instanton solution on the sphere S^4 depends. But since most of our reasoning is valid for an arbitrary compact Riemannian manifold M, we start with this general situation.

First we repeat some of the reasoning of the previous chapter using coordinate-free notation. Let M be a four-dimensional compact oriented Riemannian manifold and G a simple, non-abelian, compact Lie group. A gauge field A on M is interpreted geometrically as a connection on the principal bundle $\xi = (E, M, G, p)$ with base M (T15.3). The strength $\mathcal{F}$ of the field A is a two-form on M taking values on the fibers of the associated vector bundle $\xi_\tau = (E_\tau, M, \mathcal{G}, p_\tau)$, where $\mathcal{G}$ is the Lie algebra of G, and G acts on $\mathcal{G}$ by the adjoint representation.

The operation of taking the dual of an antisymmetric tensor gives rise to a duality operation on forms, because of the correspondences between skew-symmetric tensors and forms; we denote this operation by $*$. For example, if $\mathcal{F} = \frac{1}{2}\mathcal{F}_{\alpha\beta}\, dx^\alpha \wedge dx^\beta$, we have $*\mathcal{F} = \frac{1}{2}\check{\mathcal{F}}_{\alpha\beta}\, dx^\alpha \wedge dx^\beta$, where $\check{\mathcal{F}}_{\alpha\beta} = \frac{1}{2}\sqrt{g}\,\varepsilon_{\alpha\beta\gamma\delta}\mathcal{F}^{\gamma\delta}$. If $a = a_\alpha\, dx^\alpha$ is a one-form, the dual is $*a = \frac{1}{6}b_{\alpha\beta\gamma}\, dx^\alpha \wedge dx^\beta \wedge dx^\gamma$, where $b_{\alpha\beta\gamma} = \sqrt{g}\,\varepsilon_{\alpha\beta\gamma\delta}a^\delta$. For more details, see T6.9.

The Euclidean action of a gauge field A is

$$(31.1) \qquad S[A] = \frac{1}{4g^2}(\mathcal{F}, \mathcal{F}) = \frac{1}{4g^2}\int \langle \mathcal{F}_{\alpha\beta}, \mathcal{F}^{\alpha\beta}\rangle\, dV,$$

that is, a constant times the scalar product of the field with itself, where the scalar product of two two-forms $f = \frac{1}{2}f_{\alpha\beta}\, dx^\alpha \wedge dx^\beta$ and $h = \frac{1}{2}h_{\alpha\beta}\, dx^\alpha \wedge dx^\beta$, with values on the fibers of the vector bundle, is given by $(f, h) = \int\langle f_{\alpha\beta}, h_{\alpha\beta}\rangle\, dV$. This definition works whenever there is a scalar product defined on the fibers and a Riemannian metric on the base manifold. The scalar product of two k-forms is defined similarly (T6.9).

The topological number of a gauge field (T15.4) can obviously be written as

$$(31.2) \qquad q = \frac{1}{32\pi^2}(\mathcal{F}, *\mathcal{F}).$$

Using the fact that $(*\mathcal{F}, *\mathcal{F}) = (\mathcal{F}, \mathcal{F})$ and $(\mathcal{F}, *\mathcal{F}) = (*\mathcal{F}, \mathcal{F})$, we get

$$(31.3) \qquad 0 \le (\mathcal{F} \pm *\mathcal{F}, \mathcal{F} \pm *\mathcal{F}) = 2(\mathcal{F}, \mathcal{F}) \pm 2(\mathcal{F}, *\mathcal{F}),$$

so that $(\mathcal{F}, \mathcal{F}) \geq (\mathcal{F}, *\mathcal{F})$, and equality holds if and only if

$$(31.4) \qquad \mathcal{F} = *\mathcal{F}.$$

We see that $S[A] \geq 8\pi^2 g^{-2} q$, and equality is achieved if and only if (3.14) is satisfied. Analogously, for $q < 0$, we get $S[A] \geq 8\pi^2 g^{-1}|q|$, with equality if and only if

$$(31.5) \qquad \mathcal{F} = -*\mathcal{F}.$$

Of course, these equations simply translate to coordinate-free notation the results of the previous chapter.

Let A be a gauge field satisfying the self-duality condition (31.4). We will look for solutions A' of (31.4) that lie near A. Since we are considering gauge-equivalent solutions as identical, we remove the degrees of freedom due to gauge equivalence by imposing on A' the condition

$$(31.6) \qquad d_A^\dagger(A' - A) = 0,$$

where $d_A^\dagger$ is the operator dual to the covariant derivative d_A (recall that the difference between two gauge fields can be seen as a one-form with values in the fibers of the vector bundle $\xi_\tau = (E_\tau, M, \mathcal{G}, p_\tau)$; see T15.2). We can think of $d_A^\dagger$ as the covariant divergence: if $a = a_\mu \, dx^\mu$, we have

$$d_A^\dagger a = -(\nabla_\mu a^\mu + [A_\mu, a^\mu]),$$

where ∇_μ is the covariant derivative with respect to the Riemannian metric.

So, suppose the field $A' = A + a$ satisfies (31.4) and (31.6). If we set $P = \frac{1}{2}(1 - *)$, the one-form a satisfies

$$(31.7) \qquad P(d_A a + \tfrac{1}{2}[a, a]) = 0$$

and

$$(31.8) \qquad d_A^\dagger a = 0,$$

where the bracket of one-forms is defined by $[a_\lambda dx^\lambda, b_\mu dx^\mu] = [a_\lambda, b_\mu] \, dx^\lambda \wedge dx^\mu$.

We now consider Equations (31.7) and (31.8), ignoring the non-linear terms in (31.7). We introduce the linear operator $\mathcal{T} = (P d_A, d_A^\dagger)$, that is, $\mathcal{T}$ maps the one-form a to the pair (f, h), where $f = P d_A a$ is an anti-self-dual two-form $(f = -*f)$ and $h = d_A^\dagger a$ is a zero-form (all these forms take values in the fibers of ξ_τ). Then (31.7) and (31.8) are equivalent to

$$(31.9) \qquad \mathcal{T} a = 0.$$

If we denote the space of one-forms by Γ^1, the space of zero-forms by Γ^0, and the space of anti-self-dual two-forms by Γ_2, the operator $\mathcal{T}$ maps Γ_1 into $\Gamma_2 \dot{+} \Gamma_0$. We show that $\mathcal{T}$ is elliptic. In local coordinates, $\mathcal{T}$ maps $a = a_\lambda \, dx^\lambda$ to (f, h), where

$$f = \tfrac{1}{2} f_{\lambda\nu} \, dx^\lambda \wedge dx^\nu \quad \text{for } f_{\lambda\nu} = \tfrac{1}{2}(\partial_\lambda a_\nu - \partial_\nu a_\lambda) - \tfrac{1}{2}\sqrt{g}\, \varepsilon_{\lambda\nu\rho\sigma} \partial^\rho a^\sigma + \cdots ,$$

$$h = \partial^\mu a_\mu + \cdots ;$$

the omitted terms do not contain derivatives, and have no effect on the principal symbol of T. Thus, the principal symbol of T is the matrix linear operator that maps a_λ to

$$(\tfrac{1}{2} i (p_\lambda a_\nu - p_\nu a_\lambda - \sqrt{g}\, \varepsilon_{\lambda\nu\rho\sigma} p^\rho a^\sigma), i p^\mu a_\mu).$$

Both the domain and range of this operator have dimension $4 \dim G$. Indeed a_λ, for each value of $\lambda = 1, 2, 3, 4$, takes values in the Lie algebra of G; while the tensor $f_{\lambda\nu}$, being skew-symmetric, has three independent components, and h supplies a fourth one. It is easy to check that the principal symbol of T is a non-degenerate linear operator, and therefore T is elliptic. (Strictly speaking, in order to talk of T being elliptic, we must consider Γ_1 and $\Gamma_2 + \Gamma_0$ as spaces of sections of certain vector bundles; we can do this by introducing local coordinates.)

As mentioned before, the Riemannian metric on M and the invariant scalar product on G allow us to the define a scalar product between forms that take values in the fibers of ξ_τ. It follows that we can introduce a scalar product on Γ_1 and $\Gamma_2 + \Gamma_0$, and consider the adjoint operator $T^\dagger$ from $\Gamma_2 + \Gamma_0$ to Γ_1. This operator takes a pair $(f, h) \in \Gamma_2 + \Gamma_0$ to the one-form $d_A^\dagger f + d_A h \in \Gamma_1$. This is shown by using the fact that

$$(P d_A a, f) + (d_A^\dagger a, h) = (a, d_A^\dagger f) + (a, d_A h).$$

We want to know the number of solutions of (31.9), that is, the number $l(T)$ of zero modes of T. A lower bound for this number is provided by the index of T, since

$$l(T) = \operatorname{index} T + l(T^\dagger) \geq \operatorname{index} T.$$

The index of T can easily be computed using well-known topological results. A more transparent, although longer, calculation can be carried out using Equation (27.12), which expresses the index of an operator in terms of Seeley coefficients. For simplicity, we consider only the case where $M = S^4$ and $G = \mathrm{SU}(2)$. Then one can show that $\operatorname{index} T = 8q - 3$, so that $l(T) \geq 8q - 3$. In fact, we will show below that in this case $l(T^\dagger) = 0$, that is, $T^\dagger$ does not have zero modes. Therefore $l(T) = 8q - 3$, that is, the linear approximation to the system of equations (31.7)–(31.8) has an $(8q - 3)$-dimensional space of solutions. From this one can show that the nonlinear system itself has an $(8q - 3)$-parameter family of solutions.

The reasoning above is based on the infinite-dimensional inverse function theorem, which applies to a differentiable nonlinear map $W : E_1 \to E_2$, where E_1 and E_2 are Banach spaces. We assume that $W(x_0) = y_0$, and seek a solution to the equation $W(x) = y$, where y is near y_0. The theorem says that, if the differential $\mathcal{W}$ of W at x_0 is an invertible linear operator, the equation $W(x) = y$ has a unique solution near x_0, for y near y_0. In other words, if the derivative of an operator is invertible at a point, the operator itself is locally invertible near that point.

From this theorem one can also derive the infinite-dimensional implicit function theorem, which provides information about the map W even when the derivative $\mathcal{W}$ is not invertible, so long as the image of $\mathcal{W}$ is the whole range E_2. That is, we assume that the equation $\mathcal{W}(x) = y$ has a solution for any y, but do not assume that the solution is unique—the space of solutions is some l-dimensional subspace of E_1, where $l = l(\mathcal{W})$ is the number of zero modes of E_1. Then the solutions of $W(x) = y$, for y near y_0, also form an l-parameter family. To show this, we take l linear functionals $r_1(x), \ldots, r_l(x)$ that are linearly independent on the space $\operatorname{Ker} \mathcal{W}$ of zero modes of $\mathcal{W}$. Setting $r(x) = (r_1(x), \ldots, r_l(x))$, we get a map $r : E_1 \to \mathbf{R}^l$, and then a mapping $W' : E_1 \to E_2 \times \mathbf{R}^l$, taking $x \in E_1$ to $(W(r), r(x))$. Our assumptions imply that the differential of W' at x_0 is invertible, so we can apply the inverse function theorem to conclude that the equation $W'(x) = (y, r)$ has a unique solution for y close to y_0 and r close to 0. The solutions of this equation, as r varies, give an l-parameter family of solutions to the equation $W(x) = y$ for y close to y_0 (and in particular for $y = y_0$).

All we have to do now is notice that the nonlinear system (31.7)–(31.8) can be written in the form of a single nonlinear equation $Ta = 0$, where T is the nonlinear operator that maps a one-form $a \in \Gamma_1$ to the pair $(f, h) \in \Gamma_2 + \Gamma_0$, where $f = P(d_A a + \frac{1}{2}[a, a])$ and $h = d_A^\dagger a$. The linear operator $\mathcal{T}$ is the differential of T at $a = 0$. From $l(\mathcal{T}^\dagger) = 0$ it follows that $\mathcal{T}$ is surjective, so that T satisfies the conditions given above for the applicability of the implicit function theorem. Therefore the number of parameters needed to describe the space of solutions of $Ta = 0$ equals $l(\mathcal{T})$.

Actually, the argument above needs to be strengthened somewhat. According to our conventions, all functions, forms, and bundle sections are taken to be smooth (differentiable infinitely many times). But under these conditions one cannot make Γ_1 and $\Gamma_2 + \Gamma_0$ into Banach spaces with the necessary properties; we must instead enlarge these spaces somewhat. One could, for example, take Γ_1 as the space of one-forms whose coefficient functions have derivatives of order up to k, and whose k-th derivatives are integrable when raised to the p-th power, for $p > 4$. (Γ_1 is an example of a *Sobolev space*, and is denoted by W_p^k.) About the elements of $\Gamma_2 + \Gamma_0$ we assume that their $(k-1)$-th derivatives are integrable when raised to the p-th power.

Instead of eliminating the gauge freedom by imposing condition (31.8), we can talk of an instanton manifold where gauge-equivalent fields are identified. Denote by N_q the infinite-dimensional set of solutions to the self-duality equation (31.4), and by R_q the set obtained from N_q by identifying gauge-equivalent instantons. In other words, R_q is the set of orbits N_q/G^∞ under the action of the group G^∞ of local gauge transformations. From the preceding discussion it follows that for $G = \mathrm{SU}(2)$, $M = S^4$ and $q \geq 1$ the set R_q has dimension $8q - 3$.

Now consider an arbitrary simple, compact, non-abelian gauge group G, and $M = S^4$ as before. Then it is already not true that $l(\mathcal{T}^\dagger) = 0$ in general. Notice first that $l(\mathcal{T}^\dagger) = l(\mathcal{T}\mathcal{T}^\dagger)$, where $\mathcal{T}\mathcal{T}^\dagger$ is the operator taking a pair

$(f, h) \in \Gamma_2 \dotplus \Gamma_0$ to $(\Delta_2^A f, \Delta_0^A h)$, for $\Delta_2^A = P d_A d_A^\dagger$ and $\Delta_0^A = d_A^\dagger d_A$. This implies that $l(T^\dagger) = l(\Delta_2^A) + l(\Delta_0^A)$.

Analogously, $l(T) = l(T^\dagger T) = l(\Delta_1^A)$, where $\Delta_1^A = d_A^\dagger P d_A + d_A d_A^\dagger$. One can show that $l(\Delta_2^A) = 0$ for any instanton on S^4. The number $l(\Delta_0^A) = l(d_A)$ equals the dimension of the subspace $\mathcal{H}^A \subset \Gamma_0$ consisting of forms ω such that $d_a \omega = 0$. To each element $\omega \in \Gamma_0$ is associated an infinitesimal gauge transformation defined by

$$\delta A = d_A \omega.$$

In other words, Γ_0 can be identified with the Lie algebra $\mathcal{G}^\infty$ of the infinite-dimensional Lie group G^∞ of all gauge transformations. Thus, $\mathcal{H}^A$ consists of infinitesimal gauge transformations that leave A invariant. Put another way, $\mathcal{H}^A$ is the Lie algebra of the Lie group H^A of gauge transformations taking A to itself. For $G = \mathrm{SU}(2)$ and instantons with topological number $q > 0$ we have $H^A = 0$, but for $q = 0$ we have $\dim H^A = \dim \mathcal{H}^A = 3$: the field $A = 0$ is fixed by any global gauge transformation, that is, any gauge transformation generated by a constant function.

When H^A is trivial, that is, $l(T^\dagger) = \dim \mathcal{H}^A = 0$, the instanton is called *irreducible*. The solutions to the self-duality equation can be studied in the neighborhood of an irreducible instanton A in the same way as in the case $G = \mathrm{SU}(2)$. The dimension of the space of solutions of (31.7)–(31.8), for A irreducible, equals $l(T) = \operatorname{index} T$. For $G = \mathrm{SU}(n)$ we have

$$\operatorname{index} T = 4nq - (n^2 - 1),$$

where q is the topological number. If $q > \frac{1}{2}n$, there exist irreducible instantons, so the dimension is $4nq - (n^2 - 1)$. This implies that

$$\dim R_q = 4nq - (n^2 - 1)$$

for $q > \frac{1}{2}n$. If, on the other hand, the instanton is reducible, we have

$$l(T) = \operatorname{index} T + \dim \mathcal{H}^A.$$

Thus, the number of zero modes of T depends on the dimension of H^A. An instanton is called *regular* if this dimension does not change under a small variation in the field A. It can be shown that the dimension of R_q is $l(T) = \operatorname{index} T + \dim \mathcal{H}^A$, where A is a regular instanton. Noting that

$$\operatorname{index} T = l(\Delta_1^A) - l(\Delta_2^A) - l(\Delta_0^A) = l(\Delta_1^A) - l(\Delta_0^A)$$

and

$$\dim \mathcal{H}^A = l(\Delta_0^A),$$

we conclude that $\dim R_q = l(\Delta_1^A)$.

32. Computation of the Instanton Contribution

As discussed earlier, instantons play an important role in the calculation of the functional integrals of type (25.3) and (25.4) by the Laplace method. Recall that this method is based on expanding the exponent of the exponential in a neighborhood of its minimum, discarding terms of degree greater than two, and computing the resulting Gaussian integral. In the case of instantons the exponent achieves its minimum not at a single point, but on an entire manifold. This requires a modification to the Laplace method, which we now describe.

Consider the integral

$$(32.1) \qquad I = \int_M \varphi(x) \exp(g^{-2}S(x)) \, dV,$$

where M is an n-dimensional Riemannian manifold and dV is its volume element. We assume that $S(x)$ takes its minimum γ on a k-dimensional *critical manifold* N, that is, $S(x) \geq \gamma$ for all $x \in M$, and $S(x) = \gamma$ for $x \in N$. We assume that S is non-degenerate on N, that is, that the second differential of S at each point of N is a quadratic form of rank $n - k$. It will be convenient to write this second differential at $x \in N$ as $d^2 S = \frac{1}{2}\langle S'' \, dx, dx \rangle$, where S'' is a self-adjoint linear operator on the tangent space to M at the point $x \in N$; the non-degeneracy condition is equivalent to the condition that S'' has k zero eigenvalues. (Clearly, every tangent vector to N is an eigenvector of S'' with eigenvalue zero, so S'' has at least k zero eigenvalues.) The leading term in the asymptotic approximation of (32.1) as $g \to 0$ is

$$(32.2) \qquad (\sqrt{2\pi}g)^{\dim M - \dim N} \exp\left(-\frac{\gamma}{g^2}\right) \int_N \varphi(x)(\det S''(x))^{-1/2} \, d\mu,$$

where $d\mu$ is the measure on N determined by the Riemannian metric inherited from M, and the determinant of the singular operator is understood as the product of its nonzero eigenvalues. To obtain (32.2), we must notice that the asymptotics of (32.1) is determined by a neighborhood of N, and that in integrating in directions orthogonal to N one can employ the usual Laplace method.

If $\varphi(x)$ and $S(x)$ are invariant under the action of a compact Lie group acting G on M, we can modify (32.2). Let $g \in G$ act by the transformation $T(g)$, which we assume is an isometry for all g. If $\omega \in \mathcal{G}$ is an element of the Lie algebra of G, denote by $T_x\omega$ the vector at $x \in M$ arising from the infinitesimal transformation ω: informally, $T(1+\omega)x = x + T_x\omega$. The linear operator T_x can be interpreted as the differential of the nonlinear map $g \mapsto T(g)x$ at $g = 1$. We normalize the invariant scalar product $\langle \, , \, \rangle$ on $\mathcal{G}$ in such a way that G has unit volume.

The manifold N where S is minimized is obviously invariant under G, and likewise the functions that appear in (32.2). Hence, as explained in Chapter 25 (see (25.24)), the integral over N can be replaced by an integral over the space of orbits $R = N/G$. The leading term in the asymptotic expansion of (32.1) as $g \to 0$ then becomes

$$(32.3) \qquad I = (\sqrt{2\pi}g)^{\dim M - \dim N} \exp\left(-\frac{\gamma}{g^2}\right) \int_R \varphi(x)\, d\mu$$

for

$$(32.4) \qquad d\mu = Z(x)V(H)^{-1}\, d\mu_0,$$

where

$$(32.5) \qquad Z(x) = (\det S''(x))^{-1/2}(\det T_x^\dagger T_x)^{1/2},$$

$d\mu_0$ is the volume element in R arising from the Riemannian metric inherited from N, and $V(H)$ is the volume of the stabilizer of G with respect to the Riemannian metric on G. (We assume that all stabilizers are conjugate to a fixed subgroup $H \subset G$, and therefore all have the same volume.)

Using the operators $\Box_0^x = S''(x) + T_x T_x^\dagger$ and $\Box_1^x = T_x^\dagger T_x$, which satisfy the relations

$$(32.6) \qquad \det \Box_0^x = \det S''(x) \det T_x T_x^\dagger = \det S'' \det T_x^\dagger T_x,$$
$$(32.7) \qquad \det \Box_1^x = \det T_x^\dagger T_x,$$

we can also write

$$(32.8) \qquad Z(x) = (\det \Box_0^x)^{-1/2} \det \Box_1^x.$$

To prove (32.6), we note that the G-invariance of S implies the invariance of the second differential of S at $x \in N$ under the transformation $dx \mapsto dx + T_x\xi$, for $\xi \in \mathcal{G}$. This implies $S''T_x = 0$, which in turn implies $S''T_xT_x^\dagger = 0$ and $T_xT_x^\dagger S'' = 0$. Thus, we can find for S'' and $T_xT_x^\dagger$ a common system of eigenfunctions, each of which has eigenvalue zero for one or the other of the operators. Since in computing the determinant of a singular operator we only consider nonzero eigenvalues, we obtain (32.6).

We now apply these results, concerning the asymptotic behavior of multiple integrals (of finite multiplicity) with infinitely many minima, to the calculation of the instanton contribution to (25.3) and (25.4). (Of course, in this case the integrals are infinite-dimensional, so our calculations cannot be considered rigorous.) The Euclidean action $S_{\text{eucl}}(A) = (4g^2)^{-1}(\mathcal{F}, \mathcal{F})$ will play the role of $g^{-2}S(x)$, and $S(x)$ will correspond to the functional $(\mathcal{F}, \mathcal{F})$. For M we take the set $\mathcal{E}_q$ of gauge fields with a fixed topological number $q > 0$, so N will be the space N_q of solutions of the self-duality equation (31.4).

To expand the Euclidean action in powers of deviations from the solution to the self-duality equation, we write it in the form

$$(32.9) \qquad S_{\mathrm{eucl}} = \frac{1}{8g^2}(2(\mathcal{F}, *\mathcal{F}) + (\mathcal{F} - *\mathcal{F}, \mathcal{F} - *\mathcal{F}))$$

and use the formula for the variation of the field strength caused by an infinitesimal variation of the gauge field,

$$(32.10) \qquad \delta\mathcal{F} = d_A\,\delta A$$

(see T15.4). This implies that $\delta(\mathcal{F} - *\mathcal{F}) = 2Pd_A\,\delta A$, where $P = \frac{1}{2}(1 - *)$. Using (32.9), we conclude that the increment in the Euclidean action caused by an infinitesimal variation of the self-dual gauge field A is

$$(32.11) \quad \Delta S_{\mathrm{eucl}} = \frac{1}{2g^2}(Pd_A\,\delta A, Pd_A\,\delta A) + \cdots = \frac{1}{2g^2}(d_A^\dagger Pd_A\,\delta A, \delta A) + \cdots,$$

the discarded terms having order three or higher in δA.

We have just calculated the second variation of the Euclidean action functional S_{eucl} at an arbitrary point of N_q. The role of S'', as we have seen, is played by $d_A^\dagger Pd_A$, acting on the space Γ_1 of one-forms with values in the fibers of the fibration $\xi_\tau = (E_\tau, M\,G, p_\tau)$. Further, S_{eucl} is invariant under the infinite-dimensional group G^∞ of local gauge transformations. The Lie algebra $\mathcal{G}^\infty$ of this group can be identified with the space Γ_0 of sections of ξ_τ. The effect of an infinitesimal gauge transformation $\omega \in \mathcal{G}^\infty = \Gamma_0$ on a gauge field A is given by

$$(32.12) \qquad \delta A = d_A\omega.$$

This means that for $\mathcal{T}_x$ in the discussion above we can take the operator d_A mapping $\Gamma_0 = \mathcal{G}^\infty$ into Γ_1, and for $\mathcal{T}_x^\dagger \mathcal{T}_x$ we take $d_A^\dagger d_A$. Therefore (32.5) becomes

$$(32.13) \qquad Z_A = (\det(d_A^\dagger Pd_A))^{-1/2}\det(d_A^\dagger d_A)^{1/2} = (\det \Delta_1^A)^{-1/2}\det \Delta_0^A,$$

where $\Delta_1^A = d_A^\dagger Pd_A + d_A d_A^\dagger$ and $\Delta_0^A = d_A^\dagger d_A$. We remark that Δ_1^A and Δ_0^A are the same as in the previous chapter.

The Lie algebra $\mathcal{H}^A$ of the stabilizer H^A consists of elements $\omega \in \mathcal{G}^\infty = \Gamma_0$ such that $d_A\omega = 0$, that is, of covariantly constant sections of ξ_τ. Thus, the dimension of H^A equals the number of zero modes of d_A or, which is the same, the number of zero modes of $\Delta_0^A = d_A^\dagger d_A$. As we have seen, Δ_0^A has no zero modes for $G = \mathrm{SU}(2)$ and $q \geq 1$; for $G = \mathrm{SU}(2)$ and $q = 0$ there are three zero modes, corresponding to global gauge transformations.

We are now ready to formally apply (32.3) to the computation of the contribution of instantons with topological number $q \geq 1$ to the functional integral (25.4). We have

$$(32.14) \qquad I_q = (\sqrt{2\pi}g)^{\dim \mathcal{E}_q - \dim N_q}\exp\left(-\frac{8\pi^2}{g^2}q\right)\int_{R_q}\Phi(A)Z_A\,d\mu_0,$$

where $d\mu_0$ is the measure on the instanton manifold R_q coming from the Riemannian metric inherited from N_q.

This metric is defined as follows: if A and $A + a$ are instantons with a small, the distance between the corresponding points in R_q is $(a + d_A\varphi, a + d_A\varphi)^{1/2}$, where φ is given by the condition $d_A^\dagger(a + d_a\varphi) = 0$. Indeed, the field $A + a + d\varphi$ is obtained from $A + a$ by an infinitesimal gauge transformation, and the vector $a + d_A\varphi$ is orthogonal to the orbit of G^∞ passing through A.

Almost all quantities in (32.14) are infinite—in particular, the determinants of the infinite-dimensional operators Δ_1^A and Δ_0^A, and the dimensions $\dim \mathcal{E}_q$ and $\dim N_q$ of, respectively, the space of all gauge fields and the set of instantons with topological charge q. We can attempt to make better sense of (32.14) by discarding infinite constant factors and by regularizing determinants (Chapter 23). Such considerations allow one to conclude that, in integrating over the instanton manifold R_q, one must use the measure $d\mu = Z_A \, d\mu_0$, where Z_A is defined in (32.13), with the determinants being assumed regularized.

It is more satisfying, however, to give (32.14) a meaning by first going over to a lattice and then taking the limit as the lattice step tends to zero (that is, lifting the cutoff in momentum).

We now show how this can be done in calculating the contribution J_q of instantons with topological number q to the Euclidean Green's functions. Recall that these functions are the quotients of (25.4) by the partition function (25.3). In computing (25.3) we assume that the main contribution comes solely from classical vacuums, that is, instantons whose topological number is zero. Therefore J_q can be represented by

$$(32.15) \qquad J_q = g^{-\sigma} \exp\left(-\frac{8\pi^2}{g^2} q\right) \int_{R_q} \Phi(A) \frac{\lambda(A)}{\lambda(0)} d\mu_0,$$

where $\lambda(A) = Z_A V(H^A)^{-1}$ and where $\lambda(0) = Z_0 V(H^0)^{-1}$.

The number σ in (32.15) equals $\dim N_q - \dim N_0$, but this definition has no rigorous meaning since both dimensions are infinite. However, we can represent $\dim N_q$ as

$$(32.16) \quad \dim N_q = \dim R_q + (\dim G^\infty - \dim H^A) = l(\Delta_1^A) - l(\Delta_0^A) + \dim G^\infty,$$

where we have used the relation $\dim R_q = l(\Delta_1^A)$, which holds for a regular instanton A: see (31.10). In particular,

$$(32.17) \qquad \dim N_0 = \dim R_0 + (\dim G^\infty - \dim H^0) = \dim G^\infty - \dim G.$$

We see that

$$(32.18) \qquad \sigma = \dim R_q - \dim R_0 - (\dim H^A - \dim H^0)$$
$$= l(\Delta_1^A) - l(\Delta_0^A) - (l(\Delta_1^0) - l(\Delta_0^0)).$$

To avoid considering infinite dimensions, and also to give an interpretation to the integrand of (32.15), we pass to a lattice. In particular, we must study the asymptotic behavior of the discrete analogues $\Delta_i^A(a)$ of Δ_i^A as $a \to 0$. However, we will not follow this path, which leads to certain (surmountable) difficulties. Instead of cutoff in momenta, we will use cutoff in proper time (Chapter 28), assuming that the asymptotic behavior is the same for $\ln \det \Delta_i^A(a)$ and $\ln \det_\varepsilon \Delta_i^A$, where $\varepsilon = \text{const } a^2$. Using (28.9), we conclude that, for $a \to 0$,

$$\ln \det \Delta_i^A(a) \approx \ln \det_\varepsilon \Delta_i^A$$

$$\approx -\tfrac{1}{2}\Phi_2(\Delta_i^A)\varepsilon^{-2} - \Phi_1(\Delta_i^A)\varepsilon^{-1} + (\Phi_0(\Delta_i^A) - l(\Delta_i^A))\ln \varepsilon.$$

In the discrete case, as discussed in Chapter 25, the analogue of a gauge field is a function b_γ defined on the oriented edges of the lattice (an orientation for each edge having been fixed). Integration over the space of gauge fields A_μ is replaced with integration over a product of copies of G, one for each edge. The volume element in this product is determined by the invariant Riemannian metric

$$(32.19) \qquad ds^2 = \sum_\gamma \langle b_\gamma^{-1}\, db_\gamma, b_\gamma^{-1}\, db_\gamma \rangle,$$

where the sum is over all the edges. In order to establish a link between the volume element $d\mu_0$ in R_q and its discrete analogue $d\mu_0^a$, we recall that we have associated to the gauge field A_μ the function $b_\gamma = \mathrm{P}\exp(-\int_\gamma A_\mu\, dx_\mu)$. If the lattice step is small, we have $b_\gamma = 1 - A_\mu a$ (assuming that γ is parallel to the x^μ-axis). Using this we get

$$(32.20) \qquad ds^2 = a^{-2}\int \langle \delta A_\mu, \delta A^\mu \rangle\, dV = a^{-2}(\delta A, \delta A).$$

It follows that

$$(32.21) \qquad d\mu_0^a \approx a^{-\dim R_q}\, d\mu_0 = a^{-l(\Delta_1^A)}\, d\mu_0.$$

The volume of the discrete analogue of the group H^A is computed using the invariant metric

$$(32.22) \qquad ds^2 = \sum_\alpha \langle \rho_\alpha^{-1}\, d\rho_\alpha, \rho_\alpha^{-1}\, d\rho_\alpha \rangle$$

in the discrete analogue of the group of gauge transformations G^∞. Here ρ_α is the element of G assigned to vertex α of the lattice, and the sum is over all vertices. For each gauge transformation with a function $\rho(x)$ there is a corresponding discrete gauge transformation; the element of length (32.22) is related to the metric in G^∞ by the formula

$$(32.23) \qquad ds^2 \approx a^{-4}\int \langle \rho^{-1}(x)\, \delta\rho(x), \rho^{-1}(x)\, \delta\rho(x) \rangle\, dV.$$

Thus, upon discretization, we must replace $V(H^A)$ by $a^{-2\dim H^A}V(H^A) = a^{-2l(\Delta_0^A)}V(H^A)$.

We are now in a position to investigate the asymptotic behavior of J_q as $a \to 0$. We note first that $\lambda(A)\, d\mu_0/\lambda(0)$ behaves like

(32.24) $\text{const} \exp(-(\rho(A) - \rho(0)) \ln a)$,

where

(32.25) $\rho(A) = \Phi_0(\Delta_1^A) - 2\Phi_0(\Delta_0^A)$.

Indeed, the difference

$$(-\tfrac{1}{2} \ln \det \Delta_1^A(a) + \ln \det \Delta_0^A(a)) - (-\tfrac{1}{2} \ln \det \Delta_1^0(a) + \ln \det \Delta_0^0(a))$$

has the asymptotic behavior

$$(-\tfrac{1}{2}(\Phi_0(\Delta_1^A) - l(\Delta_1^A)) + (\Phi_0(\Delta_0^A) - l(\Delta_0^A))) \ln \varepsilon$$
$$- (-\tfrac{1}{2}(\Phi_0(\Delta_1^0) - l(\Delta_1^0)) + (\Phi_0(\Delta_0^0) - l(\Delta_0^0))) \ln \varepsilon$$
$$= -(\rho(A) - \rho(0)) \ln a + (l(\Delta_1^A) - 2l(\Delta_0^A)) - (l(\Delta_1^0) - 2l(\Delta_0^0)) \ln a$$

as $a \to 0$; the factors that diverge linearly and quadratically in $\varepsilon = a^2$ cancel out. Taking into account the divergent factors in $d\mu_0$ and $V(H^A)^{-1}$, we obtain (32.24). We see that for (32.14) to have a finite limit as $a \to 0$, the bare coupling constant g should depend on a in such a way that

$$\frac{8\pi^2}{g^2(a)} q + (\rho(A) - \rho(0)) \ln a$$

has a finite limit. Computing the Seeley coefficients $\Phi_0(\Delta_1^A)$ and $\Phi_0(\Delta_0^A)$ (Chapter 27), we get

(32.26) $\dfrac{1}{g^2(a)} = \dfrac{1}{g^2} + \dfrac{1}{8\pi^2} \tfrac{11}{6} \alpha \ln(av)$,

where α is a coefficient that depends only on the gauge group, g should be interpreted as the physical coupling constant, and v is the renormalization point. This equation represents the standard law of variation of the bare coupling constant under renormalization in a gauge theory. If $g(a)$ is chosen to be of this form, we get the following finite expression for the instanton contribution to the Euclidean Green's function as $a \to 0$:

(32.27) $J_q \approx \text{const}\, g^{-\sigma} \exp\left(-\dfrac{8\pi^2 q}{g^2}\right) \displaystyle\int_{R_q} \Phi(A)\, d\nu$

for $d\nu = (\gamma(A)/\gamma(0))\, d\mu_0$, where

(32.28) $\gamma(A) = Z_A V(H^A)^{-1} = (\det \Delta_1^A)^{-1/2} (\det \Delta_0^A) V(H^A)^{-1}$.

We have not included in the general constant the factors $\gamma(0)$ and $V(H^A)^{-1}$, which are independent of A, because they do not change under a conformal variation of the metric, which we consider below. For $G = \mathrm{SU}(n)$ we have $\alpha = 2n$ and $\sigma = 4nq$. (In general, one can show that $\sigma = 2\alpha q$.) The infinite-dimensional determinants in (32.28) are assumed regularized as explained in Chapter 28.

The calculation of these infinite-dimensional determinants requires considerable effort. However, important information about the measure $d\nu$ on the instanton manifold R_q can be obtained without computing the determinants, simply by using the conformal invariance of the action integral (25.2) for a gauge field (clearly, (25.2) does not change if the Riemannian metric $g_{\mu\nu}(x)$ on M is replaced by a conformally equivalent metric $g'_{\mu\nu}(x) = e^{\sigma(x)}g_{\mu\nu}(x)$). But the functional integral (25.4) does change under conformal changes in the metric; as mentioned in Chapter 29, this phenomenon is termed conformal anomaly, and is related to the change in scalar product in the space of fields over which we are integrating. The measure $d\nu = \gamma(A)d\mu_0(\gamma(0))^{-1}$ changes too. However, using the results of Chapter 29, one can examine the variation in $d\nu$ caused by an infinitesimal conformal variation in the metric of M. In view of (29.28), we can interpret Z_A as the partition function of the quadratic functional $S_A(a) = (d_A^\dagger P d_A a, a)$, which is invariant under the change $a \mapsto a + d_A \omega$.

Using (29.29), we obtain an expression for the variation in Z_A caused by an infinitesimal conformal variation in the metric, $\delta g_{\mu\nu} = \sigma(x)g_{\mu\nu}(x)$:

$$(32.29) \qquad \delta \ln Z_A = \tfrac{1}{2}\beta(\hat{\sigma}|\Delta_1^A) - \tfrac{1}{2}\beta(2\hat{\sigma}|\Delta_0^A),$$

where $\hat{\sigma}$ is the operator of multiplication by $\sigma(x)$. This gives an expression for the variation $\delta\, d\mu$ in the measure $d\mu = Z_A V(H^A)^{-1}\, d\mu_0 = \gamma(A)\, d\mu_0$ caused by an infinitesimal conformal variation in the metric:

$$(32.30) \qquad \delta\, d\mu = (\tfrac{1}{2}\Psi_0(\hat{\sigma}|\Delta_1^A) - \Psi_0(\hat{\sigma}|\Delta_0^A))\, d\mu.$$

The measure $d\mu_0$ and the volume $V(H^A)$ also change under a conformal change in the metric. Their variation is governed by the zero modes of Δ_1^A and Δ_0^A. The terms corresponding to the variation in the measure $d\mu_0$ cancel out with the contribution of the zero modes to the variation of the partition function Z_A.

Similarly, we can prove that the variation of $\gamma(0)$ caused by a conformal variation in the metric is

$$(32.31) \qquad \delta\gamma(0) = (\tfrac{1}{2}\Psi(\hat{\sigma}|\Delta_1^0) - \Psi(\hat{\sigma}|\Delta_0^0))\gamma(0).$$

Combining this with (32.20), we get

$$(32.32) \qquad \delta\, d\nu = (\tfrac{1}{2}\Psi_0(\hat{\sigma}|\Delta_1^A) - \Psi_0(\hat{\sigma}|\Delta_0^A) - \tfrac{1}{2}\Psi_0(\hat{\sigma}|\Delta_1^0) - \Psi_0(\hat{\sigma}|\Delta_0^0))\, d\nu.$$

Computing the Seeley coefficients (Chapter 27), we obtain

$$(32.33) \qquad \delta d\nu = \left(\frac{11}{384\pi^4}\alpha \int \sigma(x)\langle \mathcal{F}_{\mu\nu}, \mathcal{F}^{\mu\nu}\rangle\, dV\right) d\nu.$$

(Below we will show that in studying instantons in $\mathbf{R}^4$ we can make do with the previously computed coefficients $\Phi_0(\Delta_1^A)$ and $\Phi_0(\Delta_0^A)$).

We now turn to the case where M is the sphere S^4 with the usual metric. S^4 is acted on by the 15-dimensional group Q of conformal transformations (a transformation $x' = \varphi(x)$ is called *conformal* if it gives rise to a conformal change in the metric: $ds'^2 = e^{\sigma(x)} ds^2$). Thanks to the conformal invariance of the self-duality equation, the action of Q on S^4 gives rise to an action of Q on $\mathbf{R}^4$. The measure (32.28) is not invariant under this action, but using (32.30) we can study the variation of this measure under the action.

If $q = 1$, the action of Q on R_q is transitive, that is, any instanton can be obtained from any other by some conformal transformation. Hence, starting with (32.30), we can determine the measure $d\nu$ on R_1 uniquely, to within a numerical factor. Properly speaking, the measure on R_q when $M = S^4$ is of no interest since the passage to S^4 is only a means to introduce spatial cutoff; what is important is the limit of the instanton contribution when the radius of the sphere tends to infinity, that is, when the size of the instantons considered is much smaller than the radius of the sphere. By taking this limit we obtain a measure $d\nu$ on the instanton manifold for $\mathbf{R}^4$. We now turn to the study of this measure.

As explained in Chapter 30, for $G = \mathrm{SU}(n)$ the instantons with topological number $q = 1$ are characterized by their size λ and by their center a, a four-vector. The measure $d\nu$ on R_1 can be written as $d\nu = \tau(\lambda, a)\, d\lambda\, d^4 a$. It is clear that $d\nu$ is translation-invariant, so $\tau(\lambda, a)$ does not depend on a. From (32.32) it follows that the variation of $d\nu$ under an expansion $x \mapsto (1 + \tfrac{1}{2}\sigma)x$ is

$$\delta d\nu = \sigma\big(\tfrac{1}{2}\Phi_0(\Delta_1^A) - \Phi_0(\Delta_0^A) - \tfrac{1}{2}\Phi_0(\Delta_1^0) + \Phi_0(\Delta_0^0)\big)\, d\nu.$$

The Seeley coefficients that appear in this formula have already been used above. We get

$$\delta d\nu = \tfrac{11}{12}\alpha q \sigma\, d\nu,$$

which, for $q = 1$, gives

$$d\nu = \mathrm{const}\, \exp\big(\tfrac{11}{6}\alpha \ln \lambda\big) \frac{d\lambda}{\lambda^5}\, d^4 a.$$

We have thus derived the expression for the single-instanton contribution given at the end of Chapter 30. If $q = 2$ and $G = \mathrm{SU}(2)$, the manifold R_q has dimension $8 \times 2 - 3 = 13$, and the orbits of Q in R_2 are 12-dimensional. Hence, the conformal properties of $d\nu$ enable one to establish this measure to within a function of a single variable.

33. Functional Integrals for a Theory Containing Fermion Fields

In Chapter 24 we showed that many important physical quantities in quantum mechanics and quantum field theory can be represented in the form of functional integrals. The integrands of these integrals contain the exponential of either the action functional of the classical theory, or its Euclidean analogue (obtained by passing to imaginary time). The ideas of Chapter 24, however, cannot be directly applied if the theory contains fermions. One reason is that we cannot speak of a classical fermion field. Indeed, in the boson case quantization is based on the canonical commutation relations

$$(33.1) \qquad [p_j, p_k] = [q_j, q_k] = 0, \qquad [p_j, q_k] = \frac{\hbar}{i} \delta_{jk}$$

or on the equivalent relations

$$(33.2) \qquad [a_j, a_k] = [a_j^\dagger, a_k^\dagger] = 0, \qquad [a_j, a_k^\dagger] = \hbar \delta_{jk},$$

where

$$a_k = \frac{1}{\sqrt{2}}(q_k + ip_k), \qquad a_k^\dagger = \frac{1}{\sqrt{2}}(q_k - ip_k).$$

In the limit $\hbar \to 0$ the operators p_k and q_k (or a_k and $a_k^\dagger$) commute, and can be thought of as being classical quantities. By contrast, in the fermion case one starts from anticommutation relations

$$(33.3) \qquad [a_j, a_k]_+ = [a_j^\dagger, a_k^\dagger]_+ = 0, \qquad [a_j, a_k^\dagger]_+ = \hbar \delta_{jk},$$

where $[a, b]_+ = ab + ba$ is the anticommutator of a and b. Clearly, a_k and $a_k^\dagger$ do not commute even in the limit $\hbar \to 0$. The conclusion we can draw is that a fermion field has no classical limit.

However, there is a more fruitful point of view. We can say that in the limit $\hbar \to 0$ the operators a_k and $a_k^\dagger$ become anticommuting quantities. This is an indication that in the classical limit fermion fields must be thought of as fields taking on anticommuting values. This interpretation is very fruitful because, in particular, the expression of physical quantities as functional integrals remains valid in the presence of fermions, provided that we have an appropriate procedure for integrating over fields with anticommuting values. For example, for a theory that describes the interaction of a spinor field $\psi(x)$ with a scalar field $\varphi(x)$, the generating functional of the Euclidean Green's function can be written as

$$(33.4) \qquad G[j, \chi, \chi^\dagger] = \frac{Z[j, \chi, \chi^\dagger]}{Z[0, 0, 0]},$$

where $Z[j, \chi, \chi^\dagger]$ can be interpreted as the partition function in the presence of sources j and χ:

$$(33.5) \qquad Z = \int \exp\left(-S_{\text{eucl}} + \int \left(j\varphi + \chi^\dagger\psi + \psi^\dagger\chi\right) d^4x\right) \prod d\varphi \, d\psi \, d\psi^\dagger,$$

where the integral is over fields $\psi(x)$ and $\chi(x)$ taking anticommuting values, and

$$(33.6) \qquad S_{\text{eucl}} = \int \left(\tfrac{1}{2}(\partial_\mu\varphi)^2 + \psi^\dagger \, \partial\!\!\!/ \, \psi + g\psi^\dagger\psi\varphi + \lambda\varphi^4\right) dx^4$$

($\partial\!\!\!/$ being the Euclidean Dirac operator: see Chapter 26).

A precise definition of integration over fields with anticommuting values will be given later. Here we limit ourselves to remarking that, in the theories that most interest us, the action functional depends on the fermion fields quadratically, so the resulting functional integrals over anticommuting fields are Gaussian. As in the ordinary case, the evaluation of Gaussian integrals over anticommuting fields can be reduced to the calculation of determinants. The result is somewhat different, however; while in the ordinary case the integral is $(\det A)^{-1/2}$, where A is the operator corresponding to the quadratic form in the exponent, in the case of anticommuting fields the answer is $(\det A)^{1/2}$.

We now give some rigorous statements and definitions. We consider symbols $\varepsilon^1, \ldots, \varepsilon^n$ that satisfy

$$(33.7) \qquad \varepsilon^\alpha \varepsilon^\beta = -\varepsilon^\beta \varepsilon^\alpha,$$

and look at expressions of the form

$$(33.8) \qquad \omega = \sum_k \sum_{\alpha_1, \ldots, \alpha_k} a^k_{\alpha_1 \ldots \alpha_k} \varepsilon^{\alpha_1} \ldots \varepsilon^{\alpha_k}.$$

Expressions of this form can be added and multiplied together, and multiplied by scalars, in the obvious manner (taking into account (33.7)). Therefore they form an algebra known as the *Grassmann algebra* with generators $\varepsilon^1, \ldots, \varepsilon^n$. More formally, we can say that a Grassmann algebra is an associative algebra with unity and with generators $\varepsilon^1, \ldots, \varepsilon^n$ satisfying (33.7). Every element of the Grassmann algebra can be written in the form (33.8) with the $a^k_{\alpha_1 \ldots \alpha_k}$ skew-symmetric, and this representation is unique. (Another unique representation arises by imposing instead the condition that $a^k_{\alpha_1 \ldots \alpha_k} = 0$ unless $\alpha_1 < \alpha_2 < \cdots < \alpha_k$.) The product of an even number of ε^α's is called an *even monomial*, and a linear combination of even monomials is called a *even element* of the Grassmann algebra. Odd monomials and odd elements of the algebra are defined similarly.

From (33.7) it follows that an even element of a Grassmann algebra commutes with any other element, and two odd elements anticommute. If $a^0 = 0$ in (33.8), that is, if every term contains at least one ε^α, then ω is *nilpotent*, that is, $\omega^N = 0$ for some $N > 0$.

The expression (33.8) has the form of a polynomial in $\varepsilon^1, \ldots, \varepsilon^n$. This indicates that we should consider ω as a function of the anticommuting variables $\varepsilon^1, \ldots, \varepsilon^n$, and accordingly we write $\omega = \omega(\varepsilon)$. We now show that, for functions of anticommuting variables, we can define operators analogous to the usual operators of analysis. We start with the partial derivatives $d\omega/d\varepsilon^\lambda$. If

$$（33.9） \qquad \omega = \varepsilon^\lambda \varepsilon^{\alpha_1} \ldots \varepsilon^{\alpha_s}$$

we set

$$（33.10） \qquad \frac{\partial \omega}{\partial \varepsilon^\lambda} = \varepsilon^{\alpha_1} \ldots \varepsilon^{\alpha_s}.$$

Any other monomial either coincides, up to sign, with a monomial of the form (33.9), or else does not contain ε^λ, in which case we set $\partial\omega/\partial\varepsilon^\lambda = 0$. We extend the operation of partial differentiation to arbitrary elements of the form (33.8) by linearity. Thus, in order to compute $\partial\omega/\partial\varepsilon^\lambda$, where ω is written as sum of monomials, we must put ε^λ in first place in each monomial, using the anticommutation relation, and then cross out the ε^λ.

It is easy to see that, for ω_1 an even or odd monomial,

$$（33.11） \qquad \frac{\partial}{\partial \varepsilon^\lambda} \omega_1 \omega_2 = \frac{\partial \omega_1}{\partial \varepsilon^\lambda} \omega_2 \pm \omega_1 \frac{\partial \omega_2}{\partial \varepsilon^\lambda},$$

where the $+$ sign holds for ω_1 even and the $-$ sign for ω_1 odd.

The definition of the integral of a function of anticommuting variables (*Berezin integral*) is based on the relations

$$（33.12） \qquad \int \varepsilon^\lambda \, d\varepsilon^\lambda = 1, \qquad \int d\varepsilon^\lambda = 0.$$

Multiple integration, with element $d^n\varepsilon = d\varepsilon^n \ldots d\varepsilon^1$, is defined by iteration.

From (33.12) it follows that $\int \omega \, d^n\varepsilon$ vanishes if ω is a monomial of degree less than n. The integral of $\varepsilon^{\alpha_1} \ldots \varepsilon^{\alpha_n}$ equals ± 1, depending on the parity of the permutation $\alpha_1 \ldots \alpha_n$. For an expression of the form (33.8), with the $a^k_{\alpha_1,\ldots,\alpha_k}$ skew-symmetric, we have

$$（33.13） \qquad \int \omega \, d^n = n! \, a^n_{1 \ldots n}.$$

Integration can be expressed in terms of differentiation:

$$\int \omega \, d^n \varepsilon = \frac{\partial^n \omega}{\partial \varepsilon^n \ldots \partial \varepsilon^1},$$

as can be seen from the relation

$$\frac{\partial^n}{\partial \varepsilon^n \ldots \partial \varepsilon^1}(\varepsilon^1 \ldots \varepsilon^n) = 1$$

and from the fact that after n-fold differentiation the degree drops to zero.

Integration of functions of anticommuting variables has many of the properties of ordinary integration. In particular, integration by parts takes the form

$$(33.14) \qquad \int \frac{\partial \omega_1}{\partial \varepsilon^\lambda} \omega_2 \, d^n \varepsilon = \mp \int \omega_1 \frac{\partial \omega_2}{\partial \varepsilon^\lambda} \, d^n \varepsilon,$$

where ω_1 is either even ($-$ sign) or odd ($+$ sign). This follows from the rule (33.11) for the derivative of a product and from the equation

$$(33.15) \qquad \int \frac{\partial \omega}{\partial \varepsilon^\lambda} d^n \varepsilon = 0,$$

for every ω, which in turn is a simple consequence of the fact that $\partial \omega / \partial \varepsilon^\lambda$ is a sum of monomials of degree less than n.

There is also a change-of-variable formula. For instance, in a Grassmann algebra one can replace the system of generators $\varepsilon^1, \ldots, \varepsilon^n$ by the system $\tilde{\varepsilon}^1, \ldots, \tilde{\varepsilon}^n$, where $\tilde{\varepsilon}^\alpha = A^\alpha_\beta \varepsilon^\beta$ for some nonsingular matrix A^α_β. Then (32.8) can be written in terms of the $\tilde{\varepsilon}^\alpha$ as

$$\omega = \omega(\varepsilon) = \sum \tilde{a}^k_{\alpha_1 \ldots \alpha_k} \tilde{\varepsilon}^{\alpha_1} \ldots \tilde{\varepsilon}^{\alpha_k},$$

with

$$(32.16) \qquad a^k_{\alpha_1 \ldots \alpha_k} = \tilde{a}^k_{\beta_1 \ldots \beta_k} A^{\beta_1}_{\alpha_1} \ldots A^{\beta_k}_{\alpha_k}.$$

We easily conclude from this that, if the coefficients in (32.8) are skew-symmetric,

$$(32.17) \qquad a^n_{1 \ldots n} = \det A \, \tilde{a}^n_{1 \ldots n},$$

and consequently that

$$\int \omega(\tilde{\varepsilon}) \, d^n \tilde{\varepsilon} = \int \det A^{-1} \, \omega(\varepsilon) \, d^n \varepsilon.$$

Notice that this rule differs from the change-of-variable rule for ordinary integrals: here, the determinant of the change-of-variable matrix appears raised to the power -1.

We will not discuss here more general (nonlinear) changes of variables. We will need, however, one special case: substitutions of the form $\varepsilon^\alpha \mapsto \varepsilon^\alpha + \beta^\alpha$, where the β^α anticommute. The "jacobian" for such a transformation is 1.

We now list simple examples of integrals over anticommuting variables. These formulas can easily be derived from the definitions.

For σ an even element of the Grassmann algebra, we define $\exp \sigma$ by the usual series expansion: we write $\sigma = \sigma_0 + \sigma_1$, where $\sigma_0 \in \mathbf{R}$ and σ_1 is nilpotent, and set

$$\exp \sigma = \exp \sigma_0 \left(\sum \frac{1}{n!} \sigma_1^n \right),$$

where the series has only finitely many non-zero terms. With this definition we have

$$(32.18) \qquad \int \exp(a\varepsilon^1 \varepsilon^2)\, d\varepsilon^2\, d\varepsilon^1 = a,$$

where a is any number.

We also have

$$(32.19) \qquad \int \exp\left(\sum_{\alpha=1}^{n} \chi_\alpha \varepsilon^\alpha \right) d^n \varepsilon = \chi_n \cdots \chi_1,$$

where the χ_α are anticommuting quantities (that is, the integrand is an element of the Grassmann algebra with generators $\chi_1, \ldots, \chi_n, \varepsilon^1, \ldots, \varepsilon^n$).

One often has to deal with expressions of the form (33.8) where the $a^k_{\alpha_1 \ldots \alpha_k}$ depend on variables $x^1, \ldots, x^m$. (We can consider such an expression as a function of ordinary (commuting) variables x^i as well as of anticommuting variables ε^α.) For such an expression ω the integral $\int \omega\, d^m x\, d^n \varepsilon$ is defined by iteration: assuming the $a^k_{\alpha_1 \ldots \alpha_k}$ are skew-symmetric in the α^i, we have

$$\int \omega\, d^m x\, d^n \varepsilon = n! \int a_{1 \ldots n}(x)\, d^m x.$$

The most important kind of integral over anticommuting variables is the Gaussian integral, namely, the integral of $\omega = \exp \sigma$, where $\sigma = \frac{1}{2} C_{\alpha\beta} \varepsilon^\alpha \varepsilon^\beta$ is a quadratic expression on the ε^α. Without loss of generality, we can assume that the matrix $C_{\alpha\beta}$ is skew-symmetric, and we do so from now on. We have

$$(33.20) \qquad \int \exp \sigma\, d^n \varepsilon = \int \exp(\tfrac{1}{2} C_{\alpha\beta}\, \varepsilon^\alpha \varepsilon^\beta)\, d^n \varepsilon = \sqrt{\det C_{\alpha\beta}}.$$

This follows immediately from (33.18) when $\sigma = \lambda_1 \varepsilon^1 \varepsilon^2 + \lambda_2 \varepsilon^3 \varepsilon^4 + \cdots$; the general case can be reduced to this one by a linear change of variables. A more general formula is sometimes useful:

$$(33.21) \qquad \int \exp(\tfrac{1}{2} C_{\alpha\beta}\, \varepsilon^\alpha \varepsilon^\beta + \chi_\alpha \varepsilon^\alpha)\, d^n \varepsilon = \sqrt{\det C}\, \exp(\tfrac{1}{2}\chi_a (C^{-1})^{\alpha\beta} \chi_\beta).$$

Here, as in (33.19), the χ_i are anticommuting objects; furthermore the matrix C is assumed invertible.

One often encounters Grassmann algebras equipped with an involution, interpreted as complex conjugation: $\omega \mapsto \bar{\omega}$. Usually in this case the algebra has an even number of generators that occur in complex conjugate pairs: $\varepsilon^1, \bar{\varepsilon}^1, \ldots, \varepsilon^n, \bar{\varepsilon}^n$. An element of such a Grassmann algebra is *real* if it does not change under complex conjugation; we also say that such an element is a real function of the generating variables $\varepsilon^1, \bar{\varepsilon}^1, \ldots, \varepsilon^n, \bar{\varepsilon}^n$. An important special case of (33.21) is the following: if $\sigma = C_{\alpha\beta} \varepsilon^\alpha \bar{\varepsilon}^\beta$ is a real quadratic form, that is, if the matrix $C_{\alpha\beta}$ is Hermitian, we have

$$(33.22) \qquad I = \int \exp(\sigma + \varepsilon^\alpha \bar\chi_\alpha + \chi_\alpha \bar\varepsilon^\alpha)\, d^n\bar\varepsilon\, d^n\varepsilon$$

$$= \int \exp(C_{\alpha\beta}\varepsilon^\alpha \bar\varepsilon^\beta + \varepsilon^\alpha \bar\chi_\alpha + \chi_\alpha \bar\varepsilon^\alpha)\, d^n\bar\varepsilon\, d^n\varepsilon$$

$$= \det C_{\alpha\beta} \exp(-\chi_a (C^{-1})^{\alpha\beta} \bar\chi_\beta).$$

Here, as in (33.21), C is assumed invertible. We can lift the invertibility assumption as follows: let φ_k^α be an orthonormal system of eigenvectors for the matrix $C_{\alpha\beta}$, and let λ_k be the corresponding eigenvalues. Assume that $\lambda_k = 0$ for $k = 1, \ldots, s$, and $\lambda_k \neq 0$ otherwise. Then

$$(33.23) \qquad I = \nu_1 \bar\nu_1 \ldots \nu_s \bar\nu_s \det{}' C \exp(-\chi_a (C^{-1})^{\alpha\beta} \bar\chi_\beta),$$

where $\nu_k = \chi_\alpha \varphi_k^\alpha$, $\det' C$ is the product of the nonzero eigenvalues of C, and C^{-1} denotes the operator that in the basis φ_k^α has diagonal elements λ_k^{-1} if $k > s$ and 0 if $k \leq s$. (C^{-1} is uniquely defined by $C^{-1}\Pi = 0$ and $CC^{-1} = C^{-1}C = 1 - \Pi$, where Π denotes projection onto the nullspace of C.) Equation (33.23) can be proved by the linear change of variables

$$\varepsilon^a \mapsto \bar\varphi_k^\alpha \varepsilon^k, \qquad \bar\varepsilon^\alpha \mapsto \varphi_k^\alpha \bar\varepsilon^k.$$

In addition one must use (33.21) and the equation

$$\int \exp(\chi_\alpha \bar\varepsilon^\alpha + \varepsilon^\alpha \bar\chi_\alpha)\, d^n\bar\varepsilon\, d^n\varepsilon = \chi_1 \bar\chi_1 \ldots \chi_n \bar\chi_n,$$

which is a particular case of (33.19).

Up to this point we have considered finitely generated Grassmann algebras. In quantum field theory one also encounters functions of an infinite number of anticommuting variables, that is, elements of Grassmann algebras with infinitely many generators.

Consider, for concreteness, a set of generators parametrized by a continuous parameter ξ: for example, ξ might be a real number or a point in $\mathbf{R}^n$. We denote the generators by $\varepsilon(\xi)$, and assume they satisfy

$$\varepsilon(\xi_1)\varepsilon(\xi_2) = -\varepsilon(\xi_2)\varepsilon(\xi_1).$$

The elements of the Grassmann algebra generated by the $\varepsilon(\xi)$ are given formally by

$$(33.25) \qquad \omega = \sum_k \int \omega^k(\xi_1, \ldots, \xi_n)\varepsilon(\xi_1) \ldots \varepsilon(\xi_k)\, d\xi_1 \ldots d\xi_k,$$

which is the same as (33.8) if we replace the discrete index α by the continuous index ξ and the sum by the integral. We say that ω is a function of infinitely many anticommuting variables, or that it is a functional of the anticommuting-valued function $\varepsilon(\xi)$. (Actually, the definition just given is not rigorous because we have not specified the class of functions $\omega^k(\xi_1, \ldots, \xi_n)$ that can appear in

(33.25). There are different ways in which the necessary refinements can be made, but we will not dwell on them here.)

By (33.24) we can, without loss of generality, assume that the coefficient functions in (33.25) are skew-symmetric in the ξ_i. Addition and multiplication of functionals of an anticommuting-valued function are defined in the obvious way. One can also define differentiation and integration. As with functionals of ordinary functions, the integral in this case can be defined as a limit of approximating multiple integrals. More precisely, in (33.22) we must replace integration with respect to the continuous parameter ξ by summation over a finite lattice, then evaluate the integral of the resulting function of a finite number of anticommuting variables, and finally take the limit. We will not dwell here on the choice of approximating multiple integrals or on the passage to the limit, since it is essentially the same process as in Chapter 23.

Formulas (33.20)–(33.23) for Gaussian integrals can also be applied to the infinite-dimensional case. For example, (33.20) becomes

$$(33.26) \qquad \int \exp\left(\frac{1}{2}\int C(\xi,\xi')\varepsilon(\xi)\varepsilon(\xi')\,d\xi\,d\xi'\right)\prod d\varepsilon(\xi) = (\det \hat{C})^{1/2},$$

where $\hat{C}$ is the operator whose matrix has entries $C(\xi,\xi')$.

In fermion theories one can express physical quantities as functional integrals if one allows integration over anticommuting-valued functions. More exactly, the important physical quantities can be represented as functional integrals whose integrands contain either $\exp(iS)$, where S is the action, or $\exp(-S_{\text{eucl}})$, where S_{eucl} is the Euclidean action. The only difference between this case and the ones studied so far is that here the fermion fields must be assumed to take on anticommuting values; accordingly, integration over these fields follows the rules laid out in this chapter.

For example, the partition function of a fermion field in an external gauge field (which we think of as classical) is

$$(33.27) \qquad Z = \int \exp(-S_{\text{eucl}})\prod d\psi^\dagger\, d\psi = \det(\slashed{\nabla} - im),$$

where we have assumed that the Euclidean action is given by (26.7). To remove infrared divergences we must introduce infrared cutoff by assuming, for example, that all fields are defined on a compact Riemannian manifold. Ultraviolet divergences are removed by regularizing determinants, as explained in Chapter 28. The Gaussian functional integral is evaluated using (33.26), and the Green's functions of the fermion field in the presence of an external gauge field A are

$$(33.28) \quad G_n(x_1,\ldots,x_n;y_1,\ldots,y_n \mid A)$$
$$= Z^{-1}\int \psi(x_1)\ldots\psi(x_n)\psi^\dagger(y_1)\ldots\psi^\dagger(y_n)\exp(-S_{\text{eucl}})\prod d\psi^\dagger\,d\psi$$

(this is the formula for a $2n$-point Green's function).

Instead of these Green's functions it is often convenient to consider their generating functional

$$(33.29) \qquad G[\eta, \eta^\dagger] = Z^{-1} \int \exp\left(-S_{\text{eucl}} + \int (\eta^\dagger \psi + \psi^\dagger \eta)\, dx\right) \prod d\psi^\dagger\, d\psi.$$

Evaluating this functional integral by means of the infinite-dimensional analogue of (33.22), we obtain

$$(33.30) \qquad G[\eta, \eta^\dagger] = \exp\left(\int \eta^\dagger(x) D(x, x' \mid A)\eta(x')\, dx\, dx'\right),$$

where $D(x, x' \mid A) = \langle x|(\slashed{\nabla} - im)^{-1}|x'\rangle$.

So far we have assumed that the Dirac operator $\slashed{\nabla} - im$ has no zero modes. If this is not the case, we obtain $Z = 0$. But even in the presence of zero modes the functional integral

$$\int \psi(x_1) \ldots \psi(x_n)\psi^\dagger(y_1) \ldots \psi^\dagger(y_n) \exp(-S_{\text{eucl}}) \prod d\psi^\dagger\, d\psi$$

can be non-zero. It is therefore still convenient to define the Green's functions of a fermion field by means of (33.28), where we replace Z by $\det(\slashed{\nabla} - im)$ and disregard zero eigenvalues when computing the determinant (see Chapter 28). Then the generating functional of the Green's functions can be defined, as before, by (33.29). Applying the infinite-dimensional version of (33.23), we get

$$(33.31) \quad G[\eta, \eta^\dagger] = \chi_1^\dagger \chi_1 \ldots \chi_s^\dagger \chi_s \exp\left(\int \eta^\dagger(x)\langle x|(\slashed{\nabla} - im)^{-1}|x'\rangle\eta(x')\, dx\, dx'\right),$$

where $\chi_i = \langle \eta, \varphi_i \rangle = \int \eta^\dagger(x)\varphi_i(x)\, dx$ and $\varphi_i(x)$ runs over the elements of an orthonormal system of zero modes for the Dirac operator. This formula readily yields an expression for the Green's functions of a fermion field in the presence of an external gauge field. in particular, we see that the first nonzero Green's function is a $2s$-point Green's function, where s is the number of zero modes. This function is given by
$$(33.32)$$
$$G_s(x_1, \ldots, x_s; y_1, \ldots, y_s \mid A) = \text{asym}(\varphi_1(x_1) \ldots \varphi_s(x_s)\varphi_1^\dagger(y_1) \ldots \varphi_s^\dagger(y_x)),$$

where asym stands for antisymmetrization in the x_i's and y_i's.

So far we have discussed fields in a space of arbitrary dimension. We now concentrate on massless fermion fields in a $2n$-dimensional manifold. In this case the Euclidean action is invariant under the substitution

$$(33.33) \qquad \psi(x) \mapsto \psi'(x) = \exp(\beta\gamma_{2n+1})\psi(x),$$

where $\gamma_{2n+1} = -i^n\gamma_1\gamma_2 \ldots \gamma_{2n}$. But the Green's functions are not generally invariant under this transformation. Indeed, without loss of generality we can assume that the zero modes of the Dirac operator in the expression for the Green's function have a well-defined parity (that is, are either left zero modes or right zero modes). The difference $n_- - n_+$, where n_- and n_+ are the numbers of left and right zero modes, respectively, is the index of the Dirac operator considered as acting on the space of left spinors (Chapter 27). The preceding formulas show that under the transformation (33.33) the Green's function is multiplied by $\exp(-2\beta(n_- - n_+))$, and by (27.25) this equals to $\exp(-2\beta\alpha_T q(A))$, where $q(A)$ is the topological number of the field:

$$(33.34) \qquad G'_n(x_1, \ldots, x_n) = G_n(x_1, \ldots, x_n) \exp(-2\beta\alpha_T q(A)).$$

We have just computed the so-called *axial anomaly* for the particular case of axial (chiral) transformations. For the sake of completeness we dwell briefly on a more general situation. Assume the gauge field on a four-dimensional manifold contains the matrix γ^5, that is, that it can be written as

$$A_\mu(x) = V_\mu(x) + i\gamma^5 W_\mu(x),$$

where the fields $V_\mu(x)$ and $W_\mu(x)$ (the vector and axial parts of the gauge field) take values in the Lie algebra of the gauge group. The Euclidean action integral (26.7) is invariant under gauge transformations of the fermion field, $\psi(x) \mapsto T(g(x))\psi(x)$, if the field $A_\mu(x)$ is gauge-transformed at the same time. If the fermion field is massless, there is also an additional invariance under axial gauge transformations. Infinitesimal axial gauge transformations have the form

$$(33.35) \qquad \psi(x) \mapsto (1 + i\gamma^5 t(\omega(x)))\psi(x),$$
$$(33.36) \qquad A_\mu(x) \mapsto A_\mu(x) - i\gamma^5 \partial_\mu \omega(x) - i\gamma^5 [A_\mu(x), \omega(x)],$$

where $\omega(x)$ is a function with values in the Lie algebra of the gauge group.

While the invariance of the Euclidean action integral under gauge transformations implies that the partition function is likewise invariant, the axial gauge invariance breaks down at the quantum level, that is, an anomaly emerges. In other words, the partition function does not change if A_μ is replaced by a gauge-equivalent field, but it does change if A_μ undergoes the transformation (33.36). As mentioned in Chapter 29, the appearance of anomalies is due to the non-invariance of the scalar product. Under gauge transformations the scalar product of fermion fields is preserved, but under the axial gauge transformation (33.35) it changes as follows:

$$(33.37) \qquad \delta(\psi_1, \psi_2) = 2(\psi_1, \gamma^5 \omega \psi_2).$$

The ideas discussed in Chapter 29 allow one to reduce the computation of the change in the determinant of $\nabla\!\!\!\!/$ under the substitution (33.36) to the calculation of the change in the same determinant under a change in the scalar product of fermion fields described by (10.5). Using (29.17), we get

$$\delta \ln Z = \delta \ln \det \nabla\!\!\!\!/ = \tfrac{1}{2}\delta \ln \det \nabla\!\!\!\!/^{\,2} = \frac{\alpha_T}{16\pi^2} \int \omega(x)\langle \mathcal{F}_{\mu\nu}, \tilde{\mathcal{F}}^{\mu\nu}\rangle \, d^4 x.$$

34. Instantons in Quantum Chromodynamics

Recall that the QCD Lagrangian is obtained from the free fermion Lagrangian

$$(34.1) \qquad \mathcal{L} = \sum_f \bar{\psi}_f \, \partial\!\!\!/ \, \psi_f - i \sum_f m_f \bar{\psi}_f \psi_f$$

by localizing the SU(3)-symmetry. Here, for f fixed, $\psi_f = (\psi_f^1, \psi_f^2, \psi_f^3)$ is a three-dimensional vector of bispinors. In other words, when considering the quark fields ψ_f^k, where k specifies the color and f the flavor, we discard the color index k, a consider ψ_f as a vector in color space.

The Euclidean action associated with (34.1) is

$$(34.2) \qquad S = \int \sum_f (\langle \psi_f, \partial\!\!\!/ \, \psi_f \rangle - i m_f \langle \psi_f, \psi_f \rangle) \, dV,$$

where the components ψ_f^k of ψ_f take values in the space of the spinor representation of SO(4), and $\partial\!\!\!/$ denotes the Euclidean Dirac operator. The Euclidean QCD action functional, obtained by localizing the SU(3)-symmetry in (34.2), is

$$(34.3) \qquad S = \int \sum_f (\langle \psi_f, \nabla\!\!\!\!/ \, \psi_f \rangle - i m_f \langle \psi_f, \psi_f \rangle) \, dV + S_{\mathrm{YM}},$$

where S_{YM} is the Euclidean action functional for the gauge field A_μ that takes values in the Lie algebra of SU(3) (namely, the algebra of anti-Hermitian traceless matrices), and $\nabla\!\!\!\!/ = i\gamma^\mu \nabla_\mu = i\gamma^\mu(\partial_\mu + A_\mu)$. (Here the action of A_μ on ψ_f is by matrix multiplication, ψ_f being understood as a column vector in color space.)

In order to express the QCD Green's functions in terms of functional integrals, we must think of the fermion fields in (34.1)–(34.3) as fields with anti-commuting values. For example, for a given gauge-invariant functional $\Phi[A, \varphi]$ of the fields A_μ and ψ_f, the corresponding Euclidean Green's function is Z_Φ/Z, where

$$(34.4) \qquad Z_\Phi = \int \Phi[A, \psi] e^{-S} \prod dA \prod d\psi \, d\psi^\dagger$$

and the partition function Z is given by he same formula with $\Phi[A, \psi]$ replaced by 1. Take, for example,

$$(34.5) \qquad \Phi[A, \psi] = \langle b_\Gamma \psi(x_1), \psi(x_2) \rangle,$$

where $b_\Gamma = \mathrm{P}\exp(-\int_\Gamma A_\mu \, dx^\mu)$ is the path-ordered integral over a curve Γ joining x_1 and x_2. This functional is gauge-invariant and the corresponding Green's function is called the *two-point fermion Green's function*. Of course, for the functional integral (34.4) to have a meaning, we must introduce infrared and ultraviolet cutoff, the former by considering fields over a compact Riemannian manifold (e.g., the sphere) and the letter by passing to a lattice.

Without loss of generality we can assume that

(34.6) $$\Psi[A,\psi] = \psi(x_1)\dots\psi(x_m)\psi^\dagger(y_1)\dots\psi^\dagger(y_n)V[A];$$

any gauge-invariant functional is a linear combination of functionals of this form. The resulting Green's function is nonzero only if $m = n$ in (34.6), because the expression is invariant under the transformation $\psi \mapsto e^{i\alpha}\psi$ (which implies $\psi^\dagger \mapsto e^{-i\alpha}\psi^\dagger$). From now on we assume that $m = n$. The Green's function corresponding to (34.6) is called the *2n-point fermion Green's function*.

Note that we are restricting our attention to gauge-invariant functionals $\Phi[A,\psi]$. In particular, this excludes the possibility of setting $V[A] \equiv 1$ in (34.6). (It is often convenient to consider Green's functions coming from $V[A] \equiv 1$, but this requires fixing the gauge.)

The functional integral over the anticommuting fields $\psi, \psi^\dagger$ is Gaussian and can be expressed by means of the formulas of the previous chapter. For example, applying (33.27) we get for the partition function:

(34.7) $$Z = \int \prod_f \det(\slashed{\nabla} - im_f)\exp(-S_{\mathrm{YM}})\prod dA.$$

For the functional integral (34.4), with $\Phi[A,\psi]$ of the form (34.6), we get (34.8)
$$Z_\Phi = \int G_n(x_1,\dots,x_n;y_1,\dots,y_n \mid A)V[A]\prod_f \det(\slashed{\nabla} - im_f)\exp(-S_{\mathrm{YM}})\prod dA,$$

where $G_n(x_1,\dots,x_n;y_1,\dots,y_n \mid A)$ is the fermion Green's function in the presence of an external gauge field A. (Here we have used (33.28), where, as discussed before, zero modes should be discarded in the computation of the determinant of the Dirac operator.) Note that we have discarded spin, color and flavor indices from (34.7) and (34.8).

The integrals (34.7) and (34.8) can be computed using the Laplace method. The contribution of fields with topological number q can be written as an integral over the manifold of instantons with this number. From Chapter 32 it follows that the contribution from q-instantons is

(34.9) $$J_q = g^{-\sigma}\exp\left(-\frac{8\pi^2}{g^2}q\right)\int_{R_q} G_n(x_1,\dots,x_n;y_1,\dots,y_n \mid A)V[A]\,d\mu,$$

where the notation is the same as in Chapter 32, the Green's functions and the functional $V[A]$ are calculated in the presence of the instanton field A, and

$$d\mu = \frac{\nu(A)}{\nu(0)}d\mu_0,$$

with $\nu(a) = \prod_f \det(\slashed{\nabla} - im_f)\gamma(A) = \prod_f \det(\slashed{\nabla} - im_f)\det^{-1/2}\Delta_1^A \det \Delta_0^A$.

If the quark masses are small, we can write

$$(34.10) \qquad \det(\slashed{\nabla} - im_f) \approx m_f^N \det \slashed{\nabla}_A,$$

where N is the number of zero modes of $\slashed{\nabla}_A$, and the zero modes are disregarded in the calculation of $\det \slashed{\nabla}_A$.

To find out the number of zero modes of the Dirac operator, we can use the index theory developed in Chapter 27. By (27.26), the difference between the numbers of left and right zero modes of $\slashed{\nabla}_A$ (that is, the index of $\slashed{\nabla}_A$ considered as acting on left spinors only) is equal to three times the topological charge q of A. (We assume that spatial cutoff is carried out by considering the fields on a sphere or torus and applying (27.26), taking into account that signature of either of these manifolds is zero, and that the Dynkin index of the vector representation of $SU(3)$ is three.) We conclude that the number N of zero modes of $\slashed{\nabla}_A$ is no less than $3|q|$.

If the field A is a solution to the self-duality equation (30.12) or of the anti-self-duality equation (30.13), the number of zero modes is exactly $3|q|$. Indeed, in this case one can show that $\slashed{\nabla}_A$ has only left zero modes (if $q > 0$) or only right zero modes (if $q < 0$). If, for example, the field is given on a torus and satisfies the self-duality equation, the square of the Dirac operator on right spinors is given by

$$(34.11) \qquad \nabla^2 = -\nabla_\mu \nabla^\mu;$$

this follows from the formula given at the end of Chapter 27 for $\slashed{\nabla}^2$, and from the fact that $\frac{1}{2}\varepsilon_{\mu\nu\rho\sigma}\gamma^\rho\gamma^\sigma = -\gamma_\mu\gamma_\nu\gamma^5$. Noting that $(\psi, -\nabla_\mu\nabla^\mu\psi) = (\nabla_\mu\psi, \nabla^\mu\psi)$, we see that any right zero mode ψ of the Dirac operator satisfies $\nabla_\mu\psi = 0$. Using the relation between $\mathcal{F}_{\mu\nu}$ and the commutator of covariant derivatives, we obtain $t(\mathcal{F}_{\mu\nu})\psi = 0$. If the self-dual field is irreducible (Chapter 31), we get $\psi \equiv 0$. We conclude that, for an instanton A,

$$(34.12) \qquad \det(\slashed{\nabla}_A - im_f) \approx m_f^{3|q|} \det \slashed{\nabla}_A$$

if the quark mass m_f is small. This allows us to write the measure on the instanton manifold as

$$(34.13) \qquad d\mu = \prod_f m_f^{3|q|} \frac{\tilde{\nu}(A)}{\tilde{\nu}(0)} d\mu_0,$$

where

$$(34.14) \qquad \tilde{\nu}(A) = \det \slashed{\nabla}_A (\det \Delta_1^A)^{-1/2} \det \Delta_0^A.$$

If $q = 1$, we get

$$(34.15) \qquad \exp\left(-\frac{8\pi^2}{g^2}\right)d\mu = \prod_f m_f^3 \exp\left(-\frac{8\pi^2}{g^2(\lambda)}\right)\frac{d\lambda\, d^4a}{\lambda^5},$$

where, as usual, a is the position of the center of the instanton, λ is its size, which we assume much smaller than the spatial cutoff parameter, and $g(\lambda)$ is the effective coupling constant:

$$(34.16) \qquad \frac{8\pi^2}{g^2(\lambda)} = \frac{8\pi^2}{g_0^2} - (11 - \tfrac{2}{3}n_f)\ln\frac{\lambda}{\lambda_0}.$$

Equation (34.15) can be derived from renorm-group considerations. Alternatively, one can use the approach from the end of Chapter 28 to compute the variation of the determinant of the Dirac operator arising form a scale change in the gauge field. These calculations give the dependence of det ∇_A on the instanton size. The other determinants appearing in (34.14) can be found from the discussion in Chapter 22. This way of computing the single-instanton measure also gives the effective coupling constant.

We now turn to the case where one of quarks is massless. The contribution of the topologically nontrivial fields to the partition function vanishes in this case: indeed, the index of the Dirac operation in a field with a nonzero topological charge is nonzero, so there are always zero modes and the determinant must vanish.

Nonetheless, a topological nontrivial field can contribute to the functional integral (34.4). The results derived at the end of Chapter 33 imply that the $2n$-point Green's function of the Dirac operator in an external gauge field can be nonzero only if $n \geq N$, where N is the number of zero modes of the Dirac operator. Hence, fields with topological charge q can contribute to the $2n$-point Green's function only if $n \geq |3q|$, and in particular topologically nontrivial fields only contribute to Green's functions with $n \geq 3$, and only instantons with $q = \pm 1$ contribute to the six-point Green's function. Equation (33.13) can be used to express this contribution in terms of the zero modes of the Dirac operator in the field of an instanton of size λ and center a.

Note that for massless quarks the QCD action is invariant not only under transformations $\psi \mapsto \exp(i\alpha)\psi$, but also with respect to the transformations $\psi \mapsto \exp(\beta\gamma^5)\psi$ (*chiral invariance*). These invariance properties also hold for the Euclidean action integral, provided that the definition of a chiral transformation is modified appropriately (Chapter 33). It is important to note that taking into account the instanton contribution violates the chiral $U(1)$-symmetry. Indeed, at the end of Chapter 33 we discovered that this invariance is violated already in the calculation of the Green's function of a fermion in an external gauge field of nonzero topological charge. It can be said that instantons provide a solution to the so-called $U(1)$-*problem*, which can be posed as follows.

Recall that in the massless quark approximation the breakdown of chiral invariance under rotations in flavor space leads to the existence of massless Goldstone bosons (Chapter 21). In the real world the first three quarks have a small finite mass, and this leads to an approximate chiral $U(1)$-invariance, whose breakdown guarantees the existence of a multiplet of eight pseudoscalar mesons of relatively small mass. By analogy, one could think that there exists a ninth light meson, related to the breakdown, discussed above, of the chiral

$U(1)$-symmetry. Such a meson does not exist in nature, and the $U(1)$-problem consists in explaining its absence. We have seen that if we calculate the Green's functions taking into account the instanton contribution, they turn out not to be invariant under the chiral group $U(1)$, and do not contain any poles corresponding to the massless boson that is connected with the breakdown of $U(1)$-symmetry.

As in the case of a purely gauge theory, we can incorporate into the QCD Lagrangian a θ-term $\frac{1}{2}i\theta \int d^4x \, \langle \mathcal{F}_{\mu\nu}, \tilde{\mathcal{F}}^{\mu\nu} \rangle$, without changing the classical equations of motion. This term again leads to a change in the contribution to the Green's function from fields with topological charge $q \neq 0$: there appears an additional factor $\exp(-16\pi^2 q\theta i)$. In the case of massless quarks this change can be eliminated by redefining the fermion fields. Indeed, as noted earlier, in this case there is chiral $U(1)$-invariance at the classical level, but this invariance is broken at the quantum level. It can easily be verified, using the formulas obtained at the end of Chapter 33, that the change in the fermion Green's functions brought about by a chiral $U(1)$-transformation can be canceled by an appropriate θ-term in the Lagrangian. In the real world of nonzero quark masses the θ-term must lead to non-conservation of parity in strong interactions. This suggests that θ is close to zero.

Part III

Mathematical Background

35. Topological Spaces

We say that a set E is a *topological space* if we have in it the notion of *open sets*, a class of subsets of E satisfying the following conditions:

1. The intersection of a finite number of open sets is an open set.

2. The union of any number of open sets is an open set.

A metric space X can be made into a topological space in a natural way, if we define as open sets all unions of *open balls* $B_\varepsilon(x) = \{y : d(x,y) < \varepsilon\}$, for $x \in X$ and $\varepsilon > 0$.

Using the notion of open sets in a topological space, one defines *neighborhoods* and *limits* in the standard way. Continuity of maps between topological spaces is defined in terms of neighborhoods exactly as in the case of metric spaces. Equivalently, we say that a map $\varphi : E \to F$ between topological spaces is *continuous* if $\varphi^{-1}(U) \subset E$ is open whenever $U \subset F$ is open. We say that φ is a *topological equivalence*, or a *homeomorphism*, if φ is one-to-one and bicontinuous (this means that both φ and φ^{-1} are continuous).

A subset $F \subset E$ of a topological space E is *closed* if its complement $E \setminus F$ is open. Clearly, the union of finitely many closed sets is closed, and the intersection of any number of closed sets is closed. If all points x_n of a convergent sequence belong to a closed set F, the limit of the sequence also belongs to F.

Every subset E' of a topological space E is itself a topological space in a natural way: a set $U \subset E'$ is defined to be open in E' if and only if there exists an open set $V \subset E$ such that $V \cap E' = U$. This topology on E' is said to be *induced* by the topology on E. In what follows we will always give subsets of topological spaces the induced topology.

Axioms 1 and 2 above are too weak to guarantee that limits and other constructions behave in reasonable ways. For example, without further conditions, a sequence might converge to two distinct limits. In order to exclude such pathologies one generally works with topological spaces that satisfy additional requirements, called *separation axioms*. One can require, for example, that the intersection of all the neighborhoods of a point consist of that point alone; this implies that every set consisting of a single point is closed.

More commonly, the separation axiom that is used is the *Hausdorff axiom*, which says that any two distinct points have disjoint neighborhoods. In a Hausdorff space a sequence can have only one limit. All concrete topological spaces encountered in this book satisfy the Hausdorff axiom, and even stronger separation conditions. For this reason we will assume (except in Chapter 38) that all spaces are Hausdorff.

A topological space is *discrete* if every one-point set is open. For instance, the set of integers, with the topology induced from the real line, is discrete. So is a finite space: by our standing assumption that the Hausdorff axiom is satisfied, every one-point set is closed; therefore the complement of a one-point set, being the union of finitely many closed sets, is closed.

A topological space E is *compact* if every open cover of E has a finite subcover. An *open cover* of E is a family $\{U_\alpha\}$ of open subsets of E whose union is all of E; compactness means that we can choose finitely many indices $\alpha_1, \ldots, \alpha_n$ such that $\{U_{\alpha_1}, \ldots, U_{\alpha_n}\}$ is still a cover for E. If E is compact, every sequence of points $x_n \in E$ has a convergent subsequence, that is, a subsequence $x_{n_1}, \ldots, x_{n_k}, \ldots$ that converges to a point of E. For metric spaces this condition is equivalent to compactness. A compact subspace E of a space E' is closed in E'. Every closed subspace of a compact space is compact. If φ is a continuous map from a compact space E into an arbitrary space E', the image $\varphi(E)$ of E is compact. If φ is a continuous and one-to-one map from a compact space E onto a space E', the inverse φ^{-1} of φ is also continuous, so φ is a homeomorphism.

The *direct product* (or simply *product*), $E_1 \times E_2$ of two sets E_1 and E_2 is the set of ordered pairs $\{(e_1, e_2) : e_1 \in E_1, e_2 \in E_2\}$. If E_1 and E_2 are topological spaces, $E_1 \times E_2$ has a natural topology, called the *product topology*: a set $U \subset E_1 \times E_2$ is open if it is the union of sets of the form $U_1 \times U_2$, for U_1 open in E_1 and U_2 open in E_2. A map $\varphi : E_1 \times E_2 \to E$, where E is another topological space, can be seen as a function $\varphi(e_1, e_2)$ of two variables, with values in E. Saying that φ is continuous implies that it is continuous jointly in both variables. The product of two compact spaces is compact.

If every point of a topological space E has a neighborhood that is homeomorphic to $\mathbf{R}^n$, we call E an *n-dimensional manifold*. Thus, an n-dimensional manifold is a space that can be covered by open sets, called *charts*, each of which is equipped with a *local coordinate system* $(x^1, \ldots, x^n)$. If, for any two overlapping charts, the transition from one local coordinate system to another on the overlap is a smooth map, we say that E is a *smooth* manifold. A two-dimensional manifold is also called a *surface*.

A *vector* at a point $e \in E$ of a n-dimensional smooth manifold is an assignment of an n-tuple of numbers to each chart that contains e, with the condition that the n-tuples $(A^1, \ldots, A^n)$ and $(\tilde{A}^1, \ldots, \tilde{A}^n)$ associated with the charts with coordinate systems $(x^1, \ldots, x^n)$ and $(\tilde{x}^1, \ldots, \tilde{x}^n)$ satisfy the relation

$$\tilde{A}^i = \frac{\partial \tilde{x}^i}{\partial x^j} A^j.$$

The set of all vectors at a point $e \in E$ is a linear space, called the *tangent space* to E at e.

36. Groups

Let G be a set with a *composition law*, that is, a rule that assigns to each pair $(a, b) \in G \times G$ an element of G, denoted ab. The composition law is often called *multiplication*. We say that an element $e \in G$ is an *identity* if $ae = ea = a$ for every $a \in G$. Any set has at most one identity. An element $b \in G$ is the *inverse* of another element $a \in G$ if $ab = ba = e$, where e is the identity. Multiplication on G is *associative* if $(ab)c = a(bc)$ for every $a, b, c \in G$. A *group* is a set G with an associative multiplication, having an identity, and in which every element has an inverse. It is easy to verify that in this case each element has exactly one inverse; we denote the inverse of $a \in G$ by a^{-1}. In a group, $ab = e$ implies $ba = e$ and therefore $b = a^{-1}$, where e is the identity. The *left translation* by an element $g \in G$ is the map $L_g : G \to G$ taking h to gh; the *right translation* R_g takes h to hg.

A map $\varphi : G \to G'$ from one group to another is a *homomorphism* if it preserves multiplication, that is, if $\varphi(ab) = \varphi(a)\varphi(b)$. An *isomorphism* is a homomorphism that is one-to-one and onto. An isomorphism of a group onto itself is also called an *automorphism*. The *inner automorphisms* of a group G are the maps α_g defined by

$$(35.1) \qquad\qquad \alpha_g(h) = ghg^{-1};$$

we also call α_g *conjugation by* g, and two elements $h \in G$ and ghg^{-1} are called *conjugate*.

A subset $H \subset G$ is a *subgroup* of G if all products of elements of H and all inverses of elements of H are still in H. A subgroup H is *normal* if it is invariant under inner automorphisms, that is, if $h \in H$ implies $ghg^{-1} \in H$ for all $g \in G$.

Two elements h and h' of G are *conjugate* if they are taken to one another by an inner automorphism, that is, if there is $g \in G$ such that $h' = ghg^{-1}$. Similarly, two subgroups H and H' of G are *conjugate* if $H' = gHg^{-1}$ for some $g \in G$. Thus, a subgroup is normal if it has no conjugates other than itself.

The *image* $\mathrm{Im}\,\varphi$ of a group homomorphism $\varphi : G \to G'$ is the set of elements $\varphi(g)$, for $g \in G$; clearly $\mathrm{Im}\,\varphi$ is a subgroup of G'. The *kernel* of φ is the set of elements that map to the identity $e \in G$, that is, $\mathrm{Ker}\,\varphi = \varphi^{-1}(e)$; it is easy to check that $\mathrm{Ker}\,\varphi$ is a normal subgroup of G. A homomorphism is injective if and only if its kernel is the *trivial group*, that is, the group that consists only of the identity element.

Two elements $g, g' \in G$ are said to *commute* if $gg' = g'g$. A group G is called *commutative* or *abelian* if all its elements commute with one another.

Often for a commutative group the composition law is called *addition* instead of multiplication, and the identity element is called *zero*; accordingly we write $a+b$ instead of ab and 0 instead of e. The set $\mathbf{R}$ of real numbers is an abelian group under the usual addition, and the set $\mathbf{R}_+$ of positive real numbers is an abelian group under the usual multiplication. The map $x \mapsto e^x$ is an isomorphism between these two groups.

The totality of the *transformations* of a set X—that is, of the one-to-one maps from X onto itself—forms a group, with multiplication given by composition of maps:

$$(fg)(x) = (f \circ g)(x) = f(g(x)).$$

This is called the *full transformation group* of X, and any of its subgroups is also called a *transformation group*. For transformations groups one generally writes 1 for the identity.

The set of invertible $n \times n$ real matrices (those with nonzero determinant) forms a group under the usual matrix multiplication; we denote this group by $\mathrm{GL}(n, \mathbf{R})$, or simply $\mathrm{GL}(n)$. We define $\mathrm{GL}(n, \mathbf{C})$ similarly. A homomorphism from a group G into $\mathrm{GL}(n, \mathbf{R})$ or $\mathrm{GL}(n, \mathbf{C})$ is called a (real or complex) *representation* of G (see also Chapter 39).

The subgroup of $\mathrm{GL}(n, \mathbf{R})$ consisting of orthogonal matrices is denoted $O(n)$, and the subgroup of $\mathrm{GL}(n, \mathbf{C})$ consisting of unitary matrices is denoted $U(n)$. The subgroups of $O(n)$ and $U(n)$ consisting of matrices of unit determinant are denoted by $\mathrm{SO}(n)$ and $\mathrm{SU}(n)$. A *matrix group* is one of $\mathrm{GL}(n, \mathbf{R})$, $\mathrm{GL}(n, \mathbf{C})$ or their subgroups. We can regard matrix groups as groups of linear transformations, since linear transformations are in one-to-one correspondence with invertible matrices, the correspondence being established by the choice of a basis.

The group of linear transformations of four-dimensional space that leave invariant the space-time element

$$ds^2 = (dx^0)^2 - (dx^1)^2 - (dx^2)^2 - (dx^3)^2 = g_{ik}\, dx^i\, dx^k$$

is called the *full Lorentz group*. The *Lorentz group* is the subgroup of the full Lorentz group consisting of transformations that preserve the direction of time and the orientation of space; in terms of the entries a_k^i of the transformation's matrix, this means that $a_0^0 > 0$ and $\det a_k^i > 0$.

A *topological group* is a group G that is also a topological space, and such that multiplication and inversion are continuous maps (naturally, multiplication is seen as a map $G \times G \to G$). A subgroup of a topological group is also a topological group. Since $\mathrm{GL}(n, \mathbf{C})$ is a topological group (being an open subset of the space of $n \times n$ complex matrices, which can be identified with $\mathbf{R}^{2n^2}$), every subgroup of $\mathrm{GL}(n, \mathbf{C})$ is also a topological group.

The *direct product* of two groups G_1 and G_2 is the product of the sets G_1 and G_2, with componentwise multiplication: $(g_1, g_2)(g_1', g_2') = (g_1 g_1', g_2 g_2')$. If G_1 and G_2 are topological groups, so is $G_1 \times G_2$, with the product topology. If G_1 and G_2 are abelian groups with group law denoted by addition, we generally talk about the *direct sum* $G_1 \dotplus G_2$ of G_1 and G_2, instead of their direct product. We then write $(g_1, g_2) + (g'_1, g'_2) = (g_1 + g'_1, g_2 + g'_2)$.

A topological group G is an *n-dimensional Lie group* if, as a topological space, it is an n-dimensional manifold (see the end of Chapter 35). One can show that in this case the multiplication and inversion maps are smooth, not just continuous, for appropriately chosen local coordinate systems on G. Therefore G can be considered a smooth manifold. $GL(n, \mathbf{R})$, $GL(n, \mathbf{C})$, $U(n)$, $O(n)$, $SU(n)$ and $SO(n)$ are all examples of Lie groups.

A (real) *Lie algebra* is a real vector space $\mathcal{A}$ equipped with a *bracket oper-ation* $(a, b) \mapsto [a, b]$ that is distributive (that is, $[\lambda a + \mu b, c] = \lambda[a, c] + \mu[b, c]$ for $\lambda, \mu \in \mathbf{R}$ and $a, b, c \in \mathcal{A}$), anticommutative (that is, $[a, b] = -[b, a]$ for $a, b \in \mathcal{A}$), and satisfies the *Jacobi identity*

$$[a, [b, c]] + [b, [c, a]] + [c, [a, b]] = 0$$

for $a, b, c \in \mathcal{A}$. A complex Lie algebra is defined analogously. As an example of a Lie algebra, take the vector space of all real $n \times n$ matrices (with the usual addition and multiplication by scalars), and let the bracket of two elements be their commutator $[a, b] = ab - ba$.

Lie subalgebras, homomorphisms of Lie algebras, and the direct sum of two Lie algebras are defined like their group counterparts. For example, a map $t : \mathcal{A} \to \mathcal{A}'$ between Lie algebras is a homomorphism if it is linear and preserves the bracket operation, that is, $t([a, b]) = [t(a), t(b)]$ for $a, b \in \mathcal{A}$.

Every Lie group G has an associated Lie algebra, which we denote by $\mathcal{G}$. As a linear space, $\mathcal{G}$ is simply the tangent space to G at the identity element. If G is a matrix Lie group (a subgroup of $GL(n, \mathbf{R})$), the tangent space can be seen as a linear subspace of the space of all $n \times n$ matrices, and the bracket operation on $\mathcal{G}$ is given by the commutator $[a, b] = ab - ba$; one can show that the commutator of two matrices in $\mathcal{G}$ also lies in $\mathcal{G}$.

The Lie algebras of $GL(n, \mathbf{C})$ and $GL(n, \mathbf{R})$ are denoted by $\mathfrak{gl}(n, \mathbf{C})$ and $\mathfrak{gl}(n, \mathbf{R})$, and they consist, respectively, of all complex and all real $n \times n$-matrices. The Lie algebra of $U(n)$ is the space $\mathfrak{u}(n)$ of complex anti-Hermitian matrices: indeed, if $1 + a \in U(n)$ differs infinitesimally from the identity, the unitariness condition $(1 + a)(1 + a^\dagger) = 1$ becomes $a + a^\dagger = 0$. The Lie algebra of $SU(n)$ is the space $\mathfrak{su}(n)$ of traceless anti-Hermitian matrices, and the Lie algebra of $SO(n)$ is the space $\mathfrak{so}(n)$ of real skew-symmetric matrices.

Representations of Lie groups are closely related to representations of Lie algebras. An n-dimensional complex representation of a Lie algebra $\mathcal{G}$ is a homo-morphism from $\mathcal{G}$ into $\mathfrak{gl}(n, \mathbf{C})$, and real representations are defined analogously. If T is a representation of a Lie group G, there is a corresponding representa-tion of the Lie algebra $\mathcal{G}$. More generally, given a Lie group homomorphism $\varphi : G \to G'$, there is a corresponding Lie algebra homomorphism $\mathcal{G} \to \mathcal{G}'$, which is the differential (T4.1) of the map φ at the identity (recall that $\mathcal{G}$ is identified with the tangent space to G at the identity, and likewise for $\mathcal{G}'$).

Two topological groups G and G' are *locally isomorphic* if there is a one-to-one correspondence between neighborhoods of the identity in G and G' that preserves multiplication. Local isomorphism is denoted by $G \approx G'$; isomorphism is denoted by $G = G'$. Two Lie groups having the same Lie algebra $\mathcal{G}$ are locally isomorphic. Among Lie groups with Lie algebra $\mathcal{G}$ there is a simply connected Lie group G_0, unique up to isomorphism; for any Lie group G with the same Lie algebra there is a homomorphism $G_0 \to G$ whose kernel is a discrete, central subgroup of G_0 (*central* means each element of this subgroup commutes with all elements of G). We can consider G_0 the universal cover of G (T3.3).

For example, $U(n)$ is locally isomorphic to $\mathbf{R}_+ \times SU(n)$. To define the local isomorphism, we associate with each pair $(e^\alpha, u) \in \mathbf{R}_+ \times SU(n) \to U(n)$ the element $e^{i\alpha u} \in U(n)$. This gives a homomorphism, whose kernel is a subgroup of $\mathbf{R}_+ \times SU(n)$ isomorphic to $\mathbf{Z}$. As a special case, $U(1) \approx \mathbf{R}_+$, since $SU(1) = \{1\}$. It follows that

$$U(n) \approx \mathbf{R}_+ \times SU(n) \approx U(1) \times SU(n).$$

Every compact topological group G can be given an invariant measure, unique up to a constant factor. Therefore one can integrate over G, and one has the equalities

$$\int_G f(gh)\,dg = \int f(hg)\,dg = \int f(g)\,dg$$

for every continuous function f on G and every $h \in G$. In other words, we can define on the space of continuous functions on G a linear functional, the integral, which is invariant under left and right translations. One can normalize the invariant measure on a compact group so that the total volume $\int_G dg$ of the group is 1. We will not dwell on the general construction of the invariant measure, but in the particular case of a matrix group G, the construction is as follows: we give the Lie algebra $\mathcal{G}$ of G a scalar product $\langle x, y \rangle = -\operatorname{tr} xy$ that is invariant under inner automorphisms: $\langle g^{-1}xg, g^{-1}yg \rangle = \langle x, y \rangle$. This scalar product gives rise to a Riemannian metric

$$ds^2 = \langle g^{-1}\,dg, g^{-1}\,dg \rangle = -\operatorname{tr}(g^{-1}\,dg)(g^{-1}\,dg)$$

invariant under left and right translations; the volume element for this metric gives the desired measure.

37. Gluings

We now describe a process of *gluing* or *identification* through which one can obtain new topological spaces from old ones. Our presentation in this chapter will be informal, and mainly pictorial.

We start by taking a rectangle and gluing two opposite sides together. If we do the gluing without twisting (Figure 18), we obtain a *cylinder* (the cylindrical wall, not the solid). If, however, we apply a half-twist to the rectangle before gluing opposite sides, the result is the so-called *Möbius strip* (Figure 19).

If the top and bottom edges of the cylinder are glued together, we get a *torus* (Figure 20). This is the same as starting from a rectangle and gluing the top and bottom edges together, as well as the right and left edges (Figure 21).

If all the points of the circumference of a disk are identified, the result is homeomorphic (topologically equivalent) to the sphere S^2. Figure 22 shows how concentric circles in the disk are mapped to parallels of latitude of the sphere. Analogously, if we take a ball, whose boundary is a sphere, and identify all the points of the boundary, the resulting space is homeomorphic to the three-dimensional sphere S^3, the set of points in $\mathbf{R}^4$ at a fixed distance from the origin.

Now consider a square divided into vertical lines, one for each value of the x-coordinate (Figure 23). Identify together all points in each segment, so that each segment becomes a single point. The result is an interval, which we can take as the bottom edge of the square for concreteness: there is a one-to-one correspondence between vertical lines in the square and points in the bottom edge, and this correspondence is continuous in both directions (that is, segments close together correspond to points that lie close together, and vice versa).

Next, take the two-dimensional sphere S^2 and identify together pairs of diametrically opposite points. The result is the *projective plane* $\mathbf{RP}^2$. A point of

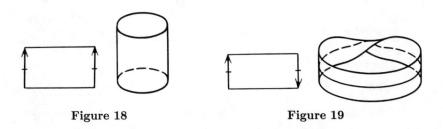

Figure 18   Figure 19

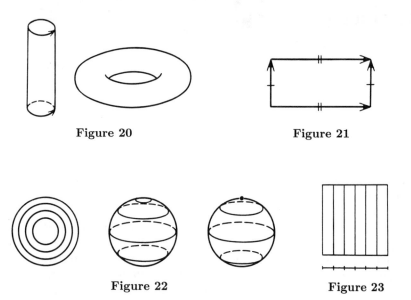

Figure 20

Figure 21

Figure 22

Figure 23

the projective plane is a set consisting of two diametrically opposite points in the sphere. To better visualize $\mathbf{RP}^2$, observe that each point below the equator is identified with one point above the equator, so we could also get the projective plane by starting with the upper hemisphere only, and identifying pairs of diametrically opposite points on the equator. Since a hemisphere is homeomorphic to a disk, the projective plane can be obtained from a disk by gluing together pairs of opposite points on the boundary.

Now concentrate on the portion of the disk bounded by two parallel chords of equal length (Figure 24). If we take the arcs that form the ends of this strip and glue them together so that opposite points on the circle match, the result is clearly the Möbius strip, whose boundary is a topological circle coming from the two chords. If, moreover, we glue together the two arcs that bound the two pieces of the strip's complement, this is the same topologically as gluing two half-disks along a diameter, so the result is homeomorphic to a disk. Thus, the projective plane can also be described as the result of gluing a Möbius strip and a disk along their boundaries.

Figure 24

Figure 25

Figure 26

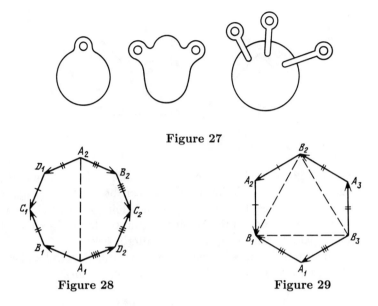

Figure 27

Figure 28   **Figure 29**

Similarly, *n-dimensional projective space* $\mathbf{RP}^n$ is defined as the n-sphere S^n with pairs of diametrically opposite points identified. It can also be thought of as the n-ball with diametrically opposite points of the bounding $(n-1)$-sphere identified.

Another way to define $\mathbf{RP}^n$ is to consider the complement of the origin in $\mathbf{R}^{n+1}$ and identify together points that have proportional coordinates (that is, that lie on the same line through the origin). To see that this construction does indeed yield $\mathbf{RP}^n$, note that any $x \neq 0$ gets identified with $x/|x|$, and this point lies on the unit sphere. So the identification process just described breaks down into two steps, one leading from $\mathbf{R}^{n+1} \setminus \{0\}$ to S^n, and the second from S^n to $\mathbf{RP}^n$.

The surface shown in Figure 25 (a *handle*) is homeomorphic to the surface obtained from a pentagon $ABCDE$ by identifying AB with DC and BC with ED (Figure 26). A handle is a surface whose boundary is a topological circle. Now take a sphere and delete k disjoint disks. This gives a *sphere with k holes*. A surface with k handles, or *surface of genus k*, is the result of gluing to the boundary of each of these deleted disks the boundary of a handle. Figure 27 shows surfaces with one, two and three handles; the first of these is topologically the torus. It is easy to verify that a sphere with k handles can be obtained form a $4k$-gon $A_1B_1C_1D_1 \ldots A_kB_kC_kD_k$ by identifying D_iC_i with A_iB_i and $A_{i+1}D_i$ with B_iC_i, where $i = 1, \ldots, k$ and $A_{k+1} = A_1$ (Figure 28).

Consider again a sphere minus k disks, but this time glue onto the bounding circles k Möbius strips (recall that the boundary of a Möbius strip is a circle). The resulting surface can be obtained from a $2k$-gon by gluing in the pattern of Figure 29: if the $2k$-gon is $A_1B_1A_2B_2 \ldots A_kB_k$, we glue A_iB_i to $B_{i-1}A_i$, where $i = 1, \ldots, k$ and $B_0 = B_k$. To show this, cut off each triangle $B_{i-1}A_iB_i$; the result is topologically a disk, of which k boundary points are identified together, so we

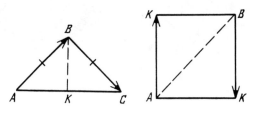

Figure 30

get a sphere with k holes. Each triangle $B_{i-1}A_iB_i$, after sides A_iB_i and $B_{i-1}A_i$ are glued together becomes a Möbius strip, as can be seen from Figure 30: cutting a triangle ABC along the altitude BK and gluing AB to BC, we arrive at the standard representation of the Möbius strip in the form of a rectangle in which one side is paired with the opposite side after a twist.

Every closed surface is homeomorphic either to a sphere with handles or a sphere with Möbius strip attached. (*Closed* in this context means the same as compact.)

38. Equivalence Relations and Quotient Spaces

In many situations in physics and mathematics it is reasonable to consider two different objects as equivalent in some sense. For example, in quantum mechanics the state of a particle or system of particles can be described by a nonzero vector in a complex Hilbert space (the state vector). But two vectors ψ and ψ' proportional to each other are physically equivalent, that is, they describe the same state. Likewise, an electromagnetic field can be described by a vector potential, but two potentials $A'_\mu(x)$ and $A_\mu(x)$ that differ by a gauge transformation (that is, that satisfy $A'_\mu(x) = A_\mu(x) + \partial_\mu \lambda(x)$) are physically equivalent.

Consider a set X and a *relation* on X, that is, a rule to decide whether one element of X is related (in some fixed sense) to another. If the relation is reflexive, symmetric and transitive (definitions follow), we say that it is an *equivalence relation*, and we write $x \sim y$ if x and y are *equivalent*, that is, related by the equivalence relation. *Reflexive* means that $x \sim x$ for every $x \in X$; *symmetric* means that $x \sim y$ implies $y \sim x$ for $x, y \in X$; and *transitive* means that $x \sim y$ and $y \sim z$ imply $x \sim z$, for $x, y, z \in X$.

Given an equivalence relation on a set X, we can consider for each $x \in X$ the set N_x of elements equivalent to x; this set is called the *equivalence class* of x. It is easy to verify that the equivalence classes of two elements either coincide (if the two elements are equivalent) or are disjoint (if not). Thus, an equivalence relation on X gives rise to a partition of X into pairwise disjoint subsets. Conversely, given a partition of a set into pairwise disjoint subsets, we can define an associated equivalence relation, under which two elements are equivalent if and only if they belong to the same subset.

If we have an equivalence relation on a set X, we can form a new set $\tilde{X}$ by replacing each equivalence class by a single point; we call this process *identifying* or *gluing together* equivalent elements. We call $\tilde{X}$ the *quotient* of X by the equivalence relation. We saw examples of identifications in Chapter 37. For example, in the space of Figure 23, we identified points in the square $[0, 1] \times [0, 1]$ that have the same horizontal coordinate, obtaining a line segment. The equivalence classes in this case are the vertical segments. The identification of opposite points on a sphere to give the projective plane is another example: each equivalence class here has two points, x and $-x$.

There exists a natural map π from X into $\tilde{X}$, taking each point x to its equivalence class N_x. This is known as the *identification map*. A map g from

$\tilde{X}$ into some set Z gives rise to a map $\tilde{g} = g\pi$ from X into Z, which clearly takes equivalent points into a single point. Conversely, if a map maps assigns the same image to all points in each equivalence class, it factors into a composition $f = h\pi = \tilde{h}$, for some $h : \tilde{X} \rightarrow Z$. Returning to the example of opposite-point identification on a sphere: any even function on the n-dimensional sphere gives rise to a function on n-dimensional projective space.

A map p from a set X into a set Y defines naturally an equivalence relation on X, where two points are equivalent if and only if they have the same image. The equivalence classes are the nonempty inverse images of points $y \in Y$. If the map is onto, there is a one-to-one correspondence between $\tilde{X}$ and Y.

If X is a topological space with an equivalence relation, the quotient $\tilde{X}$ can be given the *quotient* topology, under which a map $g : \tilde{X} \rightarrow Z$ is continuous if and only if $\tilde{g} = g\pi : X \rightarrow Z$ is continuous, where Z is an arbitrary topological space. (In particular, if $Z = \tilde{X}$ and g is the identity map, we see that $\pi : X \rightarrow \tilde{X}$ is continuous.) The quotient topology can also be characterized as follows: a set $U \subset \tilde{X}$ is open in the quotient topology if and only if its inverse image $\pi^{-1}(U) \subset X$ is open in X. However, the quotient topology may not satisfy any separation axioms. For example, one-point subsets in $\tilde{X}$ are closed if and only if equivalence classes are closed in X. Usually, however, the quotient topology is well-behaved in the cases of interest in physics.

39. Group Representations

Suppose we associate with each element of a group G a linear transformation T_g of a vector space E, in such a way that to the product of elements of G is associated the composition of the corresponding transformations:

$$(39.1) \qquad T_{gh} = T_g T_h, \qquad \text{for } g, h \in G.$$

We then say that the correspondence $g \mapsto T_g$, also denoted T, is a *linear representation* of G. In other words, T is a homomorphism from G into the group $\mathrm{GL}(E)$ of linear transformations of E. We also say that E is the *representation space* of G (under the representation T).

A subspace $E' \subset E$ is called an *invariant subspace* of the representation T if all the operators T_g map E' into itself. Obviously, the restrictions of the T_g to E' make up a representation of G in E'. If E has no nontrivial invariant subspaces (*nontrivial* means different from E and from the space consisting of the origin only), we say that T is an *irreducible* representation. If T^1 and T^2 are representations of E in vector spaces E_1 and E_2, their *direct sum* is defined as the representation T of G in $E_1 + E_2$ given by

$$T_g(e_1, e_2) = (T_g^1 e_1, T_g^2 e_2),$$

for $e_1 \in E_1$ and $e_2 \in E_2$.

Two representations T^1 and T^2 are *equivalent* if there exists an isomorphism $\alpha : E_1 \to E_2$ such that $\alpha T_g^1 = T_g^2 \alpha$ for all $g \in G$. We say that a representation T is *unitary* or *orthogonal* if each T_g is unitary or orthogonal. (Of course, this presupposes that the representation space E is a real or complex Hilbert space.) If F is an invariant subspace of an orthogonal or unitary representation T, so is its orthogonal complement $F^\perp$; furthermore, F and $F^\perp$ inherit representations of G by restriction, and the original representation T is the direct sum of the two restrictions.

For any representation T of a compact group G on a space E one can find a scalar product on E that is invariant under T. In other words, for an appropriate choice of a scalar product on E, every representation of a compact group is orthogonal (if E is a real vector space) or unitary (if E is complex). Such an invariant scalar product is constructed as follows: one starts with any scalar product and averages its images under T_g, with respect to the invariant measure dg on G. The existence of an invariant scalar product implies that every finite-dimensional representation of a compact group is equivalent to the direct sum of irreducible representations.

The representations of the Lorentz group G play an important role in physics, because the law governing the transformations of a physical quantity when one passes form one inertial frame of reference to another is described by such a representation. More specifically, suppose a physical object, such as a point or a field, is described in one inertial frame of reference K by an n-tuple $(\varphi_1, \ldots, \varphi_n)$, and in another frame K' by $(\varphi'_1, \ldots, \varphi'_n)$. We assume that the dependence of the new n-tuple in terms of the old one is described by a linear transformation T_g, where $g \in L$ is the element of the Lorentz group taking frame K to frame K'. Equation (39.1) is obviously satisfied; it simply means that changing coordinates from frame K to frame K' and then from K' to another frame K'' is equivalent to passing directly from K to K''.

Here are some examples of Lorentz group representations (physical quantities). A scalar is a quantity that does not depend on the coordinate system—we can say that it transforms according to the trivial one-dimensional representation of L (all the T_g are the identity). A (contravariant) vector is a quantity (A^0, A^1, A^2, A^3) that transforms in the same way as the components of the coordinate system:

$$(39.2) \qquad A'^i = a^i_k A^k,$$

where (a^i_k) is the matrix describing the Lorentz transformation. A covector (or subscripted vector) is a quantity (A_0, A_1, A_2, A_3) that transforms according to the law

$$(39.3) \qquad A'_i = b^k_i A'_k,$$

where $b^k_i a^l_k = \delta^l_i$, that is, (b^k_i) is the matrix inverse to the (a^i_k). An example of a covector is the quantity $\partial\varphi/\partial x^i$, where φ is a scalar field. A (rank-two contravariant) tensor A^{ik} is a quantity that transforms as the product $B^i C^k$ of two vectors:

$$(39.4) \qquad A'^{ik} = a^i_l a^k_m A^{lm}.$$

The transformation laws for a rank-two covariant tensor and a rank-two mixed tensor are obtained similarly:

$$(39.5) \qquad A'_{ik} = b^l_i b^m_k A_{lm},$$
$$(39.6) \qquad A'^k_i = a^k_l b^m_i A^l_m.$$

These three types of rank-two tensors can be written in matrix form. In particular, in this notation the transformation laws (39.4) and (39.6) become

$$A' = aAa^T, \qquad A' = aAa^{-1}.$$

Tensors or higher rank (that is, with more indices) are defined analogously.

The invariance of the space-time line element $ds^2 = g_{ik}dx^i dx^k$ under Lorentz transformations implies that the metric tensor g_{ik} does not change when one passes from one inertial reference frame to another:

$$g_{ik} = a_i^l a_k^m g_{lm}.$$

The invariance of the four-dimensional volume element under Lorentz transformations implies the same invariance for the tensor ε_{ijkl}. (This is a totally antisymmetric tensor taking the value 1 or -1 according to whether (i, j, k, l) is an even or odd permutation; if any two indices coincide, of course, its value is zero.)

The transformation laws (39.2)–(39.6) determine the corresponding representations of the Lorentz group. These laws can be applied also in the case where a_i^j is an arbitrary matrix, not necessarily a Lorentz transformation. Thus, (39.2)–(39.6) define representations for any matrix group. In particular, (39.2) defines the vector representation of $\mathrm{GL}(n, \mathbf{R})$ if the a_i^j form an arbitrary nonsingular matrix of order n and the numbers A^j are real, for $1 \le j \le n$. Similarly, (39.3) defines the covector representation of $\mathrm{GL}(n, \mathbf{R})$. If the a_i^j form a complex nonsingular matrix and the A^j and A_j are complex numbers, (39.2) and (39.3) define the vector and covector representations of $\mathrm{GL}(n, \mathbf{C})$.

In the complex case we can also consider the transformation laws

(39.7) $$A'^{\dot{m}} = \bar{a}_{\dot{k}}^{\dot{m}} A^{\dot{k}},$$

(39.8) $$A'_{\dot{m}} = \bar{b}_{\dot{m}}^{\dot{k}} A_{\dot{k}}$$

where the bar denotes complex conjugation and $1 \le \dot{k}, \dot{m} \le n$. A quantity that transforms according to (39.7) or (39.8) is called a vector or covector with *dotted indices*.

A broader class of matrix group representations is obtained if we consider tensors of arbitrary rank; in the complex case, each index can be either ordinary or dotted.

For the Lorentz group the vector and covector representations are equivalent, and the representations defined by (39.4)–(39.6) are also equivalent. This is because upper indices can be lowered by applying the metric tensor g_{kl}; for example, to the vector A^l we can associate the covector $A_k = g_{kl}A^l$, and with the tensor A^{ls} we can associate the tensor $A_{kr} = g_{kl}g_{rs}A^{ls}$.

For $\mathrm{GL}(n, \mathbf{R})$ and $\mathrm{GL}(n, \mathbf{C})$, the representations defined by (39.2) and (39.3) are not equivalent, and likewise for (39.7) and (39.8). They are also irreducible, while the representations defined by (39.4)–(39.6) are reducible. For example, the subspaces consisting of, respectively, symmetric tensors $A^{ik} = A^{ki}$ and antisymmetric tensors $A^{ik} = -A^{ki}$ are invariant under the group of nonsingular matrices. An invariant subspace of the representation (39.6) is given by all tensors of the form $\lambda \delta_k^l$, and another by all traceless tensors (those with $A_k^k = 0$). These invariant subspaces are irreducible with respect to $\mathrm{GL}(n, \mathbf{R})$ and $\mathrm{GL}(n, \mathbf{C})$.

Every representation of the Lorentz group can be realized using tensors, that is, it is equivalent to the representation of the group in one of the invariant subspaces of a tensor representation. However, the concept of a representation can be broadened by lifting the restriction that each element g of the group be associated with a single transformation T_g. A representation in this more general sense is called *multivalued*. The Lorentz group has double-valued representations that do not come from tensor representations. The simplest of them have complex dimension two, that is, each element of the Lorentz group is associated with a complex 2×2 matrix. The matrices have unit determinant, that is, they belong to $SL(2, \mathbf{C})$. Instead of describing the double-valued map from the Lorentz group into $SL(2, \mathbf{C})$, we will construct the inverse map.

We start from the representation of $SL(2, \mathbf{C})$ built with tensors of rank two having one ordinary index and one dotted index. If we regard such tensors as matrices, the representation in question assigns to each $a \in SL(2, \mathbf{C})$ the operator T_a given by

$$(39.9) \qquad\qquad T_a(A) = aAa^\dagger,$$

for A a 2×2 complex matrix. Hermitian matrices form an invariant subspace for this representation. The space of two-dimensional Hermitian matrices is a four-dimensional real vector space, since such a matrix is determined by its (real) entries a_{11} and a_{22}, and by the entry a_{12}, whose real and imaginary parts are arbitrary. We will parametrize this space by the numbers $x^0 = \frac{1}{2}(a_{11} + a_{22})$, $x^1 = \operatorname{Re} a_{12}$, $x^2 = -\operatorname{Im} a_{12}$, and $x^3 = \frac{1}{2}(a_{11} - a_{22})$, so that a Hermitian matrix $\hat{X}$ is given in terms of the parameters by

$$(39.10) \qquad\qquad \hat{X} = \begin{pmatrix} x^0 + x^3 & x^1 - ix^2 \\ x^1 + ix^2 & x^0 - x^3 \end{pmatrix} = x^0 + \mathbf{x} \cdot \boldsymbol{\sigma},$$

where $\boldsymbol{\sigma} = (\sigma_1, \sigma_2, \sigma_3)$ has as components the *Pauli matrices*

$$\sigma^1 = \begin{pmatrix} 0 & 1 \\ 1 & 0 \end{pmatrix}, \qquad \sigma^2 = \begin{pmatrix} 0 & -i \\ i & 0 \end{pmatrix}, \qquad \sigma^3 = \begin{pmatrix} 1 & 0 \\ 0 & -1 \end{pmatrix}.$$

Note that $x^i = \frac{1}{2}\operatorname{tr}(\sigma^i X)$, with $x^0 = 1$. Transformation (39.9) can be regarded as a representation of $SL(2, \mathbf{C})$ in the space of Hermitian matrices, and it preserves the determinant function on this space:

$$\det T_a(A) = \det(aAa^\dagger) = \det a \det A \det a^\dagger = \det A.$$

The determinant function itself is a quadratic form on the space $\mathcal{H}$ of Hermitian matrices—in fact, it is connected with of the space-time interval

$$\det(x^0, x^1, x^2, x^3) = (x^0)^2 - (x^1)^2 - (x^2)^2 - (x^3)^2$$

in the parametrization of $\mathcal{H}$ described above. In other words, for each $a \in SL(2, \mathbf{C})$, the operator T_a is a Lorentz transformation on the space of Hermitian matrices considered with the determinant form. More precisely, we have shown that T_a lies in the complete Lorentz group, so we have a homomorphism from $SL(2, \mathbf{C})$ to this group. Since $SL(2, \mathbf{C})$ is connected, its image is contained in the connected component of the identity, which is exactly the Lorentz group L. Therefore, T_a lies in L. The kernel of the homomorphism $\mathrm{SL}(2, \mathbf{C}) \to L$ is the two-element group $\mathbf{Z}_2$: the identity matrix and its opposite both correspond to the identity Lorentz transformation.

Now consider the one-to-two correspondence from L into $\mathrm{SL}(2, \mathbf{C})$, inverse to the map just described: each Lorentz transformation has two associated elements of $\mathrm{SL}(2, \mathbf{C})$, differing by a sign. This correspondence can be considered a double-valued two-dimensional complex representation for the Lorentz group, known as the *spinor representation*. The two-component quantities that transform according to it are called *spinors*. Their complex conjugates are called *dotted spinors*. (One also says *left* and *right spinors* instead of spinors and dotted spinors.)

Note that under the homomorphism $\mathrm{SL}(2, \mathbf{C}) \to L$ just considered, the subgroup $\mathrm{SU}(2)$ of $\mathrm{SL}(2, \mathbf{C})$ is mapped into $\mathrm{SO}(3)$. Indeed, for a unitary matrix $a \in SL(2, \mathbf{C})$, we have $T_a(\hat{X}) = a\hat{X}a^{-1}$, and therefore $\operatorname{tr} T_a(\hat{X}) = \operatorname{tr} \hat{X}$. Under our parametrization of the space $\mathcal{H}$ of Hermitian matrices, the trace is given by $\operatorname{tr} \hat{X} = x^0$, so a Lorentz transformation corresponding to a unitary matrix a preserves x^0, and is therefore an orthogonal transformation of the complementary three-dimensional space, with coordinates (x^1, x^2, x^3). We have thus built a three-dimensional representation of $\mathrm{SU}(2)$. Its inverse can be seen as a double-valued two-dimensional real representation of $\mathrm{SO}(3)$.

The correspondence between L and $\mathrm{SL}(2, \mathbf{C})$ just discussed allows one to consider any representation of $\mathrm{SL}(2, \mathbf{C})$ as a (single- or double-valued) representation of the Lorentz group. An irreducible representation of $\mathrm{SL}(2, \mathbf{C})$ is characterized by two nonnegative integers p and q, as the rank-(p, q) tensor representation, that is, as the representation in the space of tensors with p ordinary indices and q dotted indices. (For given p and q, this representation is essentially unique. It does not matter if the indices are upper or lower, because the antisymmetric tensors $\varepsilon_{\alpha\beta}$ and $\varepsilon_{\dot{\alpha}\dot{\beta}}$, being invariant under $\mathrm{SL}(2, \mathbf{C})$, allow one to lower or raise indices at will.) If $p+q$ is even, the resulting representation of L is single-valued, and it is double-valued if $p + q$ is odd.

The concept of a tensor is closely connected with that of the *tensor product* of representations. Recall that the *tensor product* $E_1 \otimes E_2$ of two vector spaces E_1 and E_2, with bases $\{e_1^{(1)}, \dots, e_m^{(1)}\}$ and $\{e_1^{(2)}, \dots, e_n^{(2)}\}$, respectively, is the space of formal linear combinations of the symbols $e_a^{(1)} \otimes e_b^{(2)}$—"formal" means that each element of $E_1 \otimes E_2$ is expressed uniquely in the form $\sum_{a,b} c^{ab} e_a^{(1)} \otimes e_b^{(2)}$. Now suppose that T_1 is a representation of a group G in a space E_1, and that $M = (M^1, \dots, M^m)$ is in E_1 (that is, M transforms according to the representation T_1). Suppose, likewise, that $N = (N^1, \dots, N^n)$ is a quantity transforming according to a representation T_2 in a space E_2. Then, by definition, the quantity with components $M^a N^b$ transforms according to the tensor product representation $T_1 \otimes T_2$, whose space is the tensor product $E_1 \otimes E_2$. The representation $T_1 \otimes T_2$ acts in $E_1 \otimes E_2$ and changes the coordinates c^{ab} according to the law

$$c'^{ab} = (T_1(g))_k^a (T_2(g))_l^b c^{kl},$$

where $(T_1(g))^a_k$ is the matrix of $T_1(g)$ in the basis $\{e_1^{(1)}, \ldots, e_m^{(1)}\}$, and likewise for $(T_2(g))^b_l$.

Although we gave a definition involving particular bases for E_1 and E_2, the definition of the tensor product of representations (and of spaces) does not depend on the choice of bases. The rank-k tensor representation of a group is the tensor product of k copies of the vector representation.

40. Group Actions

An *action* (or *left action*) of a group G on a set E is a correspondence that associates to each element $g \in G$ a map $\varphi_g : E \to E$ in such a way that

$$(40.1) \qquad \varphi_{g_1 g_2} = \varphi_{g_1} \varphi_{g_2}.$$

In other words, a group action is a homomorphism from G into the group of transformations of E. An important special case is when this homomorphism is injective, so we get an isomorphism between G and a subgroup of the transformation group of E. In this case we say that G acts *effectively*. For example, matrix groups can be considered as groups of linear transformations of $\mathbf{R}^n$ or $\mathbf{C}^n$, so they act effectively on $\mathbf{R}^n$ or $\mathbf{C}^n$.

If E is a vector space and the transformations φ_g are linear, the action is a (linear) representation of G in E. In the general case, the term *nonlinear representation* is often used in the physics literature for a group action.

A *right action* is a correspondence that associates with $g \in G$ a map $\varphi_g : E \to E$ in such a way that

$$(40.2) \qquad \varphi_{g_1 g_2} = \varphi_{g_2} \varphi_{g_1}.$$

Given a right action of G on E, we can form an associated left action by assigning to each $g \in G$ the transformation $\lambda_g = \varphi_{g^{-1}}$; we have $\lambda_{g_1 g_2} = \lambda_{g_1} \lambda_{g_2}$ because $(g_1 g_2)^{-1} = g_2^{-1} g_1^{-1}$. This means that one can reduce the study of a right action to the study of a left action. We might not even have bothered defining right actions at all, but they prove to be natural and useful in certain contexts. Note that, unless we say otherwise, all actions are assumed to be on the left.

If G acts on the left, we often write gx for $\varphi_g(x)$, so that (40.1) becomes $(g_1 g_2)x = g_1(g_2 x)$. For a right action we can write xg for $\varphi_g(x)$, so (40.2) becomes $x(g_1 g_2) = (x g_1) g_2$.

If a topological group G acts on a topological space E, we always assume that the action is *continuous*, that is, that $\varphi_g(x)$ is continuous jointly in $g \in G$ and $x \in E$.

A group action of G on E gives rise to a natural equivalence relation on E, as follows: $x_1 \in E$ and $x_2 \in E$ are equivalent if they can be obtained from one another by the action of some element of G, that is, if $x_2 = \varphi_g(x_1)$ for $g \in G$. The equivalence class N_x of a point $x \in E$ is said to be the *orbit* of x; thus N_x is the set of points that can be obtained from x by the action of elements of G. Two orbits that are not the same must be disjoint, that is, the orbits form a partition of E. The set of orbits of G in E is denoted by E/G, and it can be given the quotient topology (see Chapter 38). Note, however, that the quotient topology may not satisfy any separation axioms; in particular, all points are closed in E/G if and only if all orbits of G are closed in E. A simple case where no separation axioms are satisfied is when $\mathrm{GL}(n)$ acts on $\mathbf{R}^n$. There are exactly two orbits, one consisting of the origin and one made up of everything else; therefore E/G has two points, only one of which (the orbit of the origin) is closed.

If there is only one orbit, any point of E can be obtained from any other point by the action of G. We then say that G acts *transitively*.

The *stabilizer* H_x of a point $x \in H$ is the set of elements of G that leave x fixed, that is, $h \in H_x$ if $\varphi_h(x) = x$. Clearly, H_x is a subgroup of G. If $h \in H_x$ we have $ghg^{-1} \in H_{\varphi_g(x)}$. Therefore $H_{\varphi_g(x)}$ is obtained form H_g by an inner automorphism, and the two subgroups are conjugate, and in particular isomorphic (Chapter 36).

The definitions in the last five paragraphs apply to right as well as to left actions.

To each three-dimensional orthogonal matrix A of determinant one we can assign a rotation $x \mapsto Ax$ of $\mathbf{R}^3$ about an axis that goes through the origin. This determines an action of $\mathrm{SO}(3)$ on $\mathbf{R}^3$. The orbits are two-dimensional spheres centered at the origin, plus one orbit containing only the origin. The space of orbits is topologically equivalent to the closed half-line $\mathbf{R}_+$. Every orthogonal matrix fixes the origin, so the stabilizer of the origin is $\mathrm{SO}(3)$ itself; by contrast, the stabilizer of a point x distinct form the origin consists of rotations about the line connecting the origin with x. It is therefore isomorphic to $\mathrm{SO}(2)$, since a rotation about an axis can also be seen as a rotation of a plane perpendicular to the axis.

Similarly, we can consider $\mathrm{SO}(n)$, the group of $n \times n$ matrices with unit determinant, as a group acting on $\mathbf{R}^n$. The orbits are $(n-1)$-dimensional spheres. The stabilizer of the point $(0, \ldots, 0, r)$, for $r \neq 0$, is the group of matrices $a_{ik} \in \mathrm{SO}(n)$ such that $a_{nn} = 1$ and $a_{nk} = a_{kn} = 0$ for all $k \neq n$. By ignoring the last row and column of such a matrix, we can think of it as an $(n-1) \times (n-1)$ orthogonal matrix of determinant one, so we see that the stabilizer of the point $(0, \ldots, 0, r)$ is $\mathrm{SO}(n-1)$. The same is true for any point $x \in \mathbf{R}^n$ distinct from the origin: any such point can be obtained from a point of the form $(0, \ldots, 0, r)$ by the action of $\mathrm{SO}(n)$, and therefore the two stabilizers are conjugate, as discussed above.

For another example, consider the action of $\mathrm{GL}(1, \mathbf{R})$, the multiplicative group of nonzero real numbers, on $\mathbf{R}^n \setminus \{0\}$ given as follows: for each $\lambda \neq 0$, the associated transformation is the map

$$(x^1, \ldots, x^n) \mapsto (\lambda x^1, \ldots, \lambda x^n).$$

The orbits are straight lines through the origin; the orbit space is the real projective plane $\mathbf{RP}^{n-1}$ (see Chapter 37). The stabilizer of any point is trivial— it includes only the identity element $\lambda = 1$.

One can define in the same way an action of $\mathrm{GL}(1, \mathbf{C})$ on $\mathbf{C}^n \setminus \{0\}$. The orbit space is obtained by identifying any point $(x^1, \ldots, x^n) \in \mathbf{C}^n \setminus \{0\}$ with the points $(cx^1, \ldots, cx^n)$, for $c \in \mathbf{C} \setminus \{0\}$. This quotient is known as $(n-1)$-dimensional complex projective space, and denoted $\mathbf{CP}^{n-1}$. All stabilizers of the action are again trivial.

Let $G_{n,k}$ denote the set of k-dimensional linear subspaces of $\mathbf{R}^n$. Define an action of $\mathrm{GL}(n)$ on $G_{n,k}$ as follows: given an $n \times n$ nonsingular matrix, the corresponding linear transformation maps k-subspaces to k-subspaces, and therefore gives a map from $G_{n,k}$ into itself. This action is transitive: any k-subspace can be taken to any other by some linear transformation. The stabilizer of an element $\alpha \in G_{n,k}$ is isomorphic to the group of matrices of the form $\begin{pmatrix} A & 0 \\ B & C \end{pmatrix}$, where A is an $(n-k) \times (n-k)$ matrix, B is a $k \times (n-k)$ matrix, and C is a $k \times k$ matrix.

Now let G be a group and $H \subset G$ a subgroup. H acts on G on the left by left translations L_h, for $h \in H$ (Chapter 36). The right action R_h is defined similarly. The orbits of the left action are the *left cosets* of H, and the orbits of the right action are the *right cosets*. We will call the space of right cosets the *quotient space* or *coset space* of G by H, and denote it by G/H; thus G/H is obtained from G by identifying elements that differ by right multiplication by elements of H. (Left and right cosets are in one-to-one correspondence: if $g_1 \in G$ and $g_2 \in G$ lie in the same left coset, g_1^{-1} and g_2^{-1} are in the same right coset, because $g_1 = hg_2$ for $h \in H$ implies $g_1^{-1} = g_2^{-1}h^{-1}$. For this reason we won't use a separate notation for the space of left cosets.)

Consider the identification map $\alpha : G \to G/H$ that takes each g to the coset where it lies. When can G/H be made into a group in such a way that α is a homomorphism? If it can, H is the kernel of the homomorphism α, and therefore H is normal in G (Chapter 36). Conversely, if H is normal, we can define the product of two cosets λ_1 and λ_2 by choosing representatives g_1 and g_2 and taking the coset containing their product. (Since H is normal, the result does not change if we use different representatives $g_1' = g_1 h_1$ and $g_2' = g_2 h_2$, with $h_1, h_2 \in H$: we have $g_1' g_2' = g_1 g_2 h$, where $h = (g_2^{-1} h_1 g_2) h_2 \in H$.) The coset space G/H with this group law is called the *quotient* of G by the normal subgroup H.

If H is a normal subgroup, its left cosets and right cosets coincide. For if $g_1, g_2 \in G$ are in the same right coset, we have $g_2 = g_1 h$ with $h \in H$, so $g_2 = (g_1 h g_1^{-1}) g_1$ is in the left coset of g_1.

As an example of a quotient group, take for G the group $\mathbf{Z}$ of the integers with addition, and for H the group $m\mathbf{Z}$ of multiples of a fixed integer m. Two integers belong to the same coset if they are congruent modulo m, that is, if their difference is a multiple of m. The quotient $\mathbf{Z}/m\mathbf{Z}$ has m elements. It is called the *cyclic group of order* m and is denoted by $\mathbf{Z}_m$.

A set E where a group G acts transitively is called a *homogeneous space*. Every orbit of a (not necessarily transitive) group action is a homogeneous space. For example, we have seen that the orbits of $\mathrm{SO}(n)$ acting on $\mathbf{R}^n$ are

$(n-1)$-dimensional spheres; each such sphere is a homogeneous space with respect to the action of $SO(n)$.

Let G act transitively on E, and fix a point $e_0 \in E$. Denote by $K(e)$ the set of group elements that take e_0 to e, that is, elements such that $\varphi_g(e_0) = e$. In particular, $K(e_0)$ is the stabilizer of e_0; we set $H = K(e_0)$. If $g \in K(e)$ and $h \in H$, we have $gh \in K(e)$, and if $g_1, g_2 \in K(e)$, we have $g_1 = g_2 h$, with $h \in H$. In other words, $K(e)$ is a right coset of H. We thus obtain a one-to-one map between E and the coset space G/H.

If H is an arbitrary subgroup of G, we can define on the right coset space G/H a left action of G by multiplication. The transformation of G/H associated with a given element $g \in G$ takes the coset of an element $g_1 \in G$ to the coset of $gg_1 \in G$. The image coset is well-defined, because if we replace g_1 by another element $g_2 = g_1 h$ in the same coset, we have $gg_2 = gg_1 h$, so gg_2 and gg_1 are also in the same coset. The resulting action of G on G/H is transitive.

Any transitive action of a group G on a set E is equivalent to one of the type just described. For, if H is the stabilizer of a point $e_0 \in H$, we can construct a map $\rho : G/H \to E$ such that $\varphi_g \rho = \rho \tilde{\varphi}_g$, where φ_g denotes the action of $g \in G$ on E and $\tilde{\varphi}_g$ denotes the action of g on G/H. If G is a topological group, the map $\rho : G/H \to E$ is one-to-one and continuous. In all cases that concern us its inverse is also continuous. In particular, the sphere S^{n-1} is topologically equivalent to the quotient $SO(n)/SO(n-1)$.

41. The Adjoint Representation of a Lie Group

Every group G acts on itself by inner automorphisms: the map associated with an element g is $h \mapsto ghg^{-1}$. If G is a Lie group, the differential of each inner automorphism determines a linear transformation on the tangent space to G at the identity element, because the identity is fixed by any inner automorphism. This gives rise to a linear representation of the group G in the Lie algebra $\mathcal{G}$ of G, the *adjoint representation* of G. If G is a matrix group, the Lie algebra $\mathcal{G}$, too, is realized by matrices, and the adjoint representation τ_g maps $x \in \mathcal{G}$ to gxg^{-1}.

As we have mentioned, if G is a compact Lie group, any representation space of G can be given an invariant scalar product. In particular, the Lie algebra $\mathcal{G}$ of a compact Lie group G has a scalar product $\langle \cdot, \cdot \rangle$ invariant under G, that is, such that $\langle \tau_g(x_1), \tau_g(x_2) \rangle = \langle x_1, x_2 \rangle$. If we consider τ_g for g infinitesimally close to the identity, that is, $g = 1 + h$, with $h \in \mathcal{G}$, we find that $\tau_{1+h}(x) = x + [h, x]$. Thus, the representation σ_h of $\mathcal{G}$ corresponding to the adjoint representation of G is given by $\sigma_h(x) = [h, x]$. If $\langle \cdot, \cdot \rangle$ is an invariant scalar product we have

$$\langle [h, x_1], x_2 \rangle + \langle x_1, [h, x_2] \rangle = 0,$$

so that the expression $\langle [x, y], z \rangle$, with $x, y, z \in \mathcal{G}$, is antisymmetric in x, y and z. This expression is therefore a G-invariant antisymmetric multilinear form (multilinear means linear in each variable). If we choose an orthonormal basis $e_1, \ldots, e_n$ for $\mathcal{G}$, and denote by x^α the coordinates of an element $x = x^\alpha e_\alpha$ in this basis, we can write

$$\langle [x, y], z \rangle = f_{\alpha\beta\gamma} x^\alpha y^\beta z^\gamma,$$

where $f_{\alpha\beta\gamma}$ is an antisymmetric G-invariant tensor. The components $f_{\alpha\beta\gamma} = \langle [e_\alpha, e_\beta], e_\gamma \rangle$ of this tensor are known as *structure constants*.

If a linear subspace $\mathcal{H}$ of a Lie algebra $\mathcal{G}$ is invariant under the adjoint representation of $\mathcal{G}$, we have $\sigma_h(x) = [h, x] \in \mathcal{H}$ for all $h \in \mathcal{G}$ and $x \in \mathcal{H}$, so $\mathcal{H}$ is an invariant subalgebra of $\mathcal{G}$.

If the adjoint representation of a Lie algebra $\mathcal{G}$ is irreducible, $\mathcal{G}$ is simple, that is, it has no nontrivial invariant subalgebras. If $\mathcal{G}$ is the Lie algebra of a compact Lie group G, it follows from the results in Chapter 39 that the adjoint representation is a direct sum of irreducible representations. Thus $\mathcal{G}$ is a sum of simple Lie algebras. A Lie group whose Lie algebra is simple is also called *simple*.

Every simple compact Lie group is locally isomorphic to one of the following:

1. the *classical groups* $SU(n)$ for $n \geq 2$, $SO(n)$ for $n \geq 2$ and $n \neq 4$, and $\mathrm{Spin}(n)$ for $n \geq 1$; or

2. or one of the so-called *exceptional groups*, which are denoted G_2, F_4, E_6, E_7 and E_8.

42. Elements of Homotopy Theory

This chapter and the next are adapted from a lecture published in the Proceedings of the Third Physics Workshop, held at the Institute of Theoretical and Experimental Physics, Moscow, in 1975. They focus on the uses of homotopy theory in physics, and especially on the calculation of homotopy groups. The next chapter summarizes briefly the ways in which homotopy theory can be applied to quantum field theory.

Throughout this chapter, a "space" will mean a topological space (see Chapter 35).

Two maps f_0 and f_1 from a space X into a space Y are *homotopic* if one can be continuously deformed into the other, that is, if there is a continuous family of maps, indexed by a parameter $t \in [0, 1]$, that connects f_0 and f_1. More precisely, there must be a continuous map $f(x, t)$ from $X \times [0, 1]$ into Y satisfying the conditions $f(x, 0) = f_0(x)$ and $f(x, 1) = f_1(x)$. We write $f_t(x)$ for $f(x, t)$, and we refer to either f or the family f_t as a *homotopy* between f_0 and f_1.

Clearly, the condition for two maps being homotopic gives rise to an equivalence relation (Chapter 38). The equivalence classes of this relation are called *homotopy classes*; all maps in the same homotopy classes are homotopic to each other.

As a simple example, let Y be the Euclidean space $\mathbf{R}^n$. Then all maps from a space X into Y are homotopic to one another; a homotopy is given by the formula

$$f_t(x) = (1 - t)f_0(x) + tf_1(x).$$

In particular, all maps into $\mathbf{R}^n$ are *null-homotopic*, that is, homotopic to a constant map. (A null-homotopic map is also called *homotopically trivial*.)

If $Y = S^k$ and $X = S^l$ are spheres, with $l < k$, all maps from S^l into S^k are again homotopic to one another. To show this, we need the fact that a continuous map $f_0 : S^l \to S^k$ can be approximated to arbitrary accuracy by a smooth map $f_1 : S^l \to S^k$, which is homotopic to f_0. Now the image of f_1 cannot cover the whole of S^k, because the dimension of the domain S^l is less than the dimension of the range S^k; thus, there is a point n on S^k that is not in the image of f_1. Then we can consider f_1 as a map into $S^k \setminus \{n\}$, and this space is topologically equivalent to $\mathbf{R}^n$ (by stereographic projection, for example: see Chapter 21). It follows that f_0 is null-homotopic, and that all maps $S^l \to S^k$ are homotopic to one another if $l < k$.

We now turn to maps from the circle S^1 into itself. A point of S^1 is parametrized by a real number φ defined up to multiples of 2π, so a map $S^1 \to S^1$ can be specified by a real-valued function $f(\varphi)$, continuous on the closed interval $[0, 2\pi]$ and satisfying the condition $f(2\pi) = f(0) + 2\pi n$, where n is an integer. The number n, called the *degree* of the map, says how many times the circle wraps around itself under the action of the map; for example, the degree of the map $f(\varphi) = n\varphi$ is n. Clearly, two maps having the same degree are homotopic, while maps of different degrees are not.

A similar assertion is true for maps from the k-sphere S^k into itself. One can define an integer, the *degree*, that completely classifies homotopy classes: two maps are homotopic if and only if they have the same degree.

Recall that a space X is *connected* if any two of its points can be joined by a continuous curve, or path. (A *path* in X is a map $x : [0, 1] \to X$; we say that the path $x(t)$ *joins* the *starting point* $x(0)$ with the *endpoint* $x(1)$.) Every space can be partitioned into connected parts, called its *connected components*.

A space X is said to be *aspherical in dimension* k if every map from S^k into X is null-homotopic. As we have seen, S^n is aspherical in dimensions lower than n. A space that is aspherical in dimension 1 is also said to be *simply connected*.

The k-*dimensional homotopy group* $\pi_k(X)$ of a space X is a measure of how far X is from being aspherical in dimension k. (In particular, $\pi_k(X)$ is the trivial group if X is aspherical in dimension k.) To define $\pi_k(X)$, we need to fix a point $x_0 \in X$, called the *basepoint*; we also need a basepoint for the sphere S^k, which we can take as the south pole s. Let a k-*spheroid* be a map from S^k into X taking s to x_0; then $\pi_k(X)$ is the set of homotopy classes of k-spheroids.

A one-spheroid can be regarded as a path beginning and ending at x_0. If f_1 and f_2 are two such paths, we define their concatenation $f = f_1 f_2$ as the path obtained by going over first f_1 and then f_2. More formally, $f(t) = f_1(2t)$ for $0 \le t \le \frac{1}{2}$ and $f(t) = f_2(2t-1)$ for $\frac{1}{2} \le t \le 1$. If we replace f_1 (or f_2) by another path in the same homotopy class, the concatenation $f_1 f_2$ also remains in the same homotopy class. Therefore the operation of concatenation is well-defined on homotopy classes, and it defines an operation on $\pi_1(X)$. With this operation $\pi_1(X)$ is a group: the identity element is the class of null-homotopic paths, and the inverse of a path is the same path traversed backwards. We call $\pi_1(X)$ the *fundamental group* of X. By the discussion above, $\pi_1(S^1)$ equals $\mathbf{Z}$, while $\pi_1(S^n)$, for $n \ge 1$, is trivial. Note that the fundamental group is not always commutative.

Giving a group structure to $\pi_k(X)$, for $k > 1$, is a bit harder. We define the sum of two k-spheroids as follows (we use the word "sum" because the result is commutative on the level of homotopy classes). Divide S^k into hemispheres E_1^k and E_2^k so that the separating sphere $S^{k-1} = E_1^k \cap E_2^k$ goes through the south pole. Define a mapping $\rho_1 : E_1^k \to S^k$ taking S^{k-1} to the south pole and mapping the open hemisphere $E_1^k \setminus S^{k-1}$ homeomorphically onto $S^k \setminus \{s\}$. Define $\rho_2 : E_2^k \to S^k$ analogously. The spheroid $f = f_1 + f_2$ is defined as the map $S^k \to X$ that coincides with $f_1\rho_1$ on E_1^k and with $f_2\rho_2$ on E_2^k. Loosely speaking, f is defined by f_1 on one hemisphere and by f_2 on the other. Once more, replacing f_1 or f_2 by homotopic maps has no effect on the homotopy class of the sum, so we have an operation defined on the set $\pi_k(X)$ of equivalence classes of k-spheroids, which makes $\pi_k(X)$ into a commutative group.

A map from the k-ball into X taking the entire boundary of the ball into the point x_0 can also be regarded as a k-spheroid, because we can collapse the boundary of the k-ball to a single point to obtain a k-sphere. We can also think of a spheroid as a map from the k-dimensional cube $[0,1]^k$ into X such that the entire boundary of the cube has image x_0: this follows from the fact that a cube and a ball are topologically equivalent. In terms of maps on the cube it is easy to write a formula for the sum of two spheroids: we set

$$ f(t_1, \ldots, t_k) = \begin{cases} f_1(2t_1, t_2, \ldots, t_k) & \text{for } 0 \le t_1 \le \frac{1}{2}, \\ f_2(2t_1 - 1, t_2, \ldots, t_k) & \text{for } \frac{1}{2} \le t_1 \le 1. \end{cases} $$

Strictly speaking, the homotopy groups $\pi_k(X)$ depend not only on X but also on the basepoint x_0, so it would be more precise to denote them by $\pi_k(X, x_0)$. But, if X is connected, the groups $\pi_k(X, x_0)$ for different choices of x_0 are isomorphic. An isomorphism from $\pi_k(X, x_0)$ to $\pi_k(X, x_1)$ can be constructed by fixing a path joining x_0 and x_1. If X is simply connected, the isomorphism does not depend on the path, and the identification between $\pi_k(X, x_0)$ is $\pi_k(X, x_1)$ canonical, so we can write $\pi_k(X)$ in good conscience. If X is not simply connected, there is no distinguished isomorphism between $\pi_k(X, x_0)$ and $\pi_k(X, x_1)$, and sometimes we have to include the basepoint in the notation.

One sometimes also needs to talk about $\pi_0(X)$, the set of homotopy classes of mappings from the zero-dimensional sphere S^0 into X. The zero-sphere consists of two points s and n; a zero-spheroid takes s to the basepoint of X and n to an arbitrary point. Two zero-spheroids are homotopic if they map n to points that can be connected by a path: therefore $\pi_0(X)$ coincides with the set of path components of X. In general, this set does not have a group structure, but it does have a "zero element", the class of zero-spheroids that take n into the path component of the basepoint.

If X is a topological group, $\pi_1(X)$ is commutative, even if X is not. More generally, if X is a group, the group structure on $\pi_k(X)$, for $k \ge 1$, has the following alternative definition, equivalent to the one above: the concatenation or sum f of two spheroids f_1 and f_2 is given by $f(x) = f_1(x)f_2(x)$. This definition also works for $k = 0$, so $\pi_0(X)$ has a (possibly noncommutative) group structure if X is a group.

We now turn to the study of *fibrations*, a powerful tool in the calculation of homotopy groups. Consider a map p from a space E into a space B. Denote by $F_b = p^{-1}(b)$ the set of points of E that have image b. If all the spaces F_b are homeomorphic, we say that p is a fibration, with *base* B, *total space* E and *fiber* F_b over the point b. (We often write F for any of the fibers F_b, or for an abstract space homeomorphic to them, and we say F is the *fiber* of the fibration.) The map p is called the *projection*. A fibration with base B, total space E, fiber F and projection p is denoted by (E, B, F, p), or just (E, B, F). One also uses the terms *fiber space* or *fiber bundle* instead of fibration. (Strictly speaking, a fiber bundle is a fibration with some additional structure.)

The simplest fibrations are products. Given B and F, we take $E = B \times F$, the set of pairs (b, f) with $b \in B$ and $f \in F$. The projection is $p : (b, f) \to b$. A fibration of this type is called *trivial*.

A fibration is *locally trivial* if every point in the base has a neighborhood over which the fibration is trivial. In symbols, for every point $b_0 \in B$ we can find a neighborhood U of b_0 and a homeomorphism $\psi_U : U \times F \to p^{-1}(U)$ such that $p\psi_U(b, f) = b$ for all $b \in U$ and all $f \in F$.

Now let φ_g, for $g \in G$, be a transitive action of a topological group G on a space X (in other words, X is a homogeneous space under G: see Chapter 40). Fix $x_0 \in X$, and let H be the stabilizer of x_0. Define a map $p : G \to X$ by setting $p(g) = \varphi_g x_0$. Then (G, X, H, p) is a fibration with base X and fiber H, since, as discussed at the end of Chapter 40, the set of g such that $gx_0 = x$ for fixed x is a coset of H. In all cases of interest to us this fibration is locally trivial; in particular, this is always the case if G is a compact Lie group.

Examples of this type of fibration are easy to construct. Consider a linear representation of a group G and a point a in the space of this representation. Since G acts transitively on the orbit X_a of a, we get a fibration (G, X_a, H_a), where H_a is the stabilizer of a. For example, take $G = \mathrm{SO}(n)$ acting on $\mathbf{R}^n$ in the usual way (that is, according to the n-dimensional vector representation). If a is not the origin, the orbit X_a is an $(n-1)$-dimensional sphere, and the stabilizer H_a is isomorphic to $\mathrm{SO}(n-1)$ (see end of Chapter 40). Thus, we have a fibration $(\mathrm{SO}(n), S^{n-1}, \mathrm{SO}(n-1))$. In exactly the same way we get a fibration $(\mathrm{SU}(n), S^{2n-1}, \mathrm{SU}(n-1))$.

From now on we assume that all fibrations are locally trivial. As mentioned before, this requirement is almost always met in the cases of interest.

How do fibrations help calculate the homotopy groups of topological spaces? One tries to decompose a complicated space into simpler ones, whose homotopy groups are known. The following results are the most important ones to keep in mind. (They are particular cases of a result we will discuss later.)

(i) If $E = B \times F$ is a product, $\pi_k(E)$ is the direct sum of $\pi_k(B)$ and $\pi_k(F)$.

To prove this, write a spheroid in E in the form $f(x) = (f_1(x), f_2(x))$, where f_1 is a spheroid in B and f_2 is one in F. If $\alpha \in \pi_k(E)$ is the homotopy class of f, we associate with α the pair (α_1, α_2), where $\alpha_1 \in \pi_k(B)$ and $\alpha_2 \in \pi_k(F)$ are the homotopy classes of f_1 and f_2. It is easy to see that this correspondence is one-to-one.

(ii) If the base space of a fibration (E, B, F) is aspherical in dimensions k and $k+1$, that is, if $\pi_k(B) = \pi_{k+1}(B) = 0$, then $\pi_k(E) = \pi_k(F)$.

To construct the isomorphism, note that the inclusion $F \subset E$ gives a homomorphism $\pi_k(F) \to \pi_k(E)$, since every spheroid in F can be regarded as a spheroid in E. If $\pi_k(B) = 0$, any k-spheroid in E is homotopic to one in F, that is, the homomorphism $\pi_k(F) \to \pi_k(E)$ is onto. If $\pi_{k+1}(B) = 0$, any k-spheroid in F that is null-homotopic in E is null-homotopic in F as well, that is, the homomorphism $\pi_k(F) \to \pi_k(E)$ is injective. Therefore it is an isomorphism.

(iii) If the fiber F of the fibration (E, B, F) is aspherical in dimensions k and $k+1$, that is, if $\pi_k(F) = \pi_{k+1}(F) = 0$, then $\pi_k(E) = \pi_k(B)$.

This is because the fibration projection $p : E \to B$ gives rise to a homomorphism $\pi_k(E) \to \pi_k(B)$. Asphericity of F in dimensions k and $k+1$ implies that this homomorphism is one-to-one and onto.

(iv) If the total space of a fibration (E, B, F) is aspherical in dimensions k and $k-1$, that is, if $\pi_k(E) = \pi_{k-1}(E) = 0$, then $\pi_k(B) = \pi_{k-1}(F)$.

Consider a $(k-1)$-spheroid f in the fiber F. Because E is aspherical in dimension $k-1$, this spheroid is null-homotopic in E, so there is a map $g : S^k \to E$ that coincides with f on the boundary S^{k-1} of S^k. The map pg, where p is the fibration projection, takes S^k into B, with the boundary mapped to a single point; therefore pg can be seen as a k-spheroid in B. Using the asphericity of E in dimension k, one can show that the homotopy class of pg depends only on the homotopy class of f. We have, therefore, a homomorphism $\pi_{k-1}(F) \to \pi_k(B)$; it can be proved that this is an isomorphism.

Note that this result also applies when $k = 1$, in which case it can be rephrased as follows: if E is connected and simply connected, $\pi_0(F) = \pi_1(B)$. (The equality is between sets, since $\pi_0(F)$ generally does not have a group structure.)

(v) If G is a simply connected group and H is a discrete subgroup of G, we have $\pi_1(G/H) = H$ and $\pi_k(G/H) = \pi_k(G)$ for $k > 1$. (H is discrete with respect to the topology induced from G if any point in H has a neighborhood in G that contains no other point of H. In particular, H is closed.)

To show the first equality, consider a path in G connecting the identity element 1 to some element $h \in H$. Its image in G/H is a closed path, and so determines an element of $\pi_1(G/H)$; we associate this element with H, defining a map $H \to \pi_1(G/H)$, easily seen to be a homomorphism. Because H is discrete, it is identical with $\pi_0(H)$ (each point is a connected component). On the other hand, recall that the identification map $G \to G/H$ is the projection map of a fibration $(G, H, G/H)$. By (iv) above, the map from $H = \pi_0(H)$ to $\pi_1(G/H)$ is one-to-one and onto, and therefore it is an isomorphism.

The equality $\pi_k(G/H) = \pi_k(G)$ for $k > 1$ follows from (iii) above, since $\pi_k(H)$ is trivial for $k \geq 1$.

It follows from (v) that locally isomorphic Lie groups have the same homotopy groups in dimensions greater than 1. For, as we saw at the end of Chapter 36, two (connected) Lie groups are locally isomorphic if and only if they can be written as G/H_1 and G/H_2, where G is a simply connected Lie group and H_1, H_2 are discrete subgroups of G.

We now turn to some examples. Let $X = S^1$. Then $X = \mathbf{R}^1/\mathbf{Z}$, where $\mathbf{R}^1$ is the group of real numbers under addition and $\mathbf{Z}$ is the subgroup of the integers. $\mathbf{Z}$ is discrete in $\mathbf{R}^1$, so (v) implies that $\pi_1(S^1) = \mathbf{Z}$ (which we had already proved otherwise) and $\pi_k(S^1) = 0$ for $k > 1$. We also have $\pi_1(SO(2)) = \pi_1(U(1)) = \mathbf{Z}$, since $SO(2)$ and $U(1)$ are homeomorphic to S^1.

Near the beginning of this chapter we showed that every map $S^l \to S^k$ is null-homotopic if $l < k$; therefore $\pi_l(S^k) = 0$ for $l < k$. In particular, since $SU(2)$ is homeomorphic to S^3, we have $\pi_1(SU(2)) = \pi_2(SU(2)) = 0$. Moreover, $\pi_k(S^k) = \mathbf{Z}$ for every k; therefore $\pi_3(SU(2)) = \mathbf{Z}$. Since $SO(3) = SU(2)/\mathbf{Z}_2$, where $\mathbf{Z}_2$ is the order-two group, we get $\pi_1(SO(3)) = \mathbf{Z}_2$ and $\pi_k(SO(3)) = \pi_k(SU(2))$ for $k > 1$. In particular, $\pi_2(SO(3)) = 0$ and $\pi_3(SO(3)) = \mathbf{Z}$.

Property (ii) above, together with the fact, shown above, that $SO(n)$ fibers over S^{n-1} with fiber $SO(n-1)$, implies that

$$(43.1) \qquad\qquad \pi_k(SO(n-1)) = \pi_k(SO(n))$$

for $k < n - 2$. Thus, $\pi_1(SO(n)) = \pi_1(SO(3)) = \mathbf{Z}_2$ for $n \geq 3$. Moreover, $SO(4)$ is locally isomorphic to $SO(3) \times SO(3)$, so property (i) gives

$$\pi_k(SO(4)) = \pi_k(SO(3)) + \pi_k(SO(3)) = \pi_k(S^3) + \pi_k(S^3)$$

for $k \geq 2$. In particular, $\pi_2(SO(4)) = 0$ and $\pi_3(SO(4)) = \mathbf{Z} + \mathbf{Z}$. From (43.1) it follows that $\pi_2(SO(n)) = 0$ for all n.

Using property (ii) and the fact that $SU(n)$ fibers over S^{2n-1} with fiber $SU(n-1)$, we likewise get

$$(43.2) \qquad\qquad \pi_k(SU(n-1)) = \pi_k(SU(n))$$

for $k < 2n - 2$. Combining this with $\pi_2(SU(2)) = 0$ and $\pi_3(SU(2)) = \mathbf{Z}$ we get $\pi_2(SU(n)) = 0$ for all n and $\pi_3(SU(n)) = \mathbf{Z}$ for $n \geq 2$. It can be proved that $\pi_2(G) = 0$ for every Lie group G.

Now consider a space X and a transitive action of a compact simply connected Lie group G on X. Since $\pi_1(G) = \pi_2(G) = 0$, we can apply property (iv) to the fibration (G, X, H), where H is the stabilizer of a point. This gives $\pi_2(X) = \pi_1(H)$. This relation was used in Chapter 12.

We now turn to the computation of $\pi_3(S^2)$. Recall that $SO(3)$ acts transitively on S^2. Since $SO(3) = SU(2)/\mathbf{Z}_2$, we see that $SU(2)$ also acts transitively on S^2, with stabilizer $U(1)$. Now $SU(2)$ is homeomorphic to S^3 and $U(1)$ to S^1. We therefore get a fibration of S^3 over S^2, with fiber S^1 (the *Hopf fibration*). Keeping in mind that $\pi_k(S^1) = 0$ for $k \geq 2$ and applying property (iii), we get $\pi_k(S^3) = \pi_k(S^2)$ for $k \geq 3$; in particular, $\pi_3(S^2) = \pi_3(S^3) = \mathbf{Z}$. Every element of $\pi_3(S^2)$ is of the form $n\alpha$, where n is an integer and α is the homotopy class of the projection $S^3 \to S^2$ of the fibration just constructed.

Properties (i)–(v) are sufficient to compute many of the homotopy groups of interest in physics. We now give a general theorem that implies all of properties (i)–(v), and that sometimes allows one to compute a homotopy group that does not follow from these properties alone.

We start by defining an *exact sequence*. A sequence of groups A_n and homomorphisms α_n,

$$\cdots \xrightarrow{\alpha_{n+2}} A_{n+1} \xrightarrow{\alpha_{n+1}} A_n \xrightarrow{\alpha_n} A_{n-1} \xrightarrow{\alpha_{n-1}} \cdots,$$

is *exact* at A_n if $\operatorname{Im} \alpha_{n+1} = \operatorname{Ker} \alpha_n$, that is, if the image of α_{n+1} coincides with the kernel α_n (see Chapter 36). A sequence is *exact* if it is exact at each term.

Let (E, B, F, p) be a fibration. We have an obvious homomorphism $i_n :$ $\pi_n(F) \rightarrow \pi_n(E)$, coming from considering a spheroid in F as a spheroid in E by inclusion. Similarly, we have a homomorphism $p_n : \pi_n(E) \rightarrow \pi_n(B)$ by projection. We can also construct a *connecting homomorphisms* $\partial_n : \pi_n(B) \rightarrow$ $\pi_{n-1}(F)$ in a less obvious way, using a construction similar to the one used in the description of property (iv) above (see details below). Now we have a sequence of groups of homomorphisms

$$(43.3) \qquad \cdots \xrightarrow{\partial_{n+1}} \pi_n(F) \xrightarrow{i_n} \pi_n(E) \xrightarrow{p_n} \pi_n(B) \xrightarrow{\partial_n} \pi_{n-1}(F) \xrightarrow{i_{n-1}} \cdots,$$

called the *long homotopy sequence* of (E, B, F, p). This sequence can be shown to be exact for any fibration.

(Strictly speaking, in constructing the long homotopy sequence, we should fix basepoints $e \in F$ and $b = p(e) \in E$, and consider the homotopy groups $\pi_n(F, e)$, $\pi_n(E, e)$ and $\pi_n(B, b)$. However, for the sake of simplicity, we will ignore the dependence of the homotopy groups on the basepoint. The last three terms of the sequence, $\pi_0(F, e)$, $\pi_0(E, e)$ and $\pi_0(B, b)$, are not groups, but they do have a "zero element", as discussed above, and this allows us to talk about the kernel of ∂_1, i_0 and p_0. Therefore exactness makes sense even at the last three terms.)

Properties (ii)–(iv) follow immediately from the exact homotopy sequence. Suppose, for example, that $\pi_{k+1}(B) = 0$. Then $\operatorname{Im} \partial_{k+1} = 0$, and, by exactness, $\operatorname{Ker} i_k = 0$, that is, i_k is injective. If $\pi_k(B) = 0$, we have $\operatorname{Ker} p_k = \pi_k(E)$, and, again by exactness, $\operatorname{Im} i_k = \pi_k(E)$. Thus i_k is both injective and surjective, and therefore is an isomorphism, proving (ii). The proof of (iii) and (iv) is analogous.

To prove the exactness of the homotopy sequence (43.3), we use the *homotopy lifting property* of fibrations. If (E, B, F, p) is a fibration and X is a space, a map $f : X \rightarrow E$ is said to be a *lift* of $g : X \rightarrow B$ if $g = pf$, that is, if the image of a point x under f is in the fiber that lies above the image of x under g. Similarly, a homotopy f_t (for $t \in [0, 1]$) is a *lift* of a homotopy g_t if $g_t = pf_t$. The homotopy lifting property says that, for any fibration (E, B, F, p) and any space X, every homotopy $X \times [0, 1] \rightarrow B$ can be lifted to a homotopy $X \times [0, 1] \rightarrow E$, with a prescribed initial map. What this means is that, if $g_t : X \rightarrow B$ is a family of maps depending continuously on t, for $t \in [0, 1]$, and $h : X \rightarrow E$ is a lift of g_0, there exists a continuous family of maps f_t such that $f_0 = h$ and that f_t is a lift of g_t for all t.

For a trivial (that is, product-like) fibration the homotopy lifting property is obvious. The case of an arbitrary (locally trivial) fibration can be reduced to that of a trivial fibration.

The homotopy lifting property easily implies that any map from the cube $[0, 1]^n$ to B can be lifted to a map $[0, 1]^n \rightarrow E$ taking the origin to a specified point. We simply observe that a map $[0, 1]^n \rightarrow B$ can be considered as a homotopy between maps $[0, 1]^{n-1} \rightarrow B$, with $f_t(t_1, \ldots, t_{n-1}) = f(t_1, \ldots, t_{n-1}, t)$.

The connecting homomorphism ∂_n that appears in the long exact sequence (43.3) can be defined as follows. For a given $\alpha \in \pi_n(B, b)$, take a spheroid representing α, and regard it as a map $\varphi : [0, 1]^n \to B$ taking the boundary of the cube to the basepoint b. By the previous paragraph, we can lift φ to a map $[0, 1]^n \to E$ such that the origin is mapped to the basepoint $e \in E$ and the boundary of the cube is mapped into the fiber F_b over b. Since the boundary of $[0, 1]^n$ is homeomorphic to S^{n-1}, we have defined a spheroid $S^{n-1} \to F_b$. The homotopy class of this spheroid does not depend on how we choose the lift for φ, nor does it depend on the spheroid φ chosen to represent $\alpha \in \pi_n(B, b)$. We call this class $\partial_n \alpha$.

To prove exactness, we must show that $\operatorname{Im} i_n = \operatorname{Ker} p_n$, $\operatorname{Im} p_n = \operatorname{Ker} \partial_n$ and $\operatorname{Im} \partial_n = \operatorname{Ker} i_{n-1}$. It is immediate that $\operatorname{Im} i_n \subset \operatorname{Ker} p_n$, $\operatorname{Im} p_n \subset \operatorname{Ker} \partial_n$ and $\operatorname{Im} \partial_n \subset \operatorname{Ker} i_{n-1}$, or, equivalently,

$$(43.4) \qquad p_n i_n = 0, \qquad \partial_n p_n = 0, \qquad i_{n-1} \partial_n = 0.$$

For instance, if we think of a spheroid in F as a spheroid in E and then project it to B, we get the zero spheroid in B, and the first relation follows.

It is somewhat harder to prove the opposite inclusions,

$$(43.5) \qquad \operatorname{Ker} p_n \subset \operatorname{Im} i_n, \qquad \operatorname{Ker} \partial_n \subset \operatorname{Im} p_n, \qquad \operatorname{Ker} i_{n-1} \subset \operatorname{Im} \partial_n.$$

Suppose, for example, that $\alpha \in \operatorname{Ker} p_n$. This means that a spheroid $S^n \to E$ in class α becomes null-homotopic when projected to B; equivalently, there exists a continuous family of spheroids $g_t : S^n \to B$ with $g_0 = pf$ and g_1 the constant map with image b. By the homotopy lifting property, there exists a continuous family of spheroids $f_t : S^n \to E$ with $f_0 = f$ and f_1 a map into the fiber F_b over b. Regarding f_1 as a spheroid in F_b, and denoting its homotopy class by β, we see that $\alpha = i_n \beta$, that is, $\alpha \in \operatorname{Im} i_n$, showing the first inclusion in (43.5). The other two inclusions are proved analogously.

So far we have used fibrations as a tool for computing homotopy groups, but they are interesting for other reasons as well. We will discuss one more relevant concept, that of a *section* of a fibration. A section of the fibration (E, B, F, p) is a map $\varphi : B \to E$ such that $p\varphi$ is the identity, that is, φ takes each $x \in B$ into the fiber over x. In other words, a section is a way to select a point from each fiber, in a continuous way.

Not all fibrations have sections. If a fibration has a section, the homomorphism $p_n : \pi_n(E) \to \pi_n(B)$ is onto, because any spheroid in B can be composed with the section to give a spheroid in E. This shows that the Hopf fibration (S^3, S^2, S^1), for example, does not have a section, since $\pi_2(S^3) = 0$ and $\pi_2(S^2) \neq 0$. Using the same argument, one can show that if G is a simply connected, compact Lie group and H is a non-simply-connected subgroup of G, the fibration $(G, G/H, H)$ has no section.

In many important cases the existence of a section is a consequence of the following theorem: if the base space B has dimension n and the fiber F is aspherical in all dimensions less than n, the fibration (E, B, F) has a section.

For smooth fibrations the existence of a (continuous) section implies the existence of a smooth section. (A fibration is *smooth*, or infinitely differentiable, if the base, fiber and total spaces are smooth manifolds, and if it has local trivializations that are smooth maps and whose inverses are smooth maps.)

We conclude this chapter with a compilation of results on the homotopy groups of spheres and classical Lie groups, collected here for ease of reference.

Homotopy Groups of Spheres

$$\pi_1(S^1) = \mathbf{Z}$$
$$\pi_k(S^1) = 0 \qquad\qquad \text{for } k \geq 2$$

$$\pi_1(S^2) = 0$$
$$\pi_2(S^2) = \pi_3(S^2) = \mathbf{Z}$$
$$\pi_4(S^2) = \pi_5(S^2) = \mathbf{Z}_2$$
$$\pi_6(S^2) = \mathbf{Z}_{12}$$

$$\pi_1(S^3) = \pi_2(S^3) = 0$$
$$\pi_k(S^3) = \pi_k(S^2) \qquad\qquad \text{for } k \geq 3$$

$$\pi_i(S^n) = 0 \qquad\qquad \text{for } i < n$$
$$\pi_n(S^n) = \mathbf{Z}$$
$$\pi_{n+1}(S^n) = \pi_{n+2}(S^n) = \mathbf{Z}_2 \qquad\qquad \text{for } n \geq 3$$
$$\pi_{n+3}(S^n) = \mathbf{Z}_{24} \qquad\qquad \text{for } n \geq 5$$
$$\pi_{n+4}(S^n) = 0 \qquad\qquad \text{for } n \geq 6$$
$$\pi_{n+5}(S^n) = 0 \qquad\qquad \text{for } n \geq 7$$
$$\pi_m(S^n) = \pi_{m+1}(S^{n+1}) \qquad\qquad \text{for } m < 2n - 1$$

$$\pi_{4n-1}(S^{2n}) = \mathbf{Z} + \text{finite group}$$

All groups $\pi_m(S^n)$, except for $\pi_n(S^n)$ and $\pi_{4n-1}(S^{2n})$, are finite.

Homotopy Groups of Classical Lie Groups

$$\pi_1(\mathrm{SO}(n)) = \mathbf{Z}_2 \qquad \text{for } n > 1$$
$$\pi_1(\mathrm{SU}(n)) = \pi_1(\mathrm{Spin}(n)) = 0 \qquad \text{for } n \geq 1$$
$$\pi_1(U(n)) = \mathbf{Z} \qquad \text{for } k \geq 1$$

$$\pi_2(G) = 0 \qquad \text{for all Lie groups}$$

$$\pi_3(\mathrm{SO}(n)) = \mathbf{Z} \qquad \text{for } n = 3 \text{ and } n \geq 5$$
$$\pi_3(\mathrm{SO}(4)) = \mathbf{Z} + \mathbf{Z}$$
$$\pi_3(\mathrm{SU}(n)) = \mathbf{Z} \qquad \text{for } n \geq 2$$
$$\pi_3(\mathrm{Spin}(n)) = \mathbf{Z} \qquad \text{for } n \geq 1$$

$$\pi_4(SO(3)) = \pi_4(SU(2)) = \mathbf{Z}_2$$
$$\pi_4(SO(4)) = \mathbf{Z}_2 + \mathbf{Z}_2$$
$$\pi_4(SO(5)) = \mathbf{Z}_2$$

$$\pi_4(SO(n)) = 0 \qquad \text{for } n \geq 6$$
$$\pi_4(SU(n)) = 0 \qquad \text{for } n \geq 3$$
$$\pi_4(\text{Spin}(1)) = \mathbf{Z}_2 \qquad \text{for } n \geq 1$$

$$\pi_k(SU(1)) = 0 \qquad \text{for } k > 1$$
$$\pi_k(SO(3)) = \pi_k(SU(2)) = \pi_k(\text{Spin}(1)) = \pi_k(S^3) \quad \text{for } k > 1$$
$$\pi_k(SO(4)) = \pi_k(S^3) + \pi_k(S^3) \qquad \text{for } k > 1$$
$$\pi_k(U(n)) = \pi_k(SU(n)) \qquad \text{for } k > 1$$

The groups $\pi_k(SO(n))$, $\pi_k(U(n))$ and $\pi_k(\text{Spin}(n))$ do not depend on n for, respectively, $1 \leq k \leq n-2$, $1 \leq k \leq 2n-1$ and $1 \leq k \leq 4n+1$; the are denoted by $\pi_k(SO)$, $\pi_k(U)$ and $\pi_k(\text{Spin})$ and are known as *stable homotopy groups*. The following results allow the computation of all stable homotopy groups:

$$\pi_{k+8}(SO) = \pi_k(SO)$$
$$\pi_{k+2}(U) = \pi_k(U)$$
$$\pi_{k+8}(\text{Spin}) = \pi_k(\text{Spin}) = \pi_{k+4}(SO)$$

$$\pi_1(SO) = \mathbf{Z}_2$$
$$\pi_2(SO) = 0$$
$$\pi_3(SO) = \mathbf{Z}$$
$$\pi_4(SO) = \pi_5(SO) = \pi_6(SO) = 0$$
$$\pi_7(SO) = \mathbf{Z}$$
$$\pi_8(SO) = \mathbf{Z}_2$$

$$\pi_1(U) = \mathbf{Z}$$
$$\pi_2(U) = 0$$

The first three of these equalities constitute the *Bott periodicity theorem*.

43. Applications of Topology to Physics

We now turn briefly to applications of topology to quantum field theory, many of which were discussed in Part II.

We start with the phase space $\mathcal{E}$ of classical field theory, consisting of all fields of finite energy. We assume that zero is the energy minimum. We choose an arbitrary basepoint $e \in \mathcal{E}$. If $\mathcal{E}$ has several connected components, one can associate with each field the component that it belongs to, that is, an element of $\pi_0(\mathcal{E}, e)$. This element is an integral of motion: it cannot change as the field evolves, as we discussed in Chapter 9.

The same is still true after quantization of the classical theory. To see this, think of the quantized theory as the limit of theories with finitely many degrees of freedom, say by introducing spatial and momentum cutoff. (That is, replace the coordinate space by a lattice of points (an_1, an_2, an_3), where n_1, n_2, n_3 are integers in the range $0 \leq an_i \leq L$: see Chapter 23.) The phase space in this case is connected, but it consists of several potential wells (one for each connected component of the classical phase space), separated by potential barriers whose height tends to infinity as $a \to 0$ and $L \to \infty$. In the limit, then, there is no tunneling from one well to another. (The probability of tunneling, which can be estimated using the Feynman functional integral, depends not only on the height of the barrier but also on its width, so a rigorous argument must establish that the width does not decrease so fast as to annul the increase in the height.)

Thus, when $\mathcal{E}$ is not connected, the theory displays topological integrals of motion, or topological quantum numbers. The particles usually considered in quantum field theory and the many-particle states consisting of such particles have the same topological quantum numbers as the physical vacuum (ground state). If the theory predicts the existence of states with nontrivial quantum numbers, "unusual" particles with these quantum numbers should also exist. In the case of weak coupling, these particles correspond to the particle-like solutions of the classical equations.

To apply these ideas, one must construct theories in which the phase space $\mathcal{E}$ is disconnected. At present two approaches to this problem are known: one is used in [19, 24, 26, 61], and the other in [32, 44, 65, 66]. The first approach is based on considering fields $\varphi(x, t)$ that take values in a nonlinear manifold. An example of such a theory is chiral dynamics, where fields take values on the three-sphere S^3 [61].

Consider the space $\mathcal{F}$ of fields $\varphi(x)$ with values in a manifold M and such that $\varphi(x) \to m$ as $|x| \to \infty$, where m is a fixed point of M. When can the investigation of the phase space of a Lagrangian be reduced to the investigation of $\mathcal{F}$? Take the Lagrangian

$$(44.1) \qquad L = \tfrac{1}{2} g_{ik}(\varphi)\, \partial_\alpha \varphi^i(x)\, \partial^\alpha \varphi^k(x),$$

where $g_{ik}(m)$ is a Riemannian metric on M. The corresponding Hamiltonian has the form

$$\mathcal{H}(\pi, \varphi) = \frac{1}{2} \int g^{ik}(\varphi)\pi_i(x)\pi_k(x)\, dx + \frac{1}{2} \int g_{ik}(\varphi)(\nabla\varphi^i(x), \nabla\varphi^k(x))\, dx.$$

If $\mathcal{H}(\pi, \varphi) < \infty$, there exists a point $m \in M$ such that $\varphi(x) \to m$ as $|x| \to \infty$. Therefore the phase space is closely related to $\mathcal{F}$. (The Lagrangian (44.1) is not renormalizable. Another deficiency is that the greatest lower bound on the energy functional vanishes on all components of the phase space. This problem can easily be eliminated if we include in the Lagrangian terms containing derivatives of higher order.)

A field $\varphi(x)$ in $\mathcal{F}$ defines an element of $\pi_3(M)$. To construct this element, let σ be (for example) stereographic projection from the south pole of S^3. Then $\varphi\sigma$ is a spheroid whose class is the desired element of $\pi_3(M, m)$. (Assume that M is simply connected, so the various $\pi_3(M, m)$, as m varies over M, are canonically identified.) Two fields belong to the same connected component of $\mathcal{F}$ if and only if they determine the same element of $\pi_3(M)$.

Note that $\mathcal{F}$ may not be simply connected, for we have $\pi_1(\mathcal{F}, f) = \pi_4(M, m)$ for f an arbitrary point of M. In general, $\pi_k(\mathcal{F}, f) = \pi_{k+3}(M, m)$. In particular, $\pi_k(\mathcal{F}, f)$ does not depend on f, despite the fact that $\mathcal{F}$ is not connected.

Finkelstein and Rubinstein [26] showed that theories in which fields take values on a nonlinear manifold can describe fermions as well as bosons (cf. Chapter 33). They considered closed paths in $\mathcal{F}$ of the form

$$(44.2) \qquad \varphi_t(x) = \varphi(s_t(x)),$$

where $\varphi \in \mathcal{F}$ and s_t represents the rotation of $\mathbf{R}^3$ through an angle $t \in [0, 2\pi]$ about a fixed axis. Finkelstein and Rubinstein assume that if there are homotopically nontrivial paths of this type, the theory can contain fermions. Without getting into the merit of this assumption, we observe that the path (44.2) is null-homotopic if and only if the four-dimensional spheroid $\varphi\sigma\lambda$ is null-homotopic, where $\lambda : S^4 \to S^3$ is a homotopically nontrivial map. In other words, (44.2) is null-homotopic if and only if $\alpha\beta = 0$, where α is the element of $\pi_3(M)$ defined by φ and β is the unique nonzero element of $\pi_4(S^3)$. This assertion, whose proof we will not give here, makes it easy to analyze concrete manifolds M. For example, when $M = S^2$ or $M = S^3$, there are paths of type (44.2) that are not null-homotopic, and so, according to Finkelstein and Rubinstein's assumption, the theory can contain fermions.

The second approach to constructing theories with a disconnected phase space is based on considering Lagrangians that lead to degeneracy of the classical vacuum. In this approach, to each field of finite energy we can assign a map from S^2 into the manifold of classical vacuums. Such a map determines the asymptotic behavior of the field at infinity. This approach is described in [66], and also in Part II of this book, starting with Chapter 12.

Note, finally, that topological ideas can be applied in quantum field theory even when the topology is trivial in the corresponding classical problems. For more details, see [66].

Bibliographical Remarks

Topologically nontrivial particle-like solutions to nonlinear models were first discussed by Skyrme [60, 61]. Solutions of more general nonlinear models were studied by Finkelstein [25, 26], using homotopy theory.

Around 1974 several articles were published that aroused general interest in the role of particle-like classical solutions in quantum field theory. Faddeev and Takhtadzhian [24] wrote on the connection between solitons and quantum particles in one-dimensional models; Polyakov [44] discussing the link between quantum particles and classical solutions in the semiclassical case, as well as magnetic monopoles; 't Hooft [32] discussed magnetic monopoles; Nielsen and Olesen [41] analyzed string-like solutions; and Zeldovich et al. [80, 81] studied domain walls. Almost immediately topological methods were found to be very useful in the study of particle-like solutions [1, 43, 65].

Subsequently, many researchers investigated the link between classical solutions and quantum particles [11–13, 20–22, 63, 64]. (We do not analyze this link in detail in this book.) Particle-like classical solutions were constructed for different models in field theory: see [38] for a review.

Further momentum was provided by the discovery of instantons by Polyakov [45] and Belavin et al. [4], which were taken up by many researchers. Later on other important applications of topology to quantum field theory emerged, most of which lie outside the scope of this book (an exception being the analysis of many-valued action integrals). Mention should be made of the studies of the spin and statistics of particles corresponding to solitons, by Finkelstein and Rubinstein [26] and Witten [4] (see also [66]), the discovery by Rubakov [49] that proton decay is greatly accelerated in the presence of magnetic monopoles, and the discovery by Witten [72, 77] of a new type of anomaly, related to the topological properties of the field space (global anomalies). A very useful concept in the study of supersymmetry breaking was introduced by Witten [75], namely, the index of a supersymmetric theory (similar to the index of an elliptic operator).

The interaction of topology and physics goes both ways: ideas originating in quantum field theory have led to new topological results. It was shown in [53, 55] that physical quantities associated with appropriate action functionals can give invariants of manifolds. This idea led to the development of topological quantum field theory, where the most important results were obtained by Witten [78, 79]. In particular, Witten analyzed invariants connected with the so-called nonabelian Chern–Simons action functional, and proved that they are related to the Jones polynomial of knots. (The abelian version of this functional was studied in [53], and [58] suggested the study of the invariants connected with the nonabelian version, conjecturing that they are related to the Jones polynonimal.) A review of the results of topological quantum field theory can be found in [5].

Another area where ideas from physics are applied to topology was opened up by Donaldson [15, 16]. See [27–29] for a review of very important results in low-dimensional topology arising from this connection.

The successful application of topology to quantum field theory has stimulated the study of defects in continuous media using topological methods [62, 69–70]. The problems in this area are mathematically similar to the ones encountered in quantum field theory, which is why Part II of this book opens with a chapter on topologically stable defects of local equilibrium.

Here is a brief list of articles that served as sources for the material in some chapters, and that can be consulted for background or further study.

Chapter		References
9, 11	(1D-models, Georgi–Glashow models)	[32, 44]
10	(Abrikosov vortices)	[7, 41]
12–15	(magnetic monopoles)	[1, 43, 50, 65]
17	(symmetric gauge fields)	[2, 9, 47, 51, 72]
18	(energy of magnetic monopoles)	[7, 48]
19–20	(topologically nontrivial strings)	[56, 57, 67]
21	(nonlinear fields)	[26, 60, 61, 76]
22	(multivalued action integrals)	[14, 42, 76]
30–34	(instantons)	[3, 4, 33, 34, 45, 52, 54]

In addition, the following surveys of the applications of topology in quantum field theory can be consulted: [10 (pp. 50ff.), 18, 31, 37, 40, 46]. For basic topology material, see especially [17, 59], which are most accessible to physicists; other good references are [8, 30, 35, 39, 71]. Some textbooks on quantum field theory are [6, 23, 36, 68].

References

1. J. Arafune, P. G. O. Freund and C. J. Goebel. Topology of Higgs fields. J. Math. Phys. **16** (1975) 433.

2. F. A. Bais and Y. R. Primack. Spherically symmetric monopoles in nonabelian gauge theories. Nuclear Phys. B **123** (1977) 253.

3. A. A. Belavin and A. M. Polyakov. Quantum fluctuations of pseudoparticles. Nuclear Phys. B **123** (1977) 429.

4. A. A. Belavin, A. M. Polyakov, Yu. S. Tyupkin and A. S. Schwarz. Pseudoparticle solutions to Yang–Mills equations. Phys. Lett. B **59** (1975) 85.

5. D. Birmingham, M. Blau, M. Rakowski and G. Thompson,. Topological field theory. Phys. Rep. **209** (1991) 129.

6. N. N. Bogolyubov and D. V. Shirkov. Introduction to the Theory of Quantized Fields. Wiley, New York, 1979.

7. E. B. Bogomolny. Stability of classical solutions. Yadernaya Fiz. **24** (1976) 449.

8. V. G. Boltianskii and V. A. Efremovich. Naglyadnaya Topologia (Graphic Topology). Nauka, Moscow, 1983.

9. D. E. Burlankov. Unified spatial isotopic symmetry in Yang–Mills fields. Teoret. Mat. Fiz. **32** (1977) 326.

10. S. Coleman. The Whys of Subnuclear Physics. Plenum, New York, 1979.

11. R. F. Dashen, B. Hasslacher and A. Neveu. Non-perturbative methods and extended-hadron models in field theory. I: Semiclassical functional methods. II: Two-dimensional models and extended hadrons. III: Four-dimensional nonabelian models.. Phys. Rev. D **10** (1974) 4114, 4130, 4138.

12. R. F. Dashen, B. Hasslacher and A. Neveu. Particle spectrum in model field theories from semiclassical functional integral techniques. Phys. Rev. D **11** (1975) 3424.

13. R. F. Dashen, B. Hasslacher and A. Neveu. Semiclassical bound states in an asymptotically free theory. Phys. Rev. D **12** (1976) 2443.

14. S. Deser, R. Jackiw and S. Templeton. Three-dimensional massive gauge theories. Phys. Rev. Lett. **48** (1982) 975.

15. S. K. Donaldson. An application of gauge theory to four-dimensional topology. J. Differential Geom. **18** (1983) 279.

16. S. K. Donaldson. The geometry of four-manifolds. In Proc. Intern. Congress of Mathematicians, Berkeley, CA, 1986, edited by A. Gleason. Amer. Math. Soc., Providence, RI, 1987.

17. B. A. Dubrovin, A. T. Fomenko, S. P. Novikov. Modern Geometry: Methods and Applications. Springer, New York, 1984.

18. T. Eguchi, P. B. Gilkey and A. J. Hanson. Gravitation, gauge theories and differential geometry. Phys. Rep. C **66** (1980) 213.

19. L. D. Faddeev. Pis'ma Zh. Eksper. Teoret. Fiz.. Hadrons from Leptons? **21** (1975) 141.

20. L. D. Faddeev and V. E. Korepin. Quantization of solitons. Teoret. Mat. Fiz. **25** (1975) 147.

21. L. D. Faddeev and V. E. Korepin. About the mode problem in the quantization of solitons. Phys. Rev. Lett. B **63** (1976) 435.

22. L. D. Faddeev and V. E. Korepin. Quantum theory of solitons. Phys. Rep. C **42** (1978) 1.

23. L. D. Faddeev and A. A. Slavnov. Gauge Fields, Introduction to Quantum Theory. Addison-Wesley, Reading, MA, 1980.

24. L. D. Faddeev and L. A. Takhtadzhian. An essentially nonlinear one-dimensional model of classical field theory. Teoret. Mat. Fiz. **21** (1974) 160.

25. D. Finkelstein. Kinks. J. Math. Phys. **4** (1966) 1218.

26. D. Finkelstein and J. Rubinstein. Connection between spin, statistics, and kinks. J. Math. Phys. **9** (1968) 1762.

27. S. Donaldson and P. Kronheimer. The Geometry of Four-Manifolds. Clarendon, Oxford, 1990.

28. D. S. Freed and K. K. Uhlenbeck. Instantons and Four-manifolds. Springer, New York, 1984.

29. M. H. Freedman and F. Quinn. Topology of Four-manifolds. Princeton University Press, Princeton, NJ, 1990.

30. D. B. Fuchs and V. A. Rokhlin. Beginner's Course in Topology: Geometric Chapters. Springer, Berlin (1984),

31. P. Goddard and D. Olive. New developments in the theory of magnetic monopoles. Rep. Progr. Phys. **41** (1978) 1357.

32. G. 't Hooft. Magnetic monopoles in unified gauge theories. Nuclear Phys. B **79** (1974) 276.

33. G. 't Hooft. Symmetry breaking through Bell–Jackiw anomalies. Phys. Rev. Lett. **37** (1976) 8.

34. G. 't Hooft. Computation of the quantum effects due to a four-dimensional pseudoparticle. Phys. Rev. D **14** (1976) 3422.

35. D. Husemoller. Fiber Bundles. McGraw-Hill, New York, 1966.

36. C. Itzykson and J.-B. Zuber. Quantum Field Theory. McGraw-Hill, New York, 1980.

37. A. Jaffe and C. Taubes. Vortices and Monopoles. Birkhäuser, Boston, 1980.

38. V. G. Makhan'kov. Dynamics of classical solutions (in nonintegrable systems). Phys. Rep. C **35** (1978) 1.

39. J. W. Milnor. Topology from the Differential Viewpoint. Univ. Press of Virginia, Charlottesville, 1965.

40. C. Nash and S. Sen. Topology and Geometry for Physicists. Academic Press, London, 1983.

41. H. B. Nielsen and P. Olesen. Vortex-line models for dual strings. Nuclear Phys. B **61** (1973) 45.

42. S. P. Novikov. The Hamiltonian formalism and a many-valued analogue of Morse theory. Russian Math. Surveys **5** (1982) 1.

43. M. I. Monastyrskii and A. M. Perelomov. On the existence of monopoles in gauge field theories. Pis'ma Zh. Eksper. Teoret. Fiz. **21** 94 (1975).

44. A. M. Polyakov (1974) The particle spectrum in quantum field theory. Pis'ma Zh. Eksper. Teoret. Fiz. **20** 430.

45. A. M. Polyakov. Compact gauge fields and the infrared catastrophe. Phys. Lett. B **59** (1975) 82.

46. R. Rajaramaran. Solitons and Instantons. North-Holland, Amsterdam, 1982.

47. V. N. Romanov, Yu. S. Tyupkin and A. S. Schwarz. On spherically symmetric fieldhs in gauge theories. Nuclear Phys. B **130** (1977) 209.

48. V. N. Romanov, V. A. Fateev and A. S. Schwarz. Magnetic monopoles in unified theories of electromagnetic, weak and strong interactions. Yadernaya Fiz. **32** (1980) 1138.

49. A. Rubakov. Superheavy magnetic monopoles and proton decay. Pis'ma Zh. Eksper. Teoret. Fiz. **33** (1981) 645.

50. A. S. Schwarz. On magnetic monopoles in gauge field theories. Nuclear Phys. B **212** (1976) 358.

51. A. S. Schwarz. On symmetric gauge fields. Comm. Math. Phys. **56** (1977) 79.

52. A. S. Schwarz. Regular extremals of Euclidean action. Phys. Lett. B **67** (1977) 172.

53. A. S. Schwarz. The partition function of a degenerate quadratic functional and Ray–Singer invariants. Lett. Math. Phys. **2** (1978) 247.

54. A. S. Schwarz. Instantons and fermions in a field of instantons. Comm. Math. Phys. **64** (1979) 233.

55. A. S. Schwarz. The partition function of a degenerate functional. Comm. Math. Phys. **67** (1979) 1.

56. A. S. Schwarz. Theories with nonlocal electric charge conservation. Pis'ma Zh. Eksper. Teoret. Fiz. **34** (1981) 555.

57. A. S. Schwarz. Field theories with no local conservation of electric charge. Nuclear Phys. B **208** (1982) 141.

58. A. S. Schwarz. New topological invariants arising in the theory of quantized fields. Talk at the Baku International Topology Conference, 1987.

59. A. S. Schwarz. Topology for Physicists. Springer, Berlin, 1993.

60. T. H. R. Skyrme. A nonlinear theory of strong interactions. Proc. Roy. Soc. London A. **247** (1958) 260.

61. T. H. R. Skyrme. A unified field theory of mesons and baryons. Nuclear Phys. **31** (1962) 556.

62. G. Toulouse and M. Kleman. Principles of a classification of defects in ordered media (1976). J. Phys. Lett. **37** 149.

63. Yu. S. Tyupkin, V. A. Fateev and A. S. Schwarz. On the classical limit of the scattering matrix in quantum field theory. Dokl. Akad. Nauk SSSR **221** (1975) 70.

64. Yu. S. Tyupkin, V. A. Fateev and A. S. Schwarz. On the connection between particle-like solutions to classical equations and quantum particles (1975). Yadernaya Fiz. **22** 622.

65. Yu. S. Tyupkin, V. A. Fateev and A. S. Schwarz. On the existence of heavy particles in gauge field theories. Pis'ma Zh. Eksper. Teoret. Fiz. **21** (1975) 91.

66. Yu. S. Tyupkin, V. A. Fateev and A. S. Schwarz. Homotopy topology for physicists. In Proc. Third Physics Workshop, vol. 1 65. Institute of Theoretical and Experimental Physics, Moscow, 1975 (see Chapters 41 and 42 for a translation).

67. Yu. S. Tyupkin and A. S. Schwarz. Vortices in unified theories of weak and electromagnetic interactions. Phys. Lett. B **90** (1980) 135.

68. M. B. Voloshin and K. A. Ter-Martirosian. Teoriya Kalibrovochnykh Vzaimodeistvii Elementarnykh Chastits (Gauge Theory of Interacting Elementary Particles). Energoatomizdat, Moscow, 1984.

69. G. E. Volovik and V. P. Mineev. Pis'ma Zh. Eksper. Teoret. Fiz.. Vortices with free ends in superfluid He^3-A **23** (1976) 647.

70. G. E. Volovik and V. P. Mineev. Line and point singularities in superfluid He^3. Pis'ma Zh. Eksper. Teoret. Fiz. **24** (1976) 605.

71. A. H. Wallace. Differential Topology: First Steps. Benjamin, New York, 1968.

72. D. Wilkinson and A. S. Goldhaber. Spherically symmetric monopoles. Phys. Rev. D **16** (1977) 1221.

73. E. Witten. An SU(2)-anomaly. Phys. Lett. B **177** (1982) 324.

74. E. Witten. Supersymmetry and Morse theory. J. Diff. Geom. **17** (182) 661.

75. E. Witten. Constraints on supersymmetry breaking. Nuclear Phys. B **202** (1982) 253.

76. E. Witten. Global aspects of current algebra. Nuclear Phys. B **233** (1983) 422.

77. E. Witten. Global Gravitational Anomalies. Comm. Math. Phys. **100** (1985) 197.

78. E. Witten. Topological quantum field theory. Commun. Math. Phys. **117** (1988) 353.

79. E. Witten. Quantum field theory and the Jones polynomial. Commun. Math. Phys. **121** (1989) 351.

80. Ya. B. Zeldovich, I. Yu. Kobzarev and L. B. Okun'. Cosmological consequences of the spontaneous breakdown of discrete symmetry. Zh. Eksper. Teoret. Fiz. **64** (1974) 3.

81. Ya. B. Zeldovich, I. Yu. Kobzarev and L. B. Okun'. Spontaneous CP-violation and cosmology. Phys. Lett. B **50** (1974) 340.

82. J. Glimm and A. Jaffe. Quantum physics: A functional integral point of view. Springer, New York, 1981.

Index

Grundlehren der mathematischen Wissenschaften

A Series of Comprehensive Studies in Mathematics

A Selection